Chevrolet
Malibu
Oldsmobile
Alero and Cutlass
Pontiac
Grand Am
Automotive
Repair
Manual

**by Jay Storer
and John H Haynes**

Member of the Guild of Motoring Writers

Models covered:

Chevrolet Malibu and Oldsmobile Cutlass models
1997 through 2000

Pontiac Grand Am and Oldsmobile Alero models
1999 and 2000

(4E2 - 38026)

ABCDE
FGHIJ
KLMNO
P

Haynes Publishing Group
Sparkford Nr Yeovil
Somerset BA22 7JJ England

Haynes North America, Inc
861 Lawrence Drive
Newbury Park
California 91320 USA

Acknowledgements

Wiring diagrams provided exclusively for Haynes North America, Inc. by Valley Forge Technical Information Services. Technical consultants who contributed to this project include Bob Henderson, Jeff Kibler and Eric Godfrey.

A book in the Haynes Automotive Repair Manual Series

Printed in the U.S.A.

ISBN 1 56392 361 0

Library of Congress Catalog Card Number 00-100881

While every attempt is made to ensure that the information in this manual is correct, no liability can be accepted by the authors or publishers for loss, damage or injury caused by any errors in, or omissions from, the information given.

Contents

Haynes photographer, mechanic and author with 1999 Chevrolet Malibu

About this manual

Its purpose

The purpose of this manual is to help you get the best value from your vehicle. It can do so in several ways. It can help you decide what work must be done, even if you choose to have it done by a dealer service department or a repair shop; it provides information and procedures for routine maintenance and servicing; and it offers diagnostic and repair procedures to follow when trouble occurs.

We hope you use the manual to tackle the work yourself. For many simpler jobs, doing it yourself may be quicker than arranging an appointment to get the vehicle into a shop and making the trips to leave it and pick it up. More importantly, a lot of money can be saved by avoiding the expense the shop must pass on to you to cover its labor and overhead costs. An added benefit is the sense of satisfaction and accomplishment that you feel after doing the job yourself.

Using the manual

The manual is divided into Chapters. Each Chapter is divided into numbered Sections, which are headed in bold type between horizontal lines. Each Section consists of consecutively numbered paragraphs.

At the beginning of each numbered Section you will be referred to any illustrations which apply to the procedures in that Section. The reference numbers used in illustration captions pinpoint the pertinent Section and the Step within that Section. That is, illustration 3.2 means the illustration refers to Section 3 and Step (or paragraph) 2 within that Section.

Procedures, once described in the text, are not normally repeated. When it's necessary to refer to another Chapter, the reference will be given as Chapter and Section number. Cross references given without use of the word "Chapter" apply to Sections and/or paragraphs in the same Chapter. For example, "see Section 8" means in the same Chapter.

References to the left or right side of the vehicle assume you are sitting in the driver's seat, facing forward.

Even though we have prepared this manual with extreme care, neither the publisher nor the author can accept responsibility for any errors in, or omissions from, the information given.

NOTE

A **Note** provides information necessary to properly complete a procedure or information which will make the procedure easier to understand.

CAUTION

A **Caution** provides a special procedure or special steps which must be taken while completing the procedure where the Caution is found. Not heeding a Caution can result in damage to the assembly being worked on.

WARNING

A **Warning** provides a special procedure or special steps which must be taken while completing the procedure where the Warning is found. Not heeding a Warning can result in personal injury.

Introduction to the Chevrolet Malibu, Oldsmobile Alero and Cutlass, and Pontiac Grand Am

The Malibu and Cutlass models covered by this manual are available in four-door sedan body styles. The Oldsmobile Alero and Pontiac Grand Am are available in either two-door coupe or four-door sedan body styles.

Three engines are used in these vehicles: the 2.4-liter DOHC four-cylinder, the 3.1-liter OHV V6, and the 3.4L OHV V6 (1999 and 2000 models only). All models are equipped with Sequential Multi-port Fuel Injection (SMPFI).

The transversely mounted engine transmits power to the front wheels through an electronically controlled four-speed automatic transaxle via independent driveaxles.

The independent front suspension features coil springs with struts and lower control arms to locate the knuckle assembly at each wheel. The independent rear suspension features strut/coil spring assemblies, trailing arms and lateral link rods.

The rack-and-pinion steering unit is mounted behind the engine with power-assist as standard equipment.

The brakes are disc at the front and drums at the rear, with power assist and Anti-lock Braking System (ABS) as standard equipment.

Vehicle identification numbers

Modifications are a continuing and unpublicized part of vehicle manufacturing. Since spare parts manuals and lists are compiled on a numerical basis, the individual vehicle numbers are essential to correctly identify the component required.

Vehicle Identification Number (VIN)

This very important identification number is etched on a plate attached to the left side of the dashboard and is visible through the driver's side of the windshield (see illustration). The VIN also appears on the Vehicle Certificate of Title and Registration. It contains valuable information such as where and when the vehicle was manufactured, the model year and the body style.

VIN engine and model year codes

Two particularly important pieces of information found in the VIN are the engine code and the model year code. Counting from the left, the engine code letter designation is the 8th digit and the model year code letter designation is the 10th digit.

On the models covered by this manual the engine codes are:

T .... 2.4L four-cylinder
M .... 3.1L or 3.4L V6

On the models covered by this manual the model year codes are:

V 1997
W 1998
X 1999
Y 2000

Vehicle Safety Certification label

The Vehicle Safety Certification label is attached to the rear edge of the driver's door (see illustration). The label contains the name of the manufacturer, the month and year of production, the Gross Vehicle Weight Rating (GVWR), the Gross Axle Weight Rating (GAWR) and the certification statement.

Service Parts Identification label

Located on the underside of the spare tire cover panel, this label contains information about the options on your vehicle and the paint and trim codes (see illustration). This information is important when ordering parts or when bodywork and repainting is done.

The Vehicle Identification Number (VIN) is visible through the driver's side of the windshield

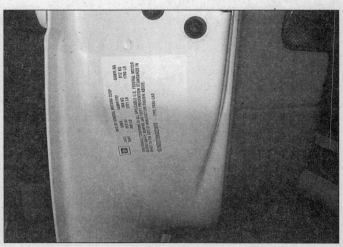

The Vehicle Safety Certification label is affixed to the driver's side door end or post

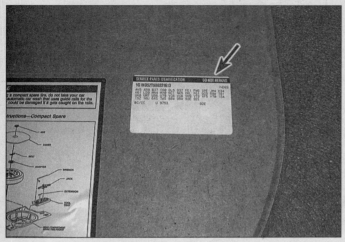

The Service Parts Identification label (arrow) contains information on options and trim/paint codes

The tire placard (label) is affixed to the rear of the driver's side rear door

Typical Engine Identification Number location (arrow) - 2.4L engine

Typical Engine Identification Number location - 3.1L and 3.4L engines

Tire placard

The tire placard (label) is attached to the rear edge of the driver's side rear door (see illustration). The label contains the tire label code, the tire sizes (front, rear and spare) tire pressures, tire speed rating, maximum vehicle capacity weight, total seating occupancy and the VIN.

Engine identification numbers

The 2.4L engine code is located on the end surface of the timing chain cover while the VIN derivative is stamped on the bottom of the engine block in front of the transaxle extension (see illustration). On the V6 engine, the codes are on the rear valve cover, just below the alternator (see illustration).

Automatic transaxle identification number

The transaxle identification information is found on a bar code label located on the front of the transaxle (see illustration).

Vehicle Emissions Control Information label

This label is found in the engine compartment. See Chapter 6 for more information on this label.

Location of the transaxle bar code label (arrow)

Buying parts

Replacement parts are available from many sources, which generally fall into one of two categories - authorized dealer parts departments and independent retail auto parts stores. Our advice concerning these parts is as follows:

Retail auto parts stores: Good auto parts stores will stock frequently needed components which wear out relatively fast, such as clutch components, exhaust systems, brake parts, tune-up parts, etc. These stores often supply new or reconditioned parts on an exchange basis, which can save a considerable amount of money. Discount auto parts stores are often very good places to buy materials and parts needed for general vehicle maintenance such as oil, grease, filters, spark plugs, belts, touch-up paint, bulbs, etc. They also usually sell tools and general accessories, have convenient hours, charge lower prices and can often be found not far from home.

Authorized dealer parts department: This is the best source for parts which are unique to the vehicle and not generally available elsewhere (such as major engine parts, transmission parts, trim pieces, etc.).

Warranty information: If the vehicle is still covered under warranty, be sure that any replacement parts purchased - regardless of the source - do not invalidate the warranty!

To be sure of obtaining the correct parts, have engine and chassis numbers available and, if possible, take the old parts along for positive identification.

Maintenance techniques, tools and working facilities

Maintenance techniques

There are a number of techniques involved in maintenance and repair that will be referred to throughout this manual. Application of these techniques will enable the home mechanic to be more efficient, better organized and capable of performing the various tasks properly, which will ensure that the repair job is thorough and complete.

Fasteners

Fasteners are nuts, bolts, studs and screws used to hold two or more parts together. There are a few things to keep in mind when working with fasteners. Almost all of them use a locking device of some type, either a lockwasher, locknut, locking tab or thread adhesive. All threaded fasteners should be clean and straight, with undamaged threads and undamaged corners on the hex head where the wrench fits. Develop the habit of replacing all damaged nuts and bolts with new ones. Special locknuts with nylon or fiber inserts can only be used once. If they are removed, they lose their locking ability and must be replaced with new ones.

Rusted nuts and bolts should be treated with a penetrating fluid to ease removal and prevent breakage. Some mechanics use turpentine in a spout-type oil can, which works quite well. After applying the rust penetrant, let it work for a few minutes before trying to loosen the nut or bolt. Badly rusted fasteners may have to be chiseled or sawed off or removed with a special nut breaker, available at tool stores.

If a bolt or stud breaks off in an assembly, it can be drilled and removed with a special tool commonly available for this purpose. Most automotive machine shops can perform this task, as well as other repair procedures, such as the repair of threaded holes that have been stripped out.

Flat washers and lockwashers, when removed from an assembly, should always be replaced exactly as removed. Replace any damaged washers with new ones. Never use a lockwasher on any soft metal surface (such as aluminum), thin sheet metal or plastic.

Fastener sizes

For a number of reasons, automobile manufacturers are making wider and wider use of metric fasteners. Therefore, it is important to be able to tell the difference between standard (sometimes called U.S. or SAE) and metric hardware, since they cannot be interchanged.

All bolts, whether standard or metric, are sized according to diameter, thread pitch and length. For example, a standard 1/2 - 13 x 1 bolt is 1/2 inch in diameter, has 13 threads per inch and is 1 inch long. An M12 - 1.75 x 25 metric bolt is 12 mm in diameter, has a thread pitch of 1.75 mm (the distance between threads) and is 25 mm long. The two bolts are nearly identical, and easily confused, but they are not interchangeable.

In addition to the differences in diameter, thread pitch and length, metric and standard bolts can also be distinguished by examining the bolt heads. To begin with, the distance across the flats on a standard bolt head is measured in inches, while the same dimension on a metric bolt is sized in millimeters (the same is true for nuts). As a result, a standard wrench should not be used on a metric bolt and a metric wrench should not be used on a standard bolt. Also, most standard bolts have slashes radiating out from the center of the head to denote the grade or strength of the bolt, which is an indication of the amount of torque that can be applied to it. The greater the number of slashes, the greater the strength of the bolt. Grades 0 through 5 are commonly used on automobiles. Metric bolts have a property class (grade) number, rather than a slash, molded into their heads to indicate bolt strength. In this case, the higher the number, the stronger the bolt. Property class numbers 8.8, 9.8 and 10.9 are commonly used on automobiles.

Strength markings can also be used to distinguish standard hex nuts from metric hex nuts. Many standard nuts have dots stamped into one side, while metric nuts are marked with a number. The greater the number of dots, or the higher the number, the greater the strength of the nut.

Metric studs are also marked on their ends according to property class (grade). Larger studs are numbered (the same as metric bolts), while smaller studs carry a geometric code to denote grade.

It should be noted that many fasteners, especially Grades 0 through 2, have no distinguishing marks on them. When such is the case, the only way to determine whether it is standard or metric is to measure the thread pitch or compare it to a known fastener of the same size.

Standard fasteners are often referred to as SAE, as opposed to metric. However, it should be noted that SAE technically refers to a non-metric fine thread fastener only. Coarse thread non-metric fasteners are referred to as USS sizes.

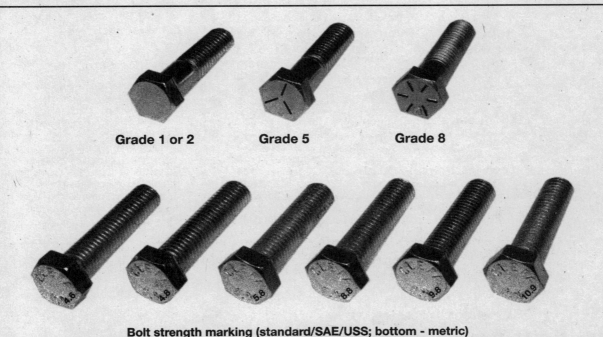

Grade 1 or 2 **Grade 5** **Grade 8**

Bolt strength marking (standard/SAE/USS; bottom - metric)

Grade	Identification	Grade	Identification
Hex Nut Grade 5	3 Dots	**Hex Nut Property Class 9**	Arabic 9
Hex Nut Grade 8	6 Dots	**Hex Nut Property Class 10**	Arabic 10

Standard hex nut strength markings   **Metric hex nut strength markings**

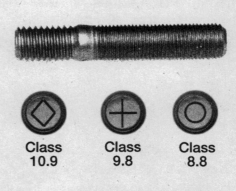

Class 10.9 **Class 9.8** **Class 8.8**

Metric stud strength markings

Since fasteners of the same size (both standard and metric) may have different strength ratings, be sure to reinstall any bolts, studs or nuts removed from your vehicle in their original locations. Also, when replacing a fastener with a new one, make sure that the new one has a strength rating equal to or greater than the original.

Tightening sequences and procedures

Most threaded fasteners should be tightened to a specific torque value (torque is the twisting force applied to a threaded com-ponent such as a nut or bolt). Overtightening the fastener can weaken it and cause it to break, while undertightening can cause it to eventually come loose. Bolts, screws and studs, depending on the material they are made of and their thread diameters, have specific torque values, many of which are noted in the Specifications at the beginning of each Chapter. Be sure to follow the torque recommendations closely. For fasteners not assigned a specific torque, a general torque value chart is presented here as a guide. These torque values are for dry (unlubricated) fasteners threaded into steel or cast iron (not aluminum). As was previously mentioned, the size and grade of a fastener determine the amount of torque that can safely be applied to it. The figures listed here are approximate for Grade 2 and Grade 3 fasteners. Higher grades can tolerate higher torque values.

Fasteners laid out in a pattern, such as cylinder head bolts, oil pan bolts, differential cover bolts, etc., must be loosened or tightened in sequence to avoid warping the component. This sequence will normally be shown in the appropriate Chapter. If a specific pattern is not given, the following procedures can be used to prevent warping.

Metric thread sizes	Ft-lbs	Nm
M-6	6 to 9	9 to 12
M-8	14 to 21	19 to 28
M-10	28 to 40	38 to 54
M-12	50 to 71	68 to 96
M-14	80 to 140	109 to 154

Pipe thread sizes	Ft-lbs	Nm
1/8	5 to 8	7 to 10
1/4	12 to 18	17 to 24
3/8	22 to 33	30 to 44
1/2	25 to 35	34 to 47

U.S. thread sizes	Ft-lbs	Nm
1/4 - 20	6 to 9	9 to 12
5/16 - 18	12 to 18	17 to 24
5/16 - 24	14 to 20	19 to 27
3/8 - 16	22 to 32	30 to 43
3/8 - 24	27 to 38	37 to 51
7/16 - 14	40 to 55	55 to 74
7/16 - 20	40 to 60	55 to 81
1/2 - 13	55 to 80	75 to 108

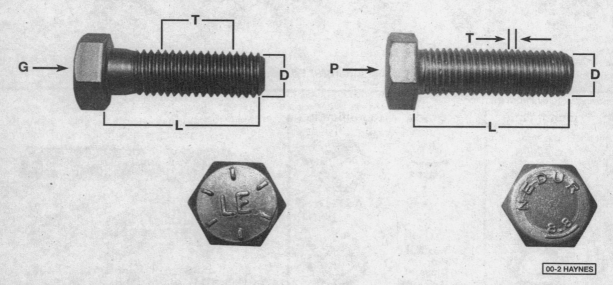

00-2 HAYNES

Standard (SAE and USS) bolt dimensions/grade marks

    G   Grade marks (bolt strength)
    L   Length (in inches)
    T   Thread pitch (number of threads per inch)
    D   Nominal diameter (in inches)

Metric bolt dimensions/grade marks

    P   Property class (bolt strength)
    L   Length (in millimeters)
    T   Thread pitch (distance between threads in millimeters)
    D   Diameter

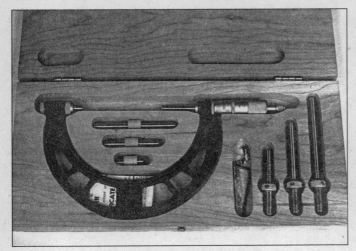

Micrometer set

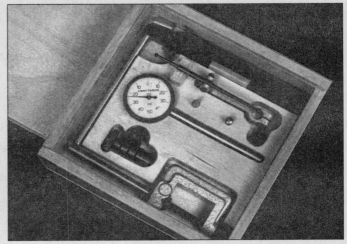

Dial indicator set

Initially, the bolts or nuts should be assembled finger-tight only. Next, they should be tightened one full turn each, in a criss-cross or diagonal pattern. After each one has been tightened one full turn, return to the first one and tighten them all one-half turn, following the same pattern. Finally, tighten each of them one-quarter turn at a time until each fastener has been tightened to the proper torque. To loosen and remove the fasteners, the procedure would be reversed.

Component disassembly

Component disassembly should be done with care and purpose to help ensure that the parts go back together properly. Always keep track of the sequence in which parts are removed. Make note of special characteristics or marks on parts that can be installed more than one way, such as a grooved thrust washer on a shaft. It is a good idea to lay the disassembled parts out on a clean surface in the order that they were removed. It may also be helpful to make sketches or take instant photos of components before removal.

When removing fasteners from a component, keep track of their locations. Sometimes threading a bolt back in a part, or putting the washers and nut back on a stud, can prevent mix-ups later. If nuts and bolts cannot be returned to their original locations, they should be kept in a compartmented box or a series of small boxes. A cupcake or muffin tin is ideal for this purpose, since each cavity can hold the bolts and nuts from a particular area (i.e. oil pan bolts, valve cover bolts, engine mount bolts, etc.). A pan of this type is especially helpful when working on assemblies with very small parts, such as the carburetor, alternator, valve train or interior dash and trim pieces. The cavities can be marked with paint or tape to identify the contents.

Whenever wiring looms, harnesses or connectors are separated, it is a good idea to identify the two halves with numbered pieces of masking tape so they can be easily reconnected.

Gasket sealing surfaces

Throughout any vehicle, gaskets are used to seal the mating surfaces between two parts and keep lubricants, fluids, vacuum or pressure contained in an assembly.

Many times these gaskets are coated with a liquid or paste-type gasket sealing compound before assembly. Age, heat and pressure can sometimes cause the two parts to stick together so tightly that they are very difficult to separate. Often, the assembly can be loosened by striking it with a soft-face hammer near the mating surfaces. A regular hammer can be used if a block of wood is placed between the hammer and the part. Do not hammer on cast parts or parts that could be easily damaged. With any particularly stubborn part, always recheck to make sure that every fastener has been removed.

Avoid using a screwdriver or bar to pry apart an assembly, as they can easily mar the gasket sealing surfaces of the parts, which must remain smooth. If prying is absolutely necessary, use an old broom handle, but keep in mind that extra clean up will be necessary if the wood splinters.

After the parts are separated, the old gasket must be carefully scraped off and the gasket surfaces cleaned. Stubborn gasket material can be soaked with rust penetrant or treated with a special chemical to soften it so it can be easily scraped off. A scraper can be fashioned from a piece of copper tubing by flattening and sharpening one end. Copper is recommended because it is usually softer than the surfaces to be scraped, which reduces the chance of gouging the part. Some gaskets can be removed with a wire brush, but regardless of the method used, the mating surfaces must be left clean and smooth. If for some reason the gasket surface is gouged, then a gasket sealer thick enough to fill scratches will have to be used during reassembly of the components. For most applications, a non-drying (or semi-drying) gasket sealer should be used.

Hose removal tips

Warning: *If the vehicle is equipped with air conditioning, do not disconnect any of the A/C hoses without first having the system depressurized by a dealer service department or a service station.*

Hose removal precautions closely parallel gasket removal precautions. Avoid scratching or gouging the surface that the hose mates against or the connection may leak. This is especially true for radiator hoses. Because of various chemical reactions, the rubber in hoses can bond itself to the metal spigot that the hose fits over. To remove a hose, first loosen the hose clamps that secure it to the spigot. Then, with slip-joint pliers, grab the hose at the clamp and rotate it around the spigot. Work it back and forth until it is completely free, then pull it off. Silicone or other lubricants will ease removal if they can be applied between the hose and the outside of the spigot. Apply the same lubricant to the inside of the hose and the outside of the spigot to simplify installation.

As a last resort (and if the hose is to be replaced with a new one anyway), the rubber can be slit with a knife and the hose peeled from the spigot. If this must be done, be careful that the metal connection is not damaged.

If a hose clamp is broken or damaged, do not reuse it. Wire-type clamps usually weaken with age, so it is a good idea to replace them with screw-type clamps whenever a hose is removed.

Tools

A selection of good tools is a basic requirement for anyone who plans to maintain and repair his or her own vehicle. For the owner who has few tools, the initial investment might seem high, but when compared to the spiraling costs of professional auto maintenance and repair, it is a wise one.

To help the owner decide which tools are needed to perform the tasks detailed in this manual, the following tool lists are offered: *Maintenance and minor repair,*

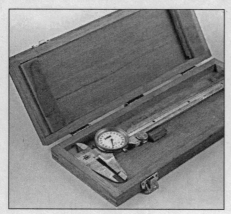

Dial caliper

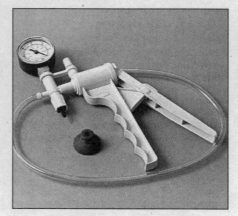

Hand-operated vacuum pump

Timing light

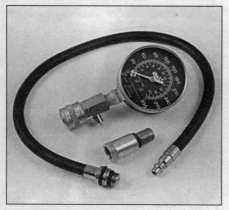

Compression gauge with spark plug hole adapter

Damper/steering wheel puller

General purpose puller

Hydraulic lifter removal tool

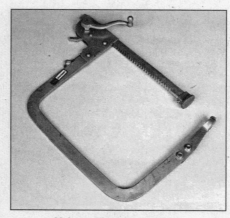

Valve spring compressor

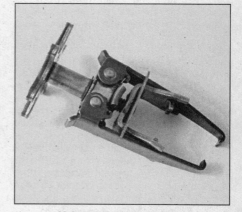

Valve spring compressor

Repair/overhaul and *Special.*

The newcomer to practical mechanics should start off with the *maintenance and minor repair* tool kit, which is adequate for the simpler jobs performed on a vehicle. Then, as confidence and experience grow, the owner can tackle more difficult tasks, buying additional tools as they are needed. Eventually the basic kit will be expanded into the *repair and overhaul* tool set. Over a period of time, the experienced do-it-yourselfer will assemble a tool set complete enough for most repair and overhaul procedures and will add tools from the special category when it is felt that the expense is justified by the frequency of use.

Maintenance and minor repair tool kit

The tools in this list should be considered the minimum required for performance of routine maintenance, servicing and minor repair work. We recommend the purchase of combination wrenches (box-end and open-end combined in one wrench). While more expensive than open end wrenches, they offer the advantages of both types of wrench.

*Combination wrench set (1/4-inch to
    1 inch or 6 mm to 19 mm)
Adjustable wrench, 8 inch
Spark plug wrench with rubber insert
Spark plug gap adjusting tool
Feeler gauge set*

Ridge reamer

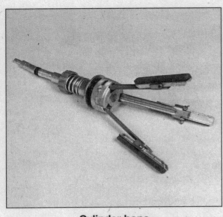

Piston ring groove cleaning tool

Ring removal/installation tool

Ring compressor

Cylinder hone

Brake hold-down spring tool

Brake bleeder wrench
Standard screwdriver (5/16-inch x
   6 inch)
Phillips screwdriver (No. 2 x 6 inch)
Combination pliers - 6 inch
Hacksaw and assortment of blades
Tire pressure gauge
Grease gun
Oil can
Fine emery cloth
Wire brush
Battery post and cable cleaning tool
Oil filter wrench
Funnel (medium size)
Safety goggles
Jackstands (2)
Drain pan

Note: *If basic tune-ups are going to be part of routine maintenance, it will be necessary to purchase a good quality stroboscopic timing light and combination tachometer/dwell meter. Although they are included in the list of special tools, it is mentioned here because they are absolutely necessary for tuning most vehicles properly.*

Repair and overhaul tool set

These tools are essential for anyone who plans to perform major repairs and are in addition to those in the maintenance and minor repair tool kit. Included is a comprehensive set of sockets which, though expensive, are invaluable because of their versatility, especially when various extensions and drives are available. We recommend the 1/2-inch drive over the 3/8-inch drive. Although the larger drive is bulky and more expensive, it has the capacity of accepting a very wide range of large sockets. Ideally, however, the mechanic should have a 3/8-inch drive set and a 1/2-inch drive set.

Socket set(s)
Reversible ratchet
Extension - 10 inch
Universal joint
Torque wrench (same size drive as
   sockets)
Ball peen hammer - 8 ounce
Soft-face hammer (plastic/rubber)
Standard screwdriver (1/4-inch x 6 inch)
Standard screwdriver (stubby -
   5/16-inch)
Phillips screwdriver (No. 3 x 8 inch)
Phillips screwdriver (stubby - No. 2)
Pliers - vise grip
Pliers - lineman's
Pliers - needle nose
Pliers - snap-ring (internal and external)
Cold chisel - 1/2-inch
Scribe

Scraper (made from flattened copper
   tubing)
Centerpunch
Pin punches (1/16, 1/8, 3/16-inch)
Steel rule/straightedge - 12 inch
Allen wrench set (1/8 to 3/8-inch or
   4 mm to 10 mm)
A selection of files
Wire brush (large)
Jackstands (second set)
Jack (scissor or hydraulic type)

Note: *Another tool which is often useful is an electric drill with a chuck capacity of 3/8-inch and a set of good quality drill bits.*

Special tools

The tools in this list include those which are not used regularly, are expensive to buy, or which need to be used in accordance with their manufacturer's instructions. Unless these tools will be used frequently, it is not very economical to purchase many of them. A consideration would be to split the cost and use between yourself and a friend or friends. In addition, most of these tools can be obtained from a tool rental shop on a temporary basis.

This list primarily contains only those tools and instruments widely available to the public, and not those special tools produced

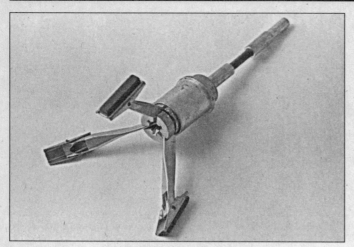

Brake cylinder hone

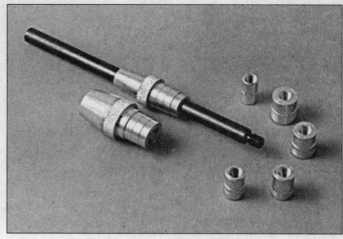

Clutch plate alignment tool

by the vehicle manufacturer for distribution to dealer service departments. Occasionally, references to the manufacturer's special tools are included in the text of this manual. Generally, an alternative method of doing the job without the special tool is offered. However, sometimes there is no alternative to their use. Where this is the case, and the tool cannot be purchased or borrowed, the work should be turned over to the dealer service department or an automotive repair shop.

Valve spring compressor
Piston ring groove cleaning tool
Piston ring compressor
Piston ring installation tool
Cylinder compression gauge
Cylinder ridge reamer
Cylinder surfacing hone
Cylinder bore gauge
Micrometers and/or dial calipers
Hydraulic lifter removal tool
Balljoint separator
Universal-type puller
Impact screwdriver
Dial indicator set
Stroboscopic timing light (inductive pick-up)

Hand operated vacuum/pressure pump
Tachometer/dwell meter
Universal electrical multimeter
Cable hoist
Brake spring removal and installation tools
Floor jack

Buying tools

For the do-it-yourselfer who is just starting to get involved in vehicle maintenance and repair, there are a number of options available when purchasing tools. If maintenance and minor repair is the extent of the work to be done, the purchase of individual tools is satisfactory. If, on the other hand, extensive work is planned, it would be a good idea to purchase a modest tool set from one of the large retail chain stores. A set can usually be bought at a substantial savings over the individual tool prices, and they often come with a tool box. As additional tools are needed, add-on sets, individual tools and a larger tool box can be purchased to expand the tool selection. Building a tool set gradually allows the cost of the tools to be spread over a longer period of time and gives the

mechanic the freedom to choose only those tools that will actually be used.

Tool stores will often be the only source of some of the special tools that are needed, but regardless of where tools are bought, try to avoid cheap ones, especially when buying screwdrivers and sockets, because they won't last very long. The expense involved in replacing cheap tools will eventually be greater than the initial cost of quality tools.

Care and maintenance of tools

Good tools are expensive, so it makes sense to treat them with respect. Keep them clean and in usable condition and store them properly when not in use. Always wipe off any dirt, grease or metal chips before putting them away. Never leave tools lying around in the work area. Upon completion of a job, always check closely under the hood for tools that may have been left there so they won't get lost during a test drive.

Some tools, such as screwdrivers, pliers, wrenches and sockets, can be hung on a panel mounted on the garage or workshop wall, while others should be kept in a tool box or tray. Measuring instruments, gauges, meters, etc. must be carefully stored where they cannot be damaged by weather or impact from other tools.

When tools are used with care and stored properly, they will last a very long time. Even with the best of care, though, tools will wear out if used frequently. When a tool is damaged or worn out, replace it. Subsequent jobs will be safer and more enjoyable if you do.

How to repair damaged threads

Sometimes, the internal threads of a nut or bolt hole can become stripped, usually from overtightening. Stripping threads is an all-too-common occurrence, especially when working with aluminum parts, because aluminum is so soft that it easily strips out.

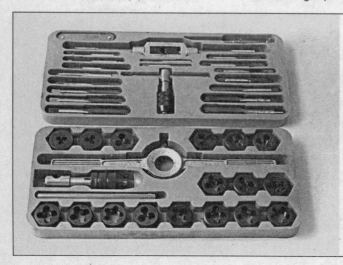

Tap and die set

Usually, external or internal threads are only partially stripped. After they've been cleaned up with a tap or die, they'll still work. Sometimes, however, threads are badly damaged. When this happens, you've got three choices:

1) Drill and tap the hole to the next suitable oversize and install a larger diameter bolt, screw or stud.

2) Drill and tap the hole to accept a threaded plug, then drill and tap the plug to the original screw size. You can also buy a plug already threaded to the original size. Then you simply drill a hole to the specified size, then run the threaded plug into the hole with a bolt and jam nut. Once the plug is fully seated, remove the jam nut and bolt.

3) The third method uses a patented thread repair kit like Heli-Coil or Slimsert. These easy-to-use kits are designed to repair damaged threads in straight-through holes and blind holes. Both are available as kits which can handle a variety of sizes and thread patterns. Drill the hole, then tap it with the special included tap. Install the Heli-Coil and the hole is back to its original diameter and thread pitch.

Regardless of which method you use, be sure to proceed calmly and carefully. A little impatience or carelessness during one of these relatively simple procedures can ruin your whole day's work and cost you a bundle if you wreck an expensive part.

Working facilities

Not to be overlooked when discussing tools is the workshop. If anything more than routine maintenance is to be carried out, some sort of suitable work area is essential.

It is understood, and appreciated, that many home mechanics do not have a good workshop or garage available, and end up removing an engine or doing major repairs outside. It is recommended, however, that the overhaul or repair be completed under the cover of a roof.

A clean, flat workbench or table of comfortable working height is an absolute necessity. The workbench should be equipped with a vise that has a jaw opening of at least four inches.

As mentioned previously, some clean, dry storage space is also required for tools, as well as the lubricants, fluids, cleaning solvents, etc. which soon become necessary.

Sometimes waste oil and fluids, drained from the engine or cooling system during normal maintenance or repairs, present a disposal problem. To avoid pouring them on the ground or into a sewage system, pour the used fluids into large containers, seal them with caps and take them to an authorized disposal site or recycling center. Plastic jugs, such as old antifreeze containers, are ideal for this purpose.

Always keep a supply of old newspapers and clean rags available. Old towels are excellent for mopping up spills. Many mechanics use rolls of paper towels for most work because they are readily available and disposable. To help keep the area under the vehicle clean, a large cardboard box can be cut open and flattened to protect the garage or shop floor.

Whenever working over a painted surface, such as when leaning over a fender to service something under the hood, always cover it with an old blanket or bedspread to protect the finish. Vinyl covered pads, made especially for this purpose, are available at auto parts stores.

Anti-theft audio system

General information

1　Some of these models are equipped with THEFTLOCK audio systems, which include an anti-theft feature that will render the stereo inoperative if stolen. If the power source to the stereo is cut with the anti-theft feature activated, the stereo will be inoperative. Even if the power source is immediately re-connected, the stereo will not function.

2　If your vehicle is equipped with this anti-theft system, do not disconnect the battery, remove the stereo or disconnect related components unless you have either turned off the feature or have the individual ID (code) number for the stereo.

Disabling the anti-theft feature

3　Press the stereo's 1 and 4 buttons at the same time for five seconds with the ignition on and the radio power off. The display will show SEC, indicating the unit is in the secure mode (anti-theft feature enabled).

4　Press the MN button. The display will show "000".

5　Press the MN button until the last two numbers are the same as your secret code.

6　Press HR until the first one or two numbers displayed match your code. The numbers will be displayed as entered.

7　Press AM/FM. If the display shows "_ _ _" you have successfully disabled the anti-theft feature. If SEC is displayed, the code you entered was incorrect and the anti-theft feature is still enabled.

Unlocking the stereo after a power loss

8　When the power is restored to the stereo, the stereo won't turn on and LOC will appear on the display. Enter your ID code as follows; pause no more than 15 seconds between Steps.

9　Turn the ignition switch to ON, but leave the stereo off.

10　Press the MN button. "000" should display.

11　Press the MN button to make the last two numbers match your code, then release the button.

12　Press the HR button until the first one or two numbers match your code.

13　Press AM/FM. SEC should appear indicating the stereo is unlocked. If LOC appears, the numbers you entered were not correct and the stereo is still inoperative.

Booster battery (jump) starting

Observe the following precautions when using a booster battery to start a vehicle:

a) *Before connecting the booster battery, make sure the ignition switch is in the Off position.*
b) *Turn off the lights, heater and other electrical loads.*
c) *Your eyes should be shielded. Safety goggles are a good idea.*
d) *Make sure the booster battery is the same voltage as the dead one in the vehicle.*
e) *The two vehicles MUST NOT TOUCH each other.*
f) *Make sure the transmission is in Neutral (manual transaxle) or Park (automatic transaxle).*
g) *If the booster battery is not a maintenance-free type, remove the vent caps and lay a cloth over the vent holes.*

Connect the red jumper cable to the positive (+) terminals of each battery.

Connect one end of the black cable to the negative (-) terminal of the booster battery. The other end of this cable should be connected to a good ground on the engine block **(see illustration)**. Make sure the cable will not come into contact with the fan, drivebelts or other moving parts of the engine.

Start the engine using the booster battery, then, with the engine running at idle speed, disconnect the jumper cables in the reverse order of connection.

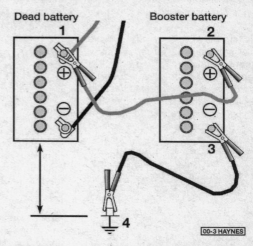

Make the booster battery cable connections in the numerical order shown (note that the negative cable of the booster battery is NOT attached to the negative terminal of the dead battery)

Jacking and towing

Jacking

Warning: *The jack supplied with the vehicle should only be used for changing a tire or placing jackstands under the frame. Never work under the vehicle or start the engine while this jack is being used as the only means of support.*

The vehicle should be on level ground. Place the shift lever in Park. Block the wheel diagonally opposite the wheel being changed. Set the parking brake.

Remove the spare tire and jack from stowage. Remove the wheel cover (on steel wheels) with the tapered end of the lug nut wrench by inserting and twisting the handle and then prying against the back of the wheel cover. On aluminum wheels, use a small screwdriver to pry out the center hub cap. Loosen the wheel lug nuts about 1/4-to-1/2 turn each.

Place the scissors-type jack under the side of the vehicle and adjust the jack height until it fits in the notch in the vertical rocker panel flange nearest the wheel to be changed. There is a front and rear jacking point on each side of the vehicle **(see illustration)**. Jacking instructions are usually located on the underside of the spare tire cover.

Turn the jack handle clockwise until the tire clears the ground. Remove the lug nuts and pull the wheel off. Replace it with the spare.

Install the lug nuts with the beveled edges facing in. Tighten them snugly. Don't attempt to tighten them completely until the vehicle is lowered or it could slip off the jack.

The jack fits over the rocker panel flange (arrows indicate the two jacking points on each side of the vehicle, indicated by a notch in the rocker panel flange)

Turn the jack handle counterclockwise to lower the vehicle. Remove the jack and tighten the lug nuts in a diagonal pattern.

Install the wheel cover (center hub cap on aluminum wheels) and be sure it's snapped into place all the way around.

Stow the tire, jack and wrench. Unblock the wheels.

Towing

As a general rule, the vehicle should be towed with the front (drive) wheels off the ground. If they can't be raised, place them on a dolly. The ignition key must be in the ACC position, since the steering lock mechanism isn't strong enough to hold the front wheels straight while towing.

The vehicle can be towed from the front only with all four wheels on the ground, provided that speeds don't exceed 30 mph and the distance is not over 40 miles. Before towing, check the transmission fluid level (see Chapter 1). If the level is below the HOT line on the dipstick, add fluid or use a towing dolly. **Caution:** *Never tow the vehicle from the rear with the front wheels on the ground.*

Equipment specifically designed for towing should be used. It should be attached to the main structural members of the vehicle, not the bumpers or brackets.

Safety is a major consideration when towing and all applicable state and local laws must be obeyed. A safety chain system must be used at all times.

Automotive chemicals and lubricants

A number of automotive chemicals and lubricants are available for use during vehicle maintenance and repair. They include a wide variety of products ranging from cleaning solvents and degreasers to lubricants and protective sprays for rubber, plastic and vinyl.

Cleaners

Carburetor cleaner and choke cleaner is a strong solvent for gum, varnish and carbon. Most carburetor cleaners leave a dry-type lubricant film which will not harden or gum up. Because of this film it is not recommended for use on electrical components.

Brake system cleaner is used to remove grease and brake fluid from the brake system, where clean surfaces are absolutely necessary. It leaves no residue and often eliminates brake squeal caused by contaminants.

Electrical cleaner removes oxidation, corrosion and carbon deposits from electrical contacts, restoring full current flow. It can also be used to clean spark plugs, carburetor jets, voltage regulators and other parts where an oil-free surface is desired.

Demoisturants remove water and moisture from electrical components such as alternators, voltage regulators, electrical connectors and fuse blocks. They are non-conductive, non-corrosive and non-flammable.

Degreasers are heavy-duty solvents used to remove grease from the outside of the engine and from chassis components. They can be sprayed or brushed on and, depending on the type, are rinsed off either with water or solvent.

Lubricants

Motor oil is the lubricant formulated for use in engines. It normally contains a wide variety of additives to prevent corrosion and reduce foaming and wear. Motor oil comes in various weights (viscosity ratings) from 0 to 50. The recommended weight of the oil depends on the season, temperature and the demands on the engine. Light oil is used in cold climates and under light load conditions. Heavy oil is used in hot climates and where high loads are encountered. Multi-viscosity oils are designed to have characteristics of both light and heavy oils and are available in a number of weights from 5W-20 to 20W-50.

Gear oil is designed to be used in differentials, manual transmissions and other areas where high-temperature lubrication is required.

Chassis and wheel bearing grease is a heavy grease used where increased loads and friction are encountered, such as for wheel bearings, balljoints, tie-rod ends and universal joints.

High-temperature wheel bearing grease is designed to withstand the extreme temperatures encountered by wheel bearings

in disc brake equipped vehicles. It usually contains molybdenum disulfide (moly), which is a dry-type lubricant.

White grease is a heavy grease for metal-to-metal applications where water is a problem. White grease stays soft under both low and high temperatures (usually from -100 to +190-degrees F), and will not wash off or dilute in the presence of water.

Assembly lube is a special extreme pressure lubricant, usually containing moly, used to lubricate high-load parts (such as main and rod bearings and cam lobes) for initial start-up of a new engine. The assembly lube lubricates the parts without being squeezed out or washed away until the engine oiling system begins to function.

Silicone lubricants are used to protect rubber, plastic, vinyl and nylon parts.

Graphite lubricants are used where oils cannot be used due to contamination problems, such as in locks. The dry graphite will lubricate metal parts while remaining uncontaminated by dirt, water, oil or acids. It is electrically conductive and will not foul electrical contacts in locks such as the ignition switch.

Moly penetrants loosen and lubricate frozen, rusted and corroded fasteners and prevent future rusting or freezing.

Heat-sink grease is a special electrically non-conductive grease that is used for mounting electronic ignition modules where it is essential that heat is transferred away from the module.

Sealants

RTV sealant is one of the most widely used gasket compounds. Made from silicone, RTV is air curing, it seals, bonds, waterproofs, fills surface irregularities, remains flexible, doesn't shrink, is relatively easy to remove, and is used as a supplementary sealer with almost all low and medium temperature gaskets.

Anaerobic sealant is much like RTV in that it can be used either to seal gaskets or to form gaskets by itself. It remains flexible, is solvent resistant and fills surface imperfections. The difference between an anaerobic sealant and an RTV-type sealant is in the curing. RTV cures when exposed to air, while an anaerobic sealant cures only in the absence of air. This means that an anaerobic sealant cures only after the assembly of parts, sealing them together.

Thread and pipe sealant is used for sealing hydraulic and pneumatic fittings and vacuum lines. It is usually made from a Teflon compound, and comes in a spray, a paint-on liquid and as a wrap-around tape.

Chemicals

Anti-seize compound prevents seizing, galling, cold welding, rust and corrosion in

fasteners. High-temperature anti-seize, usually made with copper and graphite lubricants, is used for exhaust system and exhaust manifold bolts.

Anaerobic locking compounds are used to keep fasteners from vibrating or working loose and cure only after installation, in the absence of air. Medium strength locking compound is used for small nuts, bolts and screws that may be removed later. High-strength locking compound is for large nuts, bolts and studs which aren't removed on a regular basis.

Oil additives range from viscosity index improvers to chemical treatments that claim to reduce internal engine friction. It should be noted that most oil manufacturers caution against using additives with their oils.

Gas additives perform several functions, depending on their chemical makeup. They usually contain solvents that help dissolve gum and varnish that build up on carburetor, fuel injection and intake parts. They also serve to break down carbon deposits that form on the inside surfaces of the combustion chambers. Some additives contain upper cylinder lubricants for valves and piston rings, and others contain chemicals to remove condensation from the gas tank.

Miscellaneous

Brake fluid is specially formulated hydraulic fluid that can withstand the heat and pressure encountered in brake systems. Care must be taken so this fluid does not come in contact with painted surfaces or plastics. An opened container should always be resealed to prevent contamination by water or dirt.

Weatherstrip adhesive is used to bond weatherstripping around doors, windows and trunk lids. It is sometimes used to attach trim pieces.

Undercoating is a petroleum-based, tar-like substance that is designed to protect metal surfaces on the underside of the vehicle from corrosion. It also acts as a sound-deadening agent by insulating the bottom of the vehicle.

Waxes and polishes are used to help protect painted and plated surfaces from the weather. Different types of paint may require the use of different types of wax and polish. Some polishes utilize a chemical or abrasive cleaner to help remove the top layer of oxidized (dull) paint on older vehicles. In recent years many non-wax polishes that contain a wide variety of chemicals such as polymers and silicones have been introduced. These non-wax polishes are usually easier to apply and last longer than conventional waxes and polishes.

Conversion factors

Length (distance)

Inches (in)	X	25.4	= Millimetres (mm)	X 0.0394	= Inches (in)
Feet (ft)	X	0.305	= Metres (m)	X 3.281	= Feet (ft)
Miles	X	1.609	= Kilometres (km)	X 0.621	= Miles

Volume (capacity)

Cubic inches (cu in; in³)	X	16.387	= Cubic centimetres (cc; cm³)	X 0.061	= Cubic inches (cu in; in³)
Imperial pints (Imp pt)	X	0.568	= Litres (l)	X 1.76	= Imperial pints (Imp pt)
Imperial quarts (Imp qt)	X	1.137	= Litres (l)	X 0.88	= Imperial quarts (Imp qt)
Imperial quarts (Imp qt)	X	1.201	= US quarts (US qt)	X 0.833	= Imperial quarts (Imp qt)
US quarts (US qt)	X	0.946	= Litres (l)	X 1.057	= US quarts (US qt)
Imperial gallons (Imp gal)	X	4.546	= Litres (l)	X 0.22	= Imperial gallons (Imp gal)
Imperial gallons (Imp gal)	X	1.201	= US gallons (US gal)	X 0.833	= Imperial gallons (Imp gal)
US gallons (US gal)	X	3.785	= Litres (l)	X 0.264	= US gallons (US gal)

Mass (weight)

Ounces (oz)	X	28.35	= Grams (g)	X 0.035	= Ounces (oz)
Pounds (lb)	X	0.454	= Kilograms (kg)	X 2.205	= Pounds (lb)

Force

Ounces-force (ozf; oz)	X	0.278	= Newtons (N)	X 3.6	= Ounces-force (ozf; oz)
Pounds-force (lbf; lb)	X	4.448	= Newtons (N)	X 0.225	= Pounds-force (lbf; lb)
Newtons (N)	X	0.1	= Kilograms-force (kgf; kg)	X 9.81	= Newtons (N)

Pressure

Pounds-force per square inch (psi; lbf/in²; lb/in²)	X	0.070	= Kilograms-force per square centimetre (kgf/cm²; kg/cm²)	X 14.223	= Pounds-force per square inch (psi; lbf/in²; lb/in²)
Pounds-force per square inch (psi; lbf/in²; lb/in²)	X	0.068	= Atmospheres (atm)	X 14.696	= Pounds-force per square inch (psi; lbf/in²; lb/in²)
Pounds-force per square inch (psi; lbf/in²; lb/in²)	X	0.069	= Bars	X 14.5	= Pounds-force per square inch (psi; lbf/in²; lb/in²)
Pounds-force per square inch (psi; lbf/in²; lb/in²)	X	6.895	= Kilopascals (kPa)	X 0.145	= Pounds-force per square inch (psi; lbf/in²; lb/in²)
Kilopascals (kPa)	X	0.01	= Kilograms-force per square centimetre (kgf/cm²; kg/cm²)	X 98.1	= Kilopascals (kPa)

Torque (moment of force)

Pounds-force inches (lbf in; lb in)	X	1.152	= Kilograms-force centimetre (kgf cm; kg cm)	X 0.868	= Pounds-force inches (lbf in; lb in)
Pounds-force inches (lbf in; lb in)	X	0.113	= Newton metres (Nm)	X 8.85	= Pounds-force inches (lbf in; lb in)
Pounds-force inches (lbf in; lb in)	X	0.083	= Pounds-force feet (lbf ft; lb ft)	X 12	= Pounds-force inches (lbf in; lb in)
Pounds-force feet (lbf ft; lb ft)	X	0.138	= Kilograms-force metres (kgf m; kg m)	X 7.233	= Pounds-force feet (lbf ft; lb ft)
Pounds-force feet (lbf ft; lb ft)	X	1.356	= Newton metres (Nm)	X 0.738	= Pounds-force feet (lbf ft; lb ft)
Newton metres (Nm)	X	0.102	= Kilograms-force metres (kgf m; kg m)	X 9.804	= Newton metres (Nm)

Vacuum

Inches mercury (in. Hg)	X	3.377	= Kilopascals (kPa)	X 0.2961	= Inches mercury
Inches mercury (in. Hg)	X	25.4	= Millimeters mercury (mm Hg)	X 0.0394	= Inches mercury

Power

Horsepower (hp)	X	745.7	= Watts (W)	X 0.0013	= Horsepower (hp)

Velocity (speed)

Miles per hour (miles/hr; mph)	X	1.609	= Kilometres per hour (km/hr; kph)	X 0.621	= Miles per hour (miles/hr; mph)

Fuel consumption*

Miles per gallon, Imperial (mpg)	X	0.354	= Kilometres per litre (km/l)	X 2.825	= Miles per gallon, Imperial (mpg)
Miles per gallon, US (mpg)	X	0.425	= Kilometres per litre (km/l)	X 2.352	= Miles per gallon, US (mpg)

Temperature

Degrees Fahrenheit = (°C x 1.8) + 32

Degrees Celsius (Degrees Centigrade; °C) = (°F - 32) x 0.56

*It is common practice to convert from miles per gallon (mpg) to litres/100 kilometres (l/100km), where mpg (Imperial) x l/100 km = 282 and mpg (US) x l/100 km = 235

Safety first!

Regardless of how enthusiastic you may be about getting on with the job at hand, take the time to ensure that your safety is not jeopardized. A moment's lack of attention can result in an accident, as can failure to observe certain simple safety precautions. The possibility of an accident will always exist, and the following points should not be considered a comprehensive list of all dangers. Rather, they are intended to make you aware of the risks and to encourage a safety conscious approach to all work you carry out on your vehicle.

Essential DOs and DON'Ts

DON'T rely on a jack when working under the vehicle. Always use approved jackstands to support the weight of the vehicle and place them under the recommended lift or support points.

DON'T attempt to loosen extremely tight fasteners (i.e. wheel lug nuts) while the vehicle is on a jack - it may fall.

DON'T start the engine without first making sure that the transmission is in Neutral (or Park where applicable) and the parking brake is set.

DON'T remove the radiator cap from a hot cooling system - let it cool or cover it with a cloth and release the pressure gradually.

DON'T attempt to drain the engine oil until you are sure it has cooled to the point that it will not burn you.

DON'T touch any part of the engine or exhaust system until it has cooled sufficiently to avoid burns.

DON'T siphon toxic liquids such as gasoline, antifreeze and brake fluid by mouth, or allow them to remain on your skin.

DON'T inhale brake lining dust - it is potentially hazardous (see *Asbestos* below).

DON'T allow spilled oil or grease to remain on the floor - wipe it up before someone slips on it.

DON'T use loose fitting wrenches or other tools which may slip and cause injury.

DON'T push on wrenches when loosening or tightening nuts or bolts. Always try to pull the wrench toward you. If the situation calls for pushing the wrench away, push with an open hand to avoid scraped knuckles if the wrench should slip.

DON'T attempt to lift a heavy component alone - get someone to help you.

DON'T rush or take unsafe shortcuts to finish a job.

DON'T allow children or animals in or around the vehicle while you are working on it.

DO wear eye protection when using power tools such as a drill, sander, bench grinder, etc. and when working under a vehicle.

DO keep loose clothing and long hair well out of the way of moving parts.

DO make sure that any hoist used has a safe working load rating adequate for the job.

DO get someone to check on you periodically when working alone on a vehicle.

DO carry out work in a logical sequence and make sure that everything is correctly assembled and tightened.

DO keep chemicals and fluids tightly capped and out of the reach of children and pets.

DO remember that your vehicle's safety affects that of yourself and others. If in doubt on any point, get professional advice.

Asbestos

Certain friction, insulating, sealing, and other products - such as brake linings, brake bands, clutch linings, torque converters, gaskets, etc. - may contain asbestos. Extreme care must be taken to avoid inhalation of dust from such products, since it is hazardous to health. If in doubt, assume that they do contain asbestos.

Fire

Remember at all times that gasoline is highly flammable. Never smoke or have any kind of open flame around when working on a vehicle. But the risk does not end there. A spark caused by an electrical short circuit, by two metal surfaces contacting each other, or even by static electricity built up in your body under certain conditions, can ignite gasoline vapors, which in a confined space are highly explosive. Do not, under any circumstances, use gasoline for cleaning parts. Use an approved safety solvent.

Always disconnect the battery ground (-) cable at the battery before working on any part of the fuel system or electrical system. Never risk spilling fuel on a hot engine or exhaust component. It is strongly recommended that a fire extinguisher suitable for use on fuel and electrical fires be kept handy in the garage or workshop at all times. Never try to extinguish a fuel or electrical fire with water.

Fumes

Certain fumes are highly toxic and can quickly cause unconsciousness and even death if inhaled to any extent. Gasoline vapor falls into this category, as do the vapors from some cleaning solvents. Any draining or pouring of such volatile fluids should be done in a well ventilated area.

When using cleaning fluids and solvents, read the instructions on the container carefully. Never use materials from unmarked containers.

Never run the engine in an enclosed space, such as a garage. Exhaust fumes contain carbon monoxide, which is extremely poisonous. If you need to run the engine, always do so in the open air, or at least have the rear of the vehicle outside the work area.

If you are fortunate enough to have the use of an inspection pit, never drain or pour gasoline and never run the engine while the vehicle is over the pit. The fumes, being heavier than air, will concentrate in the pit with possibly lethal results.

The battery

Never create a spark or allow a bare light bulb near a battery. They normally give off a certain amount of hydrogen gas, which is highly explosive.

Always disconnect the battery ground (-) cable at the battery before working on the fuel or electrical systems.

If possible, loosen the filler caps or cover when charging the battery from an external source (this does not apply to sealed or maintenance-free batteries). Do not charge at an excessive rate or the battery may burst.

Take care when adding water to a non maintenance-free battery and when carrying a battery. The electrolyte, even when diluted, is very corrosive and should not be allowed to contact clothing or skin.

Always wear eye protection when cleaning the battery to prevent the caustic deposits from entering your eyes.

Household current

When using an electric power tool, inspection light, etc., which operates on household current, always make sure that the tool is correctly connected to its plug and that, where necessary, it is properly grounded. Do not use such items in damp conditions and, again, do not create a spark or apply excessive heat in the vicinity of fuel or fuel vapor.

Secondary ignition system voltage

A severe electric shock can result from touching certain parts of the ignition system (such as the spark plug wires) when the engine is running or being cranked, particularly if components are damp or the insulation is defective. In the case of an electronic ignition system, the secondary system voltage is much higher and could prove fatal.

Troubleshooting

Contents

Engine and performance

1 Engine will not rotate when attempting to start

1 Battery terminal connections loose or corroded. Check the cable terminals at the battery; tighten cable clamp and/or clean off corrosion as necessary (see Chapter 1).
2 Battery discharged or faulty. If the cable ends are clean and tight on the battery connections, turn the key to the On position and switch on the headlights or windshield wipers. If they don't work, the battery is discharged.
3 Automatic transaxle not engaged in park (P) or Neutral (N).
4 Broken, loose or disconnected wires in the starting circuit. Inspect all wires and connectors at the battery, starter solenoid and ignition switch (on steering column).
5 Starter motor pinion jammed in driveplate ring gear. Remove starter (Chapter 5) and inspect pinion and driveplate (Chapter 2) at earliest convenience.
6 Starter solenoid faulty (Chapter 5).
7 Starter motor faulty (Chapter 5).
8 Ignition switch faulty (Chapter 12).
9 Engine seized. Try to turn the crankshaft with a large socket and breaker bar on the pulley bolt (see Chapter 2).

2 Engine rotates but will not start

1 Fuel tank empty.
2 Battery discharged (engine rotates slowly). Check the operation of electrical components as described in previous Section.
3 Battery terminal connections loose or corroded. See previous Section.
4 Fuel not reaching fuel injectors. Check for clogged fuel filter or lines and defective fuel pump. Also make sure the tank vent lines aren't clogged (Chapter 4).
5 Faulty ignition coil (Chapter 5).
6 Low cylinder compression. Check as described in Chapter 2.
7 Water in fuel. Drain tank and fill with new fuel.
8 Dirty or clogged fuel injectors.
9 Faulty emissions or engine control systems (Chapter 6).
10 Wet or damaged ignition components (Chapters 1 and 5).
11 Worn, faulty or incorrectly gapped spark plugs (Chapter 1).
12 Broken, loose or disconnected wires in the ignition circuit.
13 Broken, loose or disconnected wires at the ignition coils or faulty coils (Chapter 5).
14 Timing chain failure or wear affecting valve timing (Chapter 2).

3 Starter motor operates without turning engine

1 Starter pinion sticking. Remove the starter (Chapter 5) and inspect.
2 Starter pinion or driveplate teeth worn or broken. Remove the inspection cover on the left side of the engine and inspect.

4 Engine hard to start when cold

1 Battery low or discharged. Check as described in Chapter 1.
2 Fuel not reaching the fuel injectors. Check the fuel filter and lines (Chapters 1 and 4).
3 Defective spark plugs (Chapter 1).
4 Intake manifold vacuum leaks. Make sure all mounting bolts/nuts are tight and all vacuum hoses connected to the manifold are attached properly and in good condition.
5 Faulty emissions or engine control systems (Chapter 6).

5 Engine hard to start when hot

1 Air filter dirty (Chapter 1).
2 Bad engine ground connection.
3 Fuel not reaching the injectors (Chapter 4).
4 Loose connection in the ignition system (Chapter 5).
5 Faulty emissions or engine control systems (Chapter 6).

6 Starter motor noisy or engages roughly

1 Pinion or flywheel/driveplate teeth worn or broken. Remove the inspection cover and inspect.
2 Starter motor mounting bolts loose or missing.

7 Engine starts but stops immediately

1 Loose or damaged wiring in the ignition system.
2 Intake manifold vacuum leaks. Make sure all mounting bolts/nuts are tight and all vacuum hoses connected to the manifold are attached properly and in good condition.
3 Faulty emissions or engine control systems (Chapter 6).

8 Engine 'lopes' while idling or idles erratically

1 Vacuum leaks. Check mounting bolts at the intake manifold or plenum for tightness.

Make sure that all vacuum hoses are connected and in good condition. Use a stethoscope or a length of fuel hose held against your ear to listen for vacuum leaks while the engine is running. A hissing sound will be heard. A soapy water solution will also detect leaks. Check the intake manifold or plenum gasket surfaces.
2 Leaking EGR valve or plugged PCV valve (Chapter 6).
3 Clogged air filter (Chapter 1).
4 Leaking head gasket. Perform a cylinder compression check (Chapter 2).
5 Worn timing chain (Chapter 2).
6 Worn camshaft lobes (Chapter 2).
7 Valves burned or otherwise leaking (Chapter 2).
8 Ignition system not operating properly (Chapter 5).
9 Clogged or dirty injectors (Chapter 4).
10 Faulty emissions or engine control systems (Chapter 6).

9 Engine misses at idle speed

1 Spark plugs faulty or not gapped properly (Chapter 1).
2 Faulty spark plug wires (V6 only, Chapter 1).
3 Ignition components damaged or wet (Chapter 1).
4 Short circuits in spark plug wires (V6 only), ignition or coils (Chapter 5).
5 Emissions or engine control systems faulty (Chapter 6).
6 Clogged fuel filter and/or foreign matter in fuel. Replace the fuel filter (Chapter 1).
7 Vacuum leaks at intake manifold or plenum or hose connections. Check as described in Section 8.
8 Cylinder compression low or uneven. Check as described in Chapter 2.
9 Fuel injectors clogged or dirty (Chapter 4).
10 Leaky EGR valve (Chapter 6).
11 Emissions or engine control systems faulty (Chapter 6).

10 Excessively high idle speed

1 Sticking throttle linkage (Chapter 4).
2 Faulty Idle Air Control system (Chapter 4).
3 Faulty emissions or engine control systems (Chapter 6).

11 Battery will not hold a charge

1 Drivebelt defective or not adjusted properly (Chapter 1).
2 Battery cables loose or corroded (Chapter 1).
3 Alternator not charging properly (Chapter 5).
4 Loose, broken or faulty wires in the charging circuit (Chapter 5).

5   Continuous drain on the battery caused by a short circuit (Chapter 12).
6   Battery defective internally.
7   Faulty regulator (Chapter 5).

12   Alternator light stays on

1   Alternator or charging circuit fault (Chapter 5).
2   Drivebelt defective or not properly adjusted (Chapter 1).

13   Alternator light fails to come on when key is turned on

1   Faulty bulb (Chapter 12).
2   Defective alternator (Chapter 5).
3   Fault in the printed circuit, dash wiring or bulb holder (Chapter 12).

14   Engine misses throughout driving speed range

1   Fuel filter clogged and/or impurities in the fuel system. Check fuel filter (Chapter 1) or clean system (Chapter 4).
2   Faulty or incorrectly gapped spark plugs (Chapter 1).
3   Ignition system wires disconnected or damaged ignition system components (Chapter 1).
4   Defective spark plug wires (V6 only, Chapter 1).
5   Emissions or engine control system components faulty (Chapter 6).
6   Low or uneven cylinder compression pressures. Check as described in Chapter 2.
7   Ignition coils faulty or weak (Chapter 5).
8   Ignition system faulty or weak (Chapter 5).
9   Vacuum leaks at intake manifold or plenum or vacuum hoses (see Section 8).
10   Fuel injector dirty or clogged (Chapter 4).

15   Hesitation or stumble during acceleration

1   Ignition system not operating properly (Chapter 5).
2   Clogged or dirty fuel injectors (Chapter 4).
3   Fuel pressure low. Check for proper operation of the fuel pump and for restrictions in the fuel filter and lines (Chapter 4).
4   Emissions or engine control system components faulty (Chapter 6).

16   Engine stalls

1   Idle Air Control valve faulty (Chapter 6).

2   Fuel filter clogged and/or water and impurities in the fuel system (Chapter 1).
3   Wet or damaged ignition system wires or components.
4   Idle Air Control system faulty (Chapter 6).
5   Emissions or engine control system components faulty (Chapter 6).
6   Faulty or incorrectly gapped spark plugs (Chapter 1). Also check the spark plug wires on V6 engines (Chapter 1).
7   Vacuum leak at the intake manifold or plenum or vacuum hoses. Check as described in Section 8.

17   Engine lacks power

1   Check for faulty ignition wires, etc. (Chapter 1).
2   Faulty or incorrectly gapped spark plugs (Chapter 1).
3   Dirty air filter (Chapter 1).
4   Ignition coils faulty (Chapter 5).
5   Brakes binding (Chapters 1 and 9).
6   Automatic transaxle fluid level incorrect, causing slippage (Chapter 1).
7   Fuel filter clogged and/or impurities in the fuel system (Chapters 1 and 4).
8   EGR system not functioning properly (Chapter 6).
9   Use of sub-standard octane fuel. Fill tank with proper octane fuel.
10   Low or uneven cylinder compression pressures. Check as described in Chapter 2.
11   Air (vacuum) leak at intake manifold or plenum (check as described in Section 8).

18   Engine backfires

1   EGR system not functioning properly (Chapter 6).
2   Vacuum leak (refer to Section 8).
3   Damaged valve springs or sticking valves (Chapter 2).
4   Intake air (vacuum) leak (see Section 8).

19   Engine surges while holding accelerator steady

1   Intake air (vacuum) leak (see Section 8).
2   Fuel pump not working properly.
3   Idle Air Control system faulty (Chapter 6).
4   Emissions or engine control system components faulty (Chapter 6).

20   Pinging or knocking engine sounds when engine is under load

1   Use of incorrect octane fuel. Fill tank with fuel of the proper octane rating.

2   Problem in the ignition system (Chapter 5).
3   Carbon build-up in combustion chambers. Remove cylinder head(s) and clean combustion chambers (Chapter 2).
4   Incorrect spark plugs (Chapter 1).
5   Knock sensor system not functioning properly (Chapter 6).

21   Engine diesels (continues to run) after being turned off

1   Idle speed too high (Chapter 4).
2   Incorrect spark plug heat range (Chapter 1).
3   Intake air (vacuum) leak (see Section 8).
4   Carbon build-up in combustion chambers. Remove the cylinder head and clean the combustion chambers (Chapter 2).
5   Valves sticking (Chapter 2).
6   EGR system not operating properly (Chapter 6).
7   Leaking fuel injector(s) (Chapter 4).
8   Overheating. Check for causes (Section 27).

22   Low oil pressure

1   Improper oil grade.
2   Oil pump regulator valve not operating properly (Chapter 2).
3   Oil pump worn or damaged (Chapter 2).
4   Engine overheating (refer to Section 27).
5   Clogged oil filter (Chapter 1).
6   Clogged oil strainer (Chapter 2).
7   Oil pressure gauge not working properly (Chapter 2).

23   Excessive oil consumption

1   Oil pan drain plug loose.
2   Oil pan gasket or bolts loose or damaged (Chapter 2).
3   Front cover gasket or bolts loose or damaged (Chapter 2).
4   Crankshaft front or rear oil seal(s) leaking (Chapter 2).
5   Valve cover gasket bolts loose or damaged (Chapter 2).
6   Oil filter loose (Chapter 1).
7   Loose or damaged oil pressure switch (Chapter 2).
8   Pistons and cylinders worn excessively (Chapter 2).
9   Piston rings not installed correctly on pistons (Chapter 2).
10   Worn or damaged piston rings (Chapter 2).
11   Intake and/or exhaust valve oil seals worn or damaged (Chapter 2).
12   Worn valve stems.
13   Valves and or guides worn or damaged (Chapter 2).

24 Excessive fuel consumption

1   Dirty or clogged air filter element (Chapter 1).
2   Low tire pressure or incorrect tire size (Chapter 10).
3   Fuel leakage. Check all connections, lines and components in the fuel system (Chapter 4).
4   Fuel injectors clogged or dirty (Chapter 4).
5   Problem in the fuel injection system (Chapter 4).

25 Fuel odor

1   Fuel leaking out. Check all connections, lines and components in the fuel system (Chapter 4).
2   Fuel tank overfilled. Fill only to automatic shut-off.
3   Charcoal canister filter in Evaporative Emissions Control system clogged (Chapter 6).
4   Vapor leaks from Evaporative Emissions Control system lines (Chapter 6).

26 Miscellaneous engine noises

1   A strong dull noise that becomes more rapid as the engine accelerates indicates worn or damaged crankshaft bearings or an unevenly worn crankshaft. To pinpoint the trouble spot, remove the spark plug wire from one plug at a time and crank the engine over. If the noise stops, the cylinder with the removed plug wire indicates the problem area. Replace the bearing and/or service or replace the crankshaft (Chapter 2).
2   A similar (yet slightly higher pitched) noise to the crankshaft knocking described in the previous paragraph, that becomes more rapid as the engine accelerates, indicates worn or damaged connecting rod bearings (Chapter 2). The procedure for locating the problem cylinder is the same as described in Paragraph 1.
3   An overlapping metallic noise that increases in intensity as the engine speed increases, yet diminishes as the engine warms up indicates abnormal piston and cylinder wear (Chapter 2). To locate the problem cylinder, use the procedure described in Paragraph 1.
4   A rapid clicking noise that becomes faster as the engine accelerates is an indication of a worn piston pin or piston pin hole. Each time the piston hits the highest and lowest points in the stroke this sound will happen (Chapter 2). The procedure for locating the problem piston is described in Paragraph 1.
5   A metallic clicking noise coming from the water pump indicates worn or damaged water pump bearings or pump. Replace the water pump with a new one (Chapter 3).
6   A rapid tapping sound or clicking sound that becomes faster as the engine speed increases indicates "valve tapping" or stuck valve lifters. Holding one end of a section of hose to your ear and placing the other end at different spots along the rocker arm cover will help you identify this sound. The point where the sound is loudest indicates the problem valve or lifter (Chapter 2A).
7   A steady metallic rattling or rapping sound coming from the area of the timing chain cover indicates a worn, damaged or out-of-adjustment timing chain. Service or replace the chain and related components (Chapter 2).

Cooling system

27 Overheating

1   The system coolant level is low (Chapter 1).
2   Drivebelt defective or out of adjustment (Chapter 1).
3   Radiator core blocked or grille restricted (Chapter 3).
4   Thermostat faulty (Chapter 3).
5   The fan blades broken or cracked (Chapter 3).
6   Cap on the coolant reservoir is not maintaining proper pressure (Chapter 3).

28 Overcooling

The thermostat is faulty (Chapter 3).

29 External coolant leakage

1   Deteriorated/damaged hoses or loose clamps (Chapters 1 and 3).
2   Water pump seal defective (Chapters 1 and 3).
3   Leakage from radiator core or header tank (Chapter 3).
4   Engine drain or water jacket core plugs leaking (Chapter 2).
5   Leak at engine oil cooler (Chapter 3).

30 Internal coolant leakage

1   Leaking cylinder head gasket (Chapter 2).
2   The cylinder bore or cylinder head is cracked (Chapter 2).

31 Coolant loss

1   Too much coolant in system (Chapter 1).
2   Coolant boiling away because of overheating (Chapter 3).
3   The coolant reservoir cap is faulty (Chapter 3).

32 Poor coolant circulation

1   Water pump is faulty (Chapter 3).
2   Restriction in cooling system (Chapters 1 and 3).
3   Drivebelt defective or out of adjustment (Chapter 1).
4   Sticking thermostat (Chapter 3).

Manual transaxle

33 Knocking noise at low speeds

1   Worn driveaxle constant velocity (CV) joints (Chapter 8).
2   Worn side gear shaft counterbore in differential case (Chapter 7A).*

34 Noise most pronounced when turning

Differential gear noise (Chapter 7A).*

35 Clunk on acceleration or deceleration

1   Loose engine or transaxle mounts (Chapters 2 and 7A).
2   Worn differential pinion shaft in case.*
3   Worn side gear shaft counterbore in differential case (Chapter 7A).*
4   Worn or damaged inner CV joints (Chapter 8).

36 Clicking noise in turns

Worn or damaged outer CV joint (Chapter 8).

37 Vibration

1   Rough wheel bearing (Chapters 1 and 10).
2   Damaged driveaxle (Chapter 8).
3   Out of round tires (Chapter 1).
4   Tire out of balance (Chapters 1 and 10).
5   Worn CV joint (Chapter 8).

38 Noisy in neutral with engine running

1   Damaged input gear bearing (Chapter 7A).*
2   Damaged clutch release bearing (Chapter 8).

39 Noisy in one particular gear

1 Damaged or worn constant-mesh gears (Chapter 7A).*
2 Damaged or worn synchronizers (Chapter 7A).*
3 Bent reverse fork (Chapter 7A).*
4 Damaged fourth speed gear or output gear (Chapter 7A).*
5 Worn or damaged reverse idler gear or idler bushing (Chapter 7A).*

40 Noisy in all gears

1 Insufficient lubricant (Chapter 7A).
2 Damaged or worn bearings (Chapter 7A).*
3 Worn or damaged input gear shaft and/or output gear shaft (Chapter 7A).*

41 Slips out of gear

1 Worn or improperly adjusted linkage (Chapter 7A).
2 Transaxle loose on engine (Chapter 7A).
3 Shift linkage does not work freely, binds (Chapter 7A).
4 Input gear bearing retainer broken or loose (Chapter 7A).*
5 Dirt between clutch cover and engine housing (Chapter 7A).
6 Worn shift fork (Chapter 7A).*

42 Leaks lubricant

1 Driveshaft seals worn (Chapter 7A).
2 Excessive amount of lubricant in transaxle (Chapters 1 and 7A).
3 Loose or broken input gear shaft bearing retainer (Chapter 7A).*
4 Input gear bearing retainer O-ring and/or lip seal damaged (Chapter 7A).*
5 Vehicle speed sensor O-ring leaking (Chapter 7A).

43 Hard to shift

Shift linkage loose or worn (Chapter 7A).
* Although the corrective action necessary to remedy the symptoms described is beyond the scope of this manual, the above information should be helpful in isolating the cause of the condition so that the owner can communicate clearly with a professional mechanic.

Automatic transaxle

Note: Due to the complexity of the automatic transaxle, it's difficult for the home mechanic to properly diagnose and service this component. For problems other than the following, the vehicle should be taken to a dealer service department or a transmission shop.

44 Fluid leakage

1 Automatic transmission fluid is a deep red color. Fluid leaks should not be confused with engine oil, which can easily be blown by the flow of air to the transaxle.
2 Pinpoint a leak by first removing all built-up dirt and grime from the transaxle housing with degreasing agents and/or steam cleaning. Drive the vehicle at low speeds so the flow of air will not blow the leak far from its source. Raise the vehicle and determine where the leak is coming from. Common areas of leakage are:
a) Pan (Chapters 1 and 7B)
b) Filler pipe (Chapter 7B)
c) Transaxle oil lines (Chapter 7B)
d) Vehicle Speed Sensor (Chapter 6)

45 Transaxle fluid brown or has a burned smell

The transaxle has been overheated. Change the fluid (Chapter 1).

46 General shift mechanism problems

1 Chapter 7B deals with checking and adjusting the shift linkage on automatic transaxles. Common problems that may be attributed to poorly adjusted linkage are:
a) Engine starting in gears other than Park or Neutral.
b) Indicator on shifter pointing to a gear other than the one actually being used.
c) Vehicle moves when in Park.
2 Refer to Chapter 7B for the shift linkage adjustment procedure.

47 Transaxle will not downshift with accelerator pedal pressed to the floor

Pressure control valve solenoid faulty. On electronically controlled transaxles, this type of problem - which is caused by a malfunction in the control unit, a sensor or solenoid, or the circuit itself - is beyond the scope of this book. Take the vehicle to a dealer service department or a competent automatic transmission shop.

48 Engine will start in gears other than Park or Neutral

Defective or misadjusted Park/Neutral Position (PNP) switch (Chapter 7B).

49 Transaxle slips, shifts roughly, is noisy or has no drive in forward or reverse gears

There are many probable causes for the above problems, but the home mechanic should be concerned with only one possibility - fluid level. Before taking the vehicle to a repair shop, check the level and condition of the fluid as described in Chapter 1.
Correct the fluid level as necessary or change the fluid and filter if needed. If the problem persists, have a professional diagnose the probable cause.

Clutch

50 Pedal travels to floor - no pressure or very little resistance

1 Leaking clutch hydraulic release system or air in system (Chapter 8).
2 Broken release bearing or fork (Chapter 8).

51 Unable to select gears

1 Faulty transaxle (Chapter 7).
2 Faulty clutch disc or pressure plate (Chapter 8).
3 Faulty release cylinder or release bearing (Chapter 8).
4 Faulty shift lever assembly or rods (Chapter 8).

52 Clutch slips (engine speed increases with no increase in vehicle speed)

1 Clutch plate worn (Chapter 8).
2 Clutch plate is oil soaked by leaking rear main seal (Chapter 8).
3 Clutch plate not seated (Chapter 8).
4 Warped pressure plate or flywheel (Chapter 8).
5 Weak diaphragm springs (Chapter 8).
6 Clutch plate overheated. Allow to cool.
7 Faulty clutch self-adjusting mechanism (Chapter 8).

53 Grabbing (chattering) as clutch is engaged

1 Oil on clutch plate lining, burned or glazed facings (Chapter 8).
2 Worn or loose engine or transaxle mounts (Chapters 2 and 7).
3 Worn splines on clutch plate hub (Chapter 8).
4 Warped pressure plate or flywheel

(Chapter 8).
5   Burned or smeared resin on flywheel or pressure plate (Chapter 8).

54 Transaxle rattling (clicking)

1   Release fork loose (Chapter 8).
2   Low engine idle speed (Chapter 1).

55 Noise in clutch area

Faulty bearing (Chapter 8).

56 Clutch pedal stays on floor

1   Broken release bearing or fork (Chapter 8).
2   Broken or disconnected clutch cable (Chapter 8).

57 High pedal effort

1   Binding clutch cable (Chapter 8).
2   Pressure plate faulty (Chapter 8).

Driveaxles

58 Clicking noise in turns

Worn or damaged outer CV joint. Check for cut or damaged boots (Chapter 1). Repair as necessary (Chapter 8).

59 Knock or clunk when accelerating after coasting

Worn or damaged CV joint. Check for cut or damaged boots (Chapter 1). Repair as necessary (Chapter 8).

60 Shudder or vibration during acceleration

1   Worn or damaged CV joints. Repair or replace as necessary (Chapter 8).
2   Inboard joint assembly is sticking. Correct or replace as necessary (Chapter 8).

Brakes

Note: *Before assuming that a brake problem exists, make sure . . .*

a) *The tires are in good condition and properly inflated (Chapter 1).*
b) *The front-end alignment is correct (Chapter 10).*
c) *The vehicle isn't loaded with weight in an unequal manner.*

61 Vehicle pulls to one side during braking

1   Tire pressures incorrect (Chapter 1).
2   Front end out of line (have the front end aligned).
3   Unmatched tires on same axle.
4   Brake lines or hoses are restricted (Chapter 9).
5   Malfunctioning caliper or wheel cylinder (Chapter 9).
6   Suspension parts are loose (Chapter 10).
7   Brake calipers are loose (Chapter 9).
8   Brake linings contaminated (Chapters 1 and 9).

62 Noise (high-pitched squeal when the brakes are applied)

Front disc brake pads worn out. The noise comes from the wear sensor rubbing against the disc. Replace pads with new ones immediately (Chapter 9).

63 Brake roughness or chatter (pedal pulsates)

Note: *Brake pedal pulsation during operation of the Anti-Lock Brake System (ABS) is normal.*
1   Excessive front brake disc lateral runout (Chapter 9).
2   Parallelism not within specifications (Chapter 9).
3   Defective brake disc (Chapter 9).

64 Excessive pedal effort required to stop vehicle

1   Malfunctioning power brake booster (Chapter 9).
2   Partial system failure (Chapter 9).
3   Excessively worn pads (Chapter 9).
4   One or more caliper pistons or wheel cylinders seized or sticking (Chapter 9).
5   Brake pads contaminated with oil or grease (Chapter 9).
6   New pads installed and not yet seated. It will take a while for the new material to seat.

65 Excessive brake pedal travel

1   Partial brake system failure (Chapter 9).

2   Fluid level in master cylinder low (Chapters 1 and 9).
3   Air trapped in system (Chapters 1 and 9).
4   Rear shoes excessively worn (Chapter 9).

66 Dragging brakes

1   Master cylinder pistons not returning correctly (Chapter 9).
2   Restricted brakes lines or hoses (Chapters 1 and 9).
3   Parking brake adjustment incorrect (Chapter 9).
4   Pistons sticking in the calipers (Chapter 9).

67 Grabbing or uneven braking action

1   Malfunction of proportioning valves (Chapter 9).
2   Malfunction of power brake booster unit (Chapter 9).
3   Binding brake pedal mechanism (Chapter 9).
4   Pistons sticking in the calipers (Chapter 9).

68 Brake pedal feels spongy when depressed

1   Air in the hydraulic lines (Chapter 9).
2   Master cylinder mounting bolts loose (Chapter 9).
3   The master cylinder is defective (Chapter 9).

69 Brake pedal travels to the floor with little resistance

Little or no fluid in the master cylinder reservoir caused by leaking caliper or wheel cylinder pistons, loose, damaged or disconnected brake lines (Chapter 9).

70 Parking brake does not hold

Check the parking brake (Chapter 9).

Suspension and steering systems

Note: *Before attempting to diagnose the suspension and steering systems, perform the following preliminary checks:*

a) *Check the tire pressures and look for uneven wear.*
b) *Check the steering universal joints or coupling from the column to the steering gear for loose fasteners and wear.*

c) *Check the front and rear suspension and the steering gear assembly for loose and damaged parts.*

d) *Look for out-of-round or out-of-balance tires, bent rims and loose and/or rough wheel bearings.*

71   Vehicle pulls to one side

1   The tires are uneven or mismatched (Chapter 10).
2   Broken or sagging springs (Chapter 10).
3   Front wheel alignment incorrect (Chapter 10).
4   Front brakes dragging (Chapter 9).

72   Abnormal or excessive tire wear

1   Wheel alignment incorrect (Chapter 10).
2   Sagging or broken springs (Chapter 10).
3   Tire out-of-balance (Chapter 10).
4   Worn shock absorber (Chapter 10).
5   Overloaded vehicle.
6   Tires not rotated regularly.

73   Wheel makes a "thumping" noise

1   Tire blister or bump (Chapter 1).
2   Improper shock absorber action (Chapter 10).

74   Shimmy, shake or vibration

1   Tire or wheel out-of-balance or out-of-round (Chapter 10).
2   Worn or loose wheel bearings (Chapter 10).
3   Worn tie-rod ends (Chapter 10).
4   Worn balljoints (Chapter 10).
5   Wheel runout is excessive (Chapter 10).
6   Tire blister or bump (Chapter 1).

75   Hard steering

1   Worn balljoints, tie-rod ends and steering gear assembly (Chapter 10).
2   Front wheel alignment incorrect (Chapter 10).
3   Tire pressure low (Chapter 1).

76   Steering wheel does not return to center position correctly

1   Worn balljoints or tie-rod ends (Chapter 10).
2   Binding in steering column (Chapter 10).
3   Defective rack-and-pinion assembly (Chapter 10).

4   Front wheel alignment problem (Chapter 10).

77   Abnormal noise at the front end

1   Worn balljoints and tie-rod ends (Chapter 10).
2   Upper strut mount loose (Chapter 10).
3   Worn tie-rod ends (Chapter 10).
4   Stabilizer bar loose (Chapter 10).
5   Wheel lug nuts loose (Chapter 1).
6   Suspension bolts loose (Chapter 10).

78   Wander or poor steering stability

1   Tires mismatched or uneven (Chapter 10).
2   Worn balljoints or tie-rod ends (Chapters 1 and 10).
3   Shock absorbers worn or faulty (Chapter 10).
4   Stabilizer bar loose (Chapter 10).
5   Broken or sagging springs (Chapter 10).
6   Wheel alignment incorrect (Chapter 10).
7   Worn steering gear clamp bushings (Chapter 10).

79   Erratic steering when braking

1   Worn wheel bearings (Chapters 8 and 10).
2   Broken or sagging springs (Chapter 10).
3   Leaking wheel cylinder or caliper (Chapter 9).
4   Brake discs warped (Chapter 9).
5   Worn steering gear clamp bushings (Chapter 10).

80   Excessive pitching and/or rolling around corners or during braking

1   Stabilizer bar loose (Chapter 10).
2   Shock absorbers or mounts worn or damaged (Chapter 10).
3   Broken or sagging springs (Chapter 10).
4   Overloaded vehicle.

81   Suspension bottoms

1   Vehicle overloaded.
2   Shock absorbers worn or faulty (Chapter 10).
3   Incorrect, broken or sagging springs (Chapter 10).

82   Cupped tires

1   Wheel alignment incorrect (Chapter 10).
2   Shock absorbers worn or faulty (Chapter 10).
3   Worn wheel bearings (Chapters 8 and 10).
4   Tire or wheel runout excessive (Chapter 10).
5   Balljoints worn (Chapter 10).

83   Excessive tire wear on outside edge

1   Tire inflation pressures incorrect (Chapter 1).
2   Excessive speed in turns.
3   Wheel alignment incorrect (excessive toe-in or positive camber). Have professionally aligned.
4   Suspension arm bent or twisted (Chapter 10).

84   Excessive tire wear on inside edge

1   Tire pressure inflation incorrect (Chapter 1).
2   Wheel alignment incorrect (toe-out or excessive negative camber). Have professionally aligned.
3   Steering components damaged or loose (Chapter 10).

85   Tire tread worn in one place

1   Tires out-of-balance.
2   Wheel damaged or buckled. Inspect and replace if necessary.
3   Tire defective (Chapter 1).

86   Excessive play or looseness in steering system

1   Worn wheel bearings (Chapter 10).
2   Tie-rod end loose or worn (Chapter 10).
3   Steering gear loose (Chapter 10).

87   Rattling or clicking noise in rack and pinion

Steering gear mounting clamps loose (Chapter 10).

Chapter 1
Tune-up and routine maintenance

Contents

Specifications

Recommended lubricants and fluids

Note: *Listed here are manufacturer recommendations at the time this manual was written. Manufacturers occasionally upgrade their fluid and lubricant specifications, so check with your local auto parts store for current recommendations.*

Engine oil	API grade "certified for gasoline engines"
Viscosity	See accompanying chart

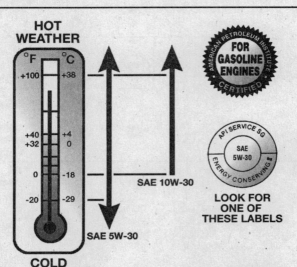

Engine oil viscosity chart - For best fuel economy and cold starting, select the lowest SAE viscosity grade for the expected temperature range

LOOK FOR ONE OF THESE LABELS

1-a3 HAYNES

Recommended lubricants and fluids

Fuel	Unleaded gasoline, 87 octane minimum
Automatic transaxle fluid	DEXRON III automatic transaxle fluid
Power steering fluid	GM power steering fluid
Brake fluid	DOT 3 brake fluid
Engine coolant	50/50 mixture of DEX-COOL coolant and de-mineralized water
Hood, door and trunk hinge lubricant	Lubriplate, aerosol spray lubricant
Door hinge and check spring grease	NLGI no. 2 multi-purpose grease
Key lock cylinder lubricant	Graphite spray
Hood latch assembly lubricant	Lubriplate, aerosol spray lubricant
Door latch lubricant	NLGI no. 2 multi-purpose grease or equivalent

Capacities*

Engine oil (including filter)	
2.4L engine	4 qts
3.1L/3.4L engine	4.5 qts
Automatic transaxle	
Fluid and filter change	7 qts
From dry, including torque converter	13 qts
Cooling system	
2.4L engine	11.3 qts
3.1L/3.4L engine	13.6 qts

All capacities approximate. Add as necessary to bring to appropriate level.

Brakes

Disc brake pad wear limit	1/8 inch
Drum brake shoe wear limit	1/16 inch

Ignition system

Spark plug type and gap	
Four-cylinder engine	
1997	AC type 41-910 platinum, or equivalent @ 0.060 inch
1998 through 2000	AC type 41-942 platinum, or equivalent @ 0.050 inch
V6 engines	AC type 41-940 platinum, or equivalent @ 0.060 inch
Firing order	
2.4L engine	1-3-4-2
3.1L/3.4L engines	1-2-3-4-5-6

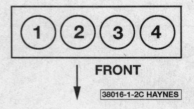

Engine cylinder identification, 2.4L

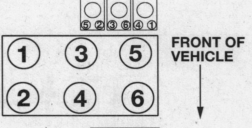

Engine cylinder identification and coil terminal locations, V6 engines

Torque specifications   Ft-lbs (unless otherwise indicated)

Spark plugs	
2.4L engine	156 in-lbs
3.1L/3.4L engines	20
Drivebelt tensioner bolt	37
Wheel lug nuts	100
Automatic transaxle	
Fluid level check plug	108 in-lbs
Pan bolts	108 in-lbs

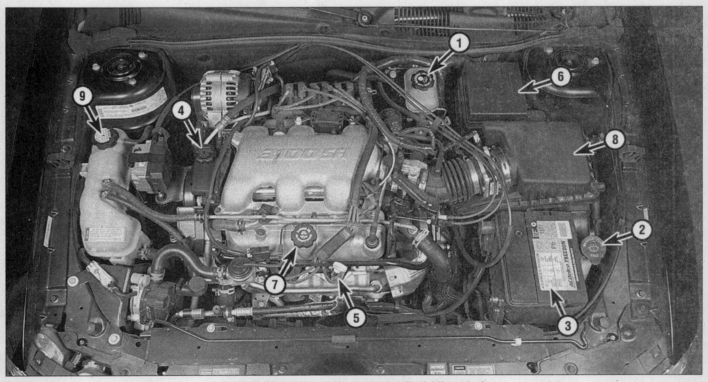

Typical engine compartment layout (2.4L four-cylinder engine)

1	Brake master cylinder reservoir	4	Power steering fluid reservoir	7	Engine oil filler cap
2	Windshield washer fluid reservoir	5	Engine oil dipstick	8	Air filter housing
3	Battery	6	Fuse and relay center	9	Coolant surge tank

Typical engine compartment layout (3.1L V6 engine)

1	Brake master cylinder reservoir	4	Power steering fluid reservoir	7	Engine oil filler cap
2	Windshield washer fluid reservoir	5	Engine oil dipstick	8	Air filter housing
3	Battery	6	Fuse and relay center	9	Coolant surge tank

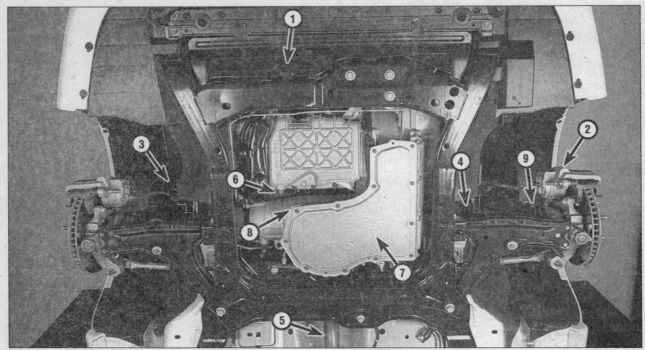

Typical engine compartment underside components

1	Engine oil filter (above plastic splash shield)	4	Inner CV joint boot	8	Automatic transaxle fluid check plug
2	Front brake caliper	5	Exhaust system	9	Outer CV joint boot
3	Strut/coil spring assembly	6	Engine oil drain plug		
		7	Automatic transaxle fluid pan		

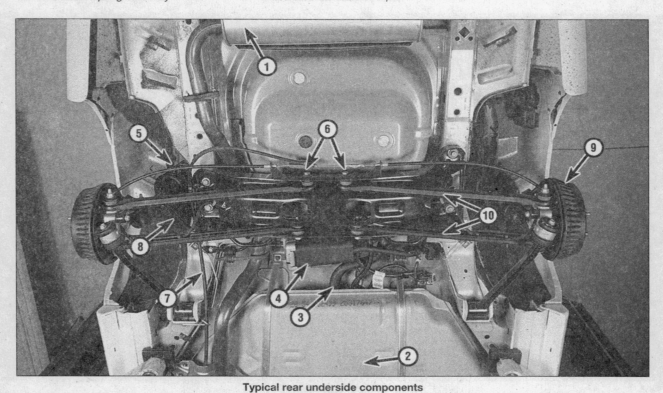

Typical rear underside components

1	Rear muffler/tailpipe	5	Strut/coil spring assembly	8	Stabilizer bar
2	Fuel tank	6	Rear wheel toe adjusters	9	Brake drum
3	Fuel filler hose	7	Parking brake cable	10	Lateral links
4	Charcoal canister				

1 Chevrolet Malibu, Oldsmobile Cutlass and Alero, and Pontiac Grand Am maintenance schedule

The following maintenance intervals are based on the assumption that the vehicle owner will be doing the maintenance or service work, as opposed to having a dealer service department do the work. Although the time/mileage intervals are loosely based on factory recommendations, most have been shortened to ensure, for example, that such items as lubricants and fluids are checked/changed at intervals that promote maximum engine/driveline service life. Also, subject to the preference of the individual owner interested in keeping his or her vehicle in peak condition at all times, and with the vehicle's ultimate resale in mind, many of the maintenance procedures may be performed more often than recommended in the following schedule. We encourage such owner initiative.

When your vehicle is new, follow the maintenance schedule to the letter, record the maintenance performed in your owners manual and keep all receipts to protect the new vehicle warranty. In many cases, the initial maintenance check is done at no cost to the owner.

Every 250 miles or weekly, whichever comes first

Check the engine oil level (Section 4)
Check the engine coolant level (Section 4)
Check the windshield washer fluid level (Section 4)
Check the brake fluid level (Section 4)
Check the tires and tire pressures (Section 5)
Check the operation of all lights
Check the horn operation

Every 3000 miles or 3 months, whichever comes first

All items listed above plus:
Check the power steering fluid level (Section 6)
Check the automatic transmission fluid level (Section 7)
Change the engine oil and filter (Section 8)

Every 6000 miles or 6 months, whichever comes first

All items listed above plus:
Check the battery (Section 9)
Check the cooling system (Section 10)
Inspect the underhood hoses (Section 11)
Check the engine drivebelt (Section 12)
Rotate the tires (Section 13)
Inspect the seat belts (Section 14)

Every 15,000 miles or 12 months, whichever comes first

Check the suspension, steering and driveaxle boots (Section 15)
Check the brakes (Section 16)*
Inspect the exhaust system (Section 17)
Inspect and replace, if necessary, the windshield wiper blades (Section 18)

Every 30,000 miles or 24 months, whichever comes first

All items listed above plus:
Replace the air filter (Section 19)
Inspect the PCV valve (Section 20)
Inspect the spark plug wires (V6 engines only) (Section 21)
Replace the spark plugs (conventional [non-platinum] spark plugs) (Section 22)
Replace the fuel filter (Section 23)
Inspect the fuel system (Section 24)
Service the cooling system (drain, flush and refill) (green-colored ethylene glycol anti-freeze only) (Section 25)
Change the brake fluid (Section 26)

Every 60,000 miles or 48 months, whichever comes first

Change the automatic transaxle fluid and filter (Section 27)**

Every 100,000 miles or 5 years, whichever comes first

Replace the spark plugs (platinum-tipped spark plugs) (Section 22)
Service the cooling system (drain, flush and refill) (orange-colored "DEX-COOL" silicate-free coolant only) (Section 25)

* *If the vehicle frequently tows a trailer, is operated primarily in stop-and-go conditions or its brakes receive severe usage for any other reason, check the brakes every 3000 miles or three months.*

** *If operated under one or more of the following conditions, change the automatic transmission fluid every 30,000 miles:*
In heavy city traffic where the outside temperature regularly reaches 90-degrees F (32-degrees C) or higher.
In hilly or mountainous terrain.
Frequent trailer pulling.

2   Introduction

This Chapter is designed to help the home mechanic maintain the Chevrolet Malibu, Oldsmobile Cutlass and Alero, and Pontiac Grand Am models with the goals of maximum performance, economy, safety and reliability in mind.

Included is a master maintenance schedule, followed by procedures dealing specifically with each item on the schedule. Visual checks, adjustments, component replacement and other helpful items are included. Refer to the accompanying illustrations of the engine compartment and the underside of the vehicle for the locations of various components.

Adhering to the mileage/time maintenance schedule and following the step-by-step procedures, which is simply a preventive maintenance program, will result in maximum reliability and vehicle service life. Keep in mind that it's a comprehensive program - maintaining some items but not others at the specified intervals will not produce the same results.

As you service the vehicle, you'll discover that many of the procedures can - and should - be grouped together because of the nature of the particular procedure you're performing or because of the close proximity of two otherwise unrelated components to one another.

For example, if the vehicle is raised, you should inspect the exhaust, suspension, steering and fuel systems while you're under the vehicle. When you're rotating the tires, it makes good sense to check the brakes, since the wheels are already removed. Finally, let's suppose you have to borrow or rent a torque wrench. Even if you only need it to tighten the spark plugs, you might as well check the torque of as many critical fasteners as time allows.

The first step in this maintenance program is to prepare before the actual work begins. Read through all the procedures you're planning to do, then gather up all the parts and tools needed. If it looks like you might run into problems during a particular job, seek advice from a mechanic or an experienced do-it-yourselfer.

Caution: *On models equipped with the Theftlock audio system, be sure the lockout feature is turned off before performing any procedure which requires disconnecting the battery (see the front of this manual).*

3   Tune-up general information

The term tune-up is used in this manual to represent a combination of individual operations rather than one specific procedure.

If, from the time the vehicle is new, the routine maintenance schedule is followed closely and frequent checks are made of fluid levels and high wear items, as suggested

throughout this manual, the engine will be kept in relatively good running condition and the need for additional work due to lack of regular maintenance will be minimized. This is even more likely if a used vehicle, which has not received regular and frequent maintenance checks, is purchased. In such cases, an engine tune-up will be needed outside of the regular routine maintenance intervals.

The first step in any tune-up or diagnostic procedure to help correct a poor running engine is a cylinder compression check. A compression check (see Chapter 2, Part C) will help determine the condition of internal engine components and should be used as a guide for tune-up and repair procedures. If, for instance, a compression check indicates serious internal engine wear, a conventional tune-up won't improve the performance of the engine and would be a waste of time and money. Because of its importance, the compression check should be done by someone with the right equipment and the knowledge to use it properly.

The following procedures are those most often needed to bring a generally poor running engine back into a proper state of tune.

Minor tune-up

Check all engine related fluids (Section 4)
Clean, inspect and test the battery (Section 9)
Check the cooling system (Section 10)
Check all underhood hoses (Section 11)
Check and adjust the drivebelt (Section 12)
Check the air filter (Section 19)
Check the PCV valve (Section 20)
Inspect the spark plug wires (V6 engines only) (Section 21)
Replace the spark plugs (Section 22)

Major tune-up

All items listed under Minor tune-up plus . . .
Replace the air filter (Section 19)
Replace the spark plug wires (Section 21)
Replace the fuel filter (Section 23)
Check the fuel system (Section 24)
Check the ignition timing (Chapter 5)
Check the charging system (Chapter 5)
Check the EGR system (Chapter 6)

4   Fluid level checks (every 250 miles or weekly)

Note: *The following are fluid level checks to be done on a 250 mile or weekly basis. Additional fluid level checks can be found in specific maintenance procedures that follow. Regardless of intervals, be alert to fluid leaks under the vehicle, which would indicate a problem to be corrected immediately.*

1   Fluids are an essential part of the lubrication, cooling, brake and windshield washer systems. Because the fluids gradually become depleted and/or contaminated during normal operation of the vehicle, they must

4.2 The engine oil dipstick (arrow) is located on the firewall side of the engine on 2.4L models (shown); it's on the radiator side on V6 models

be periodically replenished. See *Recommended lubricants and fluids* at the beginning of this Chapter before adding fluid to any of the following components. **Note:** *The vehicle must be on level ground when fluid levels are checked.*

Engine oil

Refer to illustrations 4.2, 4.4, 4.6a and 4.6b

2   The engine oil level is checked with a dipstick **(see illustration)**. The dipstick extends through a metal tube down into the oil pan.

3   The oil level should be checked before the vehicle has been driven, or about 15 minutes after the engine has been shut off. If the oil is checked immediately after driving the vehicle, some of the oil will remain in the upper part of the engine, resulting in an inaccurate reading on the dipstick.

4   Pull the dipstick from the tube and wipe all the oil from the end with a clean rag or paper towel. Insert the clean dipstick all the way back into the tube and pull it out again. Note the oil at the end of the dipstick. Add oil as necessary to keep the level above the ADD mark in the cross hatched area of the dipstick **(see illustration)**.

5   Do not overfill the engine by adding too much oil since this may result in oil fouled spark plugs, oil leaks or oil seal failures.

6   Oil is added to the engine after removing a twist-off cap located on the valve cover **(see illustrations)**. A funnel may help to reduce spills.

7   Checking the oil level is an important preventive maintenance step. A consistently low oil level indicates oil leakage through damaged seals, defective gaskets or past worn rings or valve guides. If the oil looks milky in color or has water droplets in it, the cylinder head gasket may be blown or the head or block may be cracked. The engine should be checked immediately. The condition of the oil should also be checked. When-

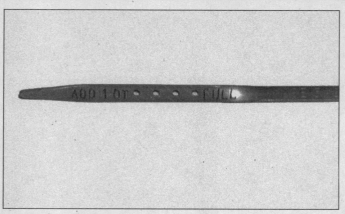

4.4 The oil level should be at or near the upper hole or in the cross-hatched area on the dipstick - if it's below the ADD line, add enough oil to bring the level into the upper hole or top of the cross-hatched area

4.6a The engine oil filler cap (arrow) is clearly marked and threads into the tube on the front valve cover on V6 models - turn it counterclockwise to remove it

4.6b The oil filler cap (arrow) on four-cylinder models is located at the right rear

4.9 The coolant surge tank is located on the right (passenger's) side of the engine compartment - the coolant level can be checked by observing it through the translucent tank

ever you check the oil level, slide your thumb and index finger up the dipstick before wiping off the oil. If you see small dirt or metal particles clinging to the dipstick, the oil should be changed (see Section 8).

Engine coolant

Refer to illustration 4.9

Warning: *Do not allow antifreeze to come in contact with your skin or painted surfaces of the vehicle. Rinse off spills immediately with plenty of water. Antifreeze is highly toxic if ingested. Never leave antifreeze lying around in an open container or in puddles on the floor; children and pets are attracted by its sweet smell and may drink it. Check with local authorities on disposing of used anti-freeze. Many communities have collection centers that will see that antifreeze is disposed of safely.*

Note: *Non-toxic antifreeze is now manufactured and available at local auto parts stores, but even this type should be disposed of properly.*

Caution: *Never mix green-colored ethylene*

glycol anti-freeze and orange-colored "DEX-COOL" silicate-free coolant because doing so will destroy the efficiency of the "DEX-COOL" coolant which is designed to last for 100,000 miles or five years.*

8   All vehicles covered by this manual are equipped with a pressurized coolant surge tank, located at the right side of the engine compartment, and connected by hoses to the radiator and cooling system.

9   The coolant level in the surge tank should be checked regularly. **Warning:** *Do not remove the pressure cap to check the coolant level when the engine is warm.* The level of coolant in the surge tank varies with the temperature of the engine. When the engine is cold, the coolant level should be at or slightly above the COLD mark on the surge tank **(see illustration)**. Once the engine has warmed up, the level should be at or near the HOT mark. If it isn't, add coolant to the surge tank. To add coolant simply twist open the cap and add a 50/50 mixture of ethylene glycol based green-colored antifreeze or orange-colored "DEX-COOL" silicate-free

coolant and water (see **Caution** above).

10 Drive the vehicle and recheck the coolant level. If only a small amount of coolant is required to bring the system up to the proper level, water can be used. However, repeated additions of water will dilute the antifreeze and water solution. In order to maintain the proper ratio of antifreeze and water, always top up the coolant level with the correct mixture. An empty plastic milk jug or bleach bottle makes an excellent container for mixing coolant. Do not use rust inhibitors or additives.

11 If the coolant level drops consistently, there may be a leak in the system. Inspect the radiator, hoses, filler cap, drain plugs and water pump (see Section 10). If no leaks are noted, have the pressure cap tested by a service station.

12 If you have to remove the pressure cap, wait until the engine has cooled completely, then wrap a thick cloth around the cap and unscrew it slowly. If coolant or steam escapes, let the engine cool down longer, then remove the cap.

4.14 Flip the windshield washer fluid cap (arrow) up to add fluid

4.18 The fluid level inside the brake fluid reservoir can easily be checked by observing the level through the translucent sides of the reservoir - if necessary, add additional fluid until it reaches the top of the MAX mark - DO NOT OVERFILL

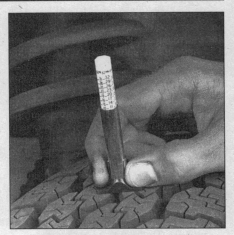

5.2 Use a tire tread depth gauge to monitor tire wear - they are available at auto parts stores and service stations and cost very little

13   Check the condition of the coolant as well. It should be relatively clear. If it is brown or rust colored, the system should be drained, flushed and refilled. Even if the coolant appears to be normal, the corrosion inhibitors wear out, so it must be replaced at the specified intervals. If the system is filled with standard green coolant/water, it must be flushed and replaced more frequently than if the original "DEX-COOL" coolant is retained.

Windshield washer fluid

Refer to illustration 4.14

14   Fluid for the windshield washer system is located in a plastic reservoir on the left side (driver's side) of the engine compartment **(see illustration)**. In milder climates, plain water can be used in the reservoir, but it should be kept no more than two-thirds full to allow for expansion if the water freezes. In colder climates, use windshield washer system antifreeze, available at any auto parts store, to lower the freezing point of the fluid. Mix the antifreeze with water in accordance with the manufacturer's directions on the container. **Caution:** *Do not use cooling system antifreeze - it will damage the vehicle's paint.*

15   To help prevent icing in cold weather, warm the windshield with the defroster before using the washer.

Battery electrolyte

16   All vehicles covered by this manual are equipped with a battery that is permanently sealed (except for vent holes) and has no filler caps. Water does not have to be added to these batteries at any time.

Brake fluid

Refer to illustration 4.18

17   The brake fluid level is checked by looking through the plastic reservoir mounted on the master cylinder. The master cylinder is mounted on the front of the power booster unit in the left (driver's side) rear corner of the

engine compartment.

18   The fluid level should be at or near the base of the reservoir filler neck **(see illustration)**. If the fluid level is low, wipe the top of the reservoir and the lid with a clean rag to prevent contamination of the system as the lid is pried off.

19   When adding fluid, pour it carefully into the reservoir to avoid spilling it on surrounding painted surfaces. Be sure the specified fluid is used, since mixing different types of brake fluid can cause damage to the system. See *Recommended lubricants and fluids* at the front of this Chapter or your owner's manual. **Warning:** *Brake fluid can harm your eyes and damage painted surfaces, so use extreme caution when handling or pouring it. Do not use brake fluid that has been standing open or is more than one year old. Brake fluid absorbs moisture from the air. Excess moisture can cause a dangerous loss of braking effectiveness.*

20   At this time the fluid and master cylinder can be inspected for contamination. The system should be drained and refilled if deposits, dirt particles or water droplets are seen in the fluid.

21   After filling the reservoir to the proper level, make sure the lid completely snaps in place to prevent fluid leakage.

22   The brake fluid level in the master cylinder will drop slightly as the pads at each wheel wear down during normal operation. If the master cylinder requires repeated replenishing to keep it at the proper level, this is an indication of leakage in the brake system, which should be corrected immediately. Check all brake lines and connections (see Section 16 for more information).

23   If, when checking the master cylinder fluid level, you discover one or both reservoirs empty or nearly empty, the brake system should be bled and inspected for leaks (see Chapter 9).

5   Tire and tire pressure checks (every 250 miles or weekly)

Refer to illustrations 5.2, 5.3, 5.4a, 5.4b and 5.8

1   Periodic inspection of the tires may spare you the inconvenience of being stranded with a flat tire. It can also provide you with vital information regarding possible problems in the steering and suspension systems before major damage occurs.

2   The original tires on this vehicle are equipped with 1/2-inch wide bands that appear when tread depth reaches 1/16-inch, indicating the tires are worn out. Tread wear can be monitored with a simple, inexpensive device known as a tread depth indicator **(see illustration)**.

3   Note any abnormal tread wear **(see illustration)**. Tread pattern irregularities such as cupping, flat spots and more wear on one side than the other are indications of front end alignment and/or balance problems. If any of these conditions are noted, take the vehicle to a tire shop or service station to correct the problem.

4   Look closely for cuts, punctures and embedded nails or tacks. Sometimes a tire will hold air pressure for short time or leak down very slowly after a nail has embedded itself in the tread. If a slow leak persists, check the valve stem core to make sure it's tight **(see illustration)**. Examine the tread for an object that may have embedded itself in the tire or for a "plug" that may have begun to leak (radial tire punctures are repaired with a plug that's installed in a puncture). If a puncture is suspected, it can be easily verified by spraying a solution of soapy water onto the suspected area **(see illustration)**. The soapy solution will bubble if there's a leak. Unless the puncture is unusually large, a tire shop or service station can usually repair the tire.

5   Carefully inspect the inner sidewall of each tire for evidence of brake fluid. If you

1

UNDERINFLATION

CUPPING

Cupping may be caused by:

- Underinflation and/or mechanical irregularities such as out-of-balance condition of wheel and/or tire, and bent or damaged wheel.
- Loose or worn steering tie-rod or steering idler arm.
- Loose, damaged or worn front suspension parts.

OVERINFLATION

INCORRECT TOE-IN OR EXTREME CAMBER

FEATHERING DUE TO MISALIGNMENT

5.3 This chart will help you determine the condition of the tires, the probable cause(s) of abnormal wear and the corrective action necessary

5.4a If a tire loses air on a steady basis, check the valve core first to make sure it's snug (special inexpensive wrenches are commonly available at auto parts stores)

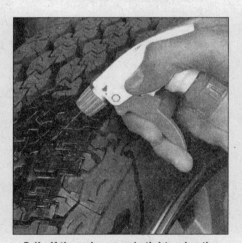

5.4b If the valve core is tight, raise the corner of the vehicle with the low tire and spray a soapy water solution onto the tread as the tire is turned slowly - leaks will cause small bubbles to appear

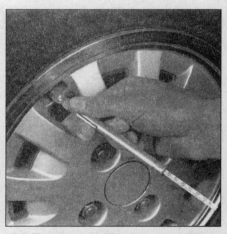

5.8 To extend the life of the tires, check the air pressure at least once a week with an accurate gauge (don't forget the spare)

see any, inspect the brakes immediately.

6   Correct air pressure adds miles to the lifespan of the tires, improves mileage and enhances overall ride quality. Tire pressure cannot be accurately estimated by looking at a tire, especially if it's a radial. A tire pressure gauge is essential. Keep an accurate gauge in the vehicle. The pressure gauges attached to the nozzles of air hoses at gas stations are often inaccurate.

7   Always check tire pressure when the tires are cold. Cold, in this case, means the vehicle has not been driven over a mile in the three hours preceding a tire pressure check. A pressure rise of four to eight pounds is not uncommon once the tires are warm.

8   Unscrew the valve cap protruding from the wheel or hubcap and push the gauge firmly onto the valve stem (see illustration). Note the reading on the gauge and compare the figure to the recommended tire pressure shown on the label attached to the inside of the glove compartment door. Be sure to reinstall the valve cap to keep dirt and moisture

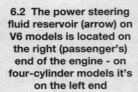

6.2 The power steering fluid reservoir (arrow) on V6 models is located on the right (passenger's) end of the engine - on four-cylinder models it's on the left end

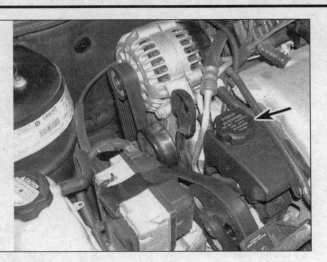

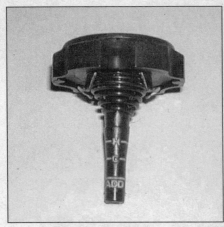

6.6 The marks on the dipstick indicate the safe fluid range

out of the valve stem mechanism. Check all four tires and, if necessary, add enough air to bring them up to the recommended pressure.

9   Don't forget to keep the spare tire inflated to the specified pressure (refer to your owner's manual or the tire sidewall).

10   Some models may have an optional Tire Pressure Monitor system, which includes a dashboard warning light to let you know if any tire is 12 psi lower than recommended. If you change tire pressures, get new tires or have yours rotated, the Check Tire Pressure light may come on. To reset it, turn the key On (engine Off), then push once on the reset button in the driver's side interior fuse panel (see illustration 8.19). Press the button again after you see the Oil Change light flashing. When the Check Tire Pressure light begins to flash, push and hold the button again until a chime is heard, then let go.

6   Power steering fluid level check (every 3000 miles or 3 months)

Refer to illustrations 6.2 and 6.6

1   The power steering system relies on fluid that may, over a period of time, require replenishing.

2   The fluid reservoir for the power steering pump is mounted on the right side of the V6 engine by the engine drivebelt (see illustration). On four-cylinder models, the power steering pump/reservoir is driven by one of the camshafts and is located at the transaxle end of the engine.

3   For the check, the front wheels should be pointed straight ahead and the engine should be off.

4   Use a clean rag to wipe off the reservoir cap and the area around the cap. This will help prevent any foreign matter from entering the reservoir during the check.

5   Twist off the cap and check the temperature of the fluid at the end of the dipstick with your finger.

6   Wipe off the fluid with a clean rag, reinsert it, then withdraw it and read the fluid

level. The level should be at the HOT mark if the fluid was hot to the touch (see illustration). It should be at the COLD mark if the fluid was cool to the touch.

7   If additional fluid is required, pour the specified type directly into the reservoir, using a funnel to prevent spills.

8   If the reservoir requires frequent fluid additions, all power steering hoses, hose connections, the power steering pump and the rack and pinion assembly should be carefully checked for leaks.

7   Automatic transaxle fluid level check

On these models, it is not necessary to check the fluid level at regular intervals. If leaks or shifting problems lead you to suspect a low fluid level, refer to Section 27.

8   Engine oil and filter change (every 3000 miles or 3 months)

Refer to illustrations 8.2, 8.7, 8.12, 8.14 and 8.19

1   Frequent oil changes are the best preventive maintenance the home mechanic can give the engine, because aging oil becomes diluted and contaminated, which leads to premature engine wear. **Note:** *Some 1999 and later models have an Oil Change Indicator light on the instrument panel. We recommend that you change your oil according to the Maintenance Schedule at the beginning of this Chapter, even if the Oil Change light has not come on.*

2   Make sure you have all the necessary tools before you begin this procedure (see illustration). You should also have plenty of rags or newspapers handy for mopping up any spills.

3   Access to the underside of the vehicle is greatly improved if the vehicle can be lifted on a hoist, driven onto ramps or supported

by jackstands. **Warning:** *Do not work under a vehicle that is supported only by a hydraulic or scissors-type jack.*

4   If this is your first oil change, get under the vehicle and familiarize yourself with the locations of the oil drain plug and the oil filter. The engine and exhaust components will be warm during the actual work, so try to anticipate any potential problems before the engine and accessories are hot.

5   Park the vehicle on a level spot. Start the engine and allow it to reach its normal operating temperature. Warm oil and sludge will flow out more easily. Turn off the engine when it's warmed up. Remove the filler cap from the valve cover.

6   Raise the vehicle and support it securely on jackstands. **Warning:** *Never get beneath the vehicle when it is supported only by a jack. The jack provided with your vehicle is designed solely for raising the vehicle to remove and replace the wheels. Always use jackstands to support the vehicle when it becomes necessary to place your body underneath the vehicle.*

7   Being careful not to touch the hot exhaust components, place the drain pan under the drain plug in the bottom of the pan and remove the plug (see illustration). You may want to wear gloves while unscrewing the plug the final few turns if the engine is hot.

8   Allow the old oil to drain into the pan. It may be necessary to move the pan farther under the engine as the oil flow slows to a trickle. Inspect the old oil for the presence of metal shavings and chips.

9   After all the oil has drained, wipe off the drain plug with a clean rag. Even minute metal particles clinging to the plug would immediately contaminate the new oil.

10   Clean the area around the drain plug opening, reinstall the plug and tighten it to the specifications listed at the beginning of this Chapter.

11   Move the drain pan into position under the oil filter.

12   Loosen the oil filter (see illustration) by turning it counterclockwise with the filter

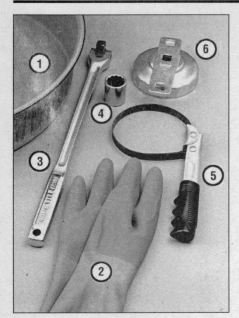

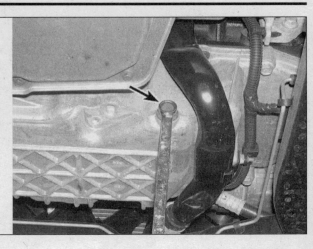

8.7 The engine oil drain plug is located at the rear of the oil pan - it is usually very tight, so use a socket or box-end wrench to avoid rounding off the hex

8.2 These tools are required when changing the engine oil and filter

1   **Drain pan** - It should be fairly shallow in depth, but wide to prevent spills
2   **Rubber gloves** - When removing the drain plug and filter, you will get oil on your hands (the gloves will prevent burns)
3   **Breaker bar** - Sometimes the oil drain plug is tight, and a long breaker bar is needed to loosen it
4   **Socket** - To be used with the breaker bar or a ratchet (must be the correct size to fit the drain plug)
5   **Filter wrench** - This is a metal band-type wrench, which requires clearance around the filter to be effective
6   **Filter wrench** - This type fits on the bottom of the filter and can be turned with a ratchet or breaker bar (different-size wrenches are available for different types of filters)

wrench. **Note:** *Oil filters on all engines are located on the front (radiator) side of the engine block.* Use a quality filter wrench of the correct size and be careful not to collapse the canister as you apply pressure. Once the filter is loose, use your hands to unscrew it from the block. Just as the filter is detached from the block, immediately tilt the open end up to prevent the oil inside the filter from spilling out. **Warning:** *The exhaust system may still be hot, so be careful.*

13   With a clean rag, wipe off the mounting surface on the block. If a residue of old oil is allowed to remain, it will smoke when the block is heated up. Also make sure that none of the old gasket remains stuck to the mounting surface. It can be removed with a scraper if necessary.

14   Compare the old filter with the new one to make sure they are the same type. Smear some clean engine oil on the rubber gasket of the new filter and screw it into place **(see illustration)**. Because overtightening the filter will damage the gasket, do not use a filter wrench to tighten the filter. Tighten it by hand until the gasket contacts the seating surface. Then seat the filter by giving it an additional 3/4-turn.

15   Remove all tools, rags, etc. from under the vehicle, being careful not to spill the oil in the drain pan, then lower the vehicle.

16   Add new oil to the engine through the oil filler cap. Use a funnel, if necessary, to prevent oil from spilling onto the top of the engine. Pour three quarts of fresh oil into the engine. Wait a few minutes to allow the oil to drain into the pan, then check the level on the oil dipstick (see Section 4 if necessary). If the oil level is at or near the upper hole on the dipstick, install the filler cap hand tight, start the engine and allow the new oil to circulate.

17   Allow the engine to run for about a minute. While the engine is running, look under the vehicle and check for leaks at the oil pan drain plug and around the oil filter. If either is leaking, stop the engine and tighten the plug or filter.

18   Wait a few minutes to allow the oil to trickle down into the pan, then recheck the level on the dipstick and, if necessary, add enough oil to bring the level to the upper hole.

19   During the first few trips after an oil change, make it a point to check frequently for leaks and proper oil level. **Note:** *If your vehicle is equipped with the Change Oil light,*

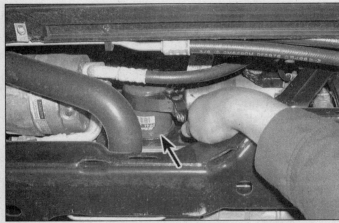

8.12 The oil filter is usually on very tight as well and will require a special wrench for removal - DO NOT use the wrench to tighten the new filter!

8.14 Lubricate the oil filter gasket with clean engine oil before installing the filter on the engine

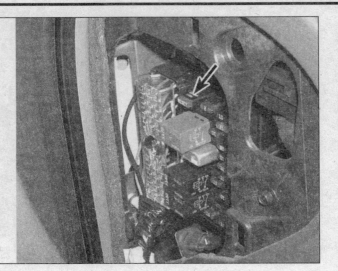

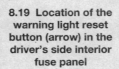

8.19 Location of the warning light reset button (arrow) in the driver's side interior fuse panel

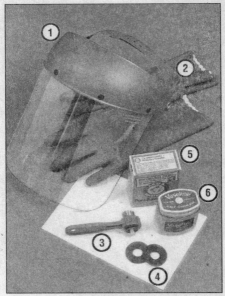

9.1 Tools and materials required for battery maintenance

1   *Face shield/safety goggles* - When removing corrosion with a brush, the acidic particles can easily fly up into your eyes
2   *Rubber gloves* - Another safety item to consider when servicing the battery - remember that's acid inside the battery!
3   *Battery terminal/cable cleaner* - This wire brush cleaning tool will remove all traces of corrosion from the battery and cable
4   *Treated felt washers* - Placing one of these on each terminal, directly under the cable end, will help prevent corrosion (be sure to get the correct type for side-terminal batteries)
5   *Baking soda* - A solution of baking soda and water can be used to neutralize corrosion
6   *Petroleum jelly* - A layer of this on the battery terminal bolts will help prevent corrosion

you must reset the light. With the Key On and engine off, depress the button once, which will start the light blinking **(see illustration)**. Now depress the button and hold it. The light should go out and a chime should sound to indicate that the reset is complete.

Terminal end corrosion or damage.

Insulation cracks.

Chafed insulation or exposed wires.

Burned or melted insulation.

9.4 Typical battery cable problems

20   The old oil drained from the engine cannot be re-used in its present state and should be recycled. Check with your local auto parts store, disposal facility or environmental agency to see whether they will accept the oil for recycling. Don't pour used oil into drains or onto the ground. After the oil has cooled, it can be drained into a suitable container (capped plastic jugs, topped bottles, milk cartons, etc.) for transport to one of these recycling sites.

9   Battery check, maintenance and charging (every 6000 miles or 6 months)

Refer to illustrations 9.1, 9.4, 9.5a, 9.5b and 9.5c
Warning: *Hydrogen gas is produced by the battery, so keep open flames and lighted tobacco away from it at all times. Always wear eye protection when working around the battery. Rinse off spilled electrolyte immediately with large amounts of water. When removing the battery cables, always detach the negative cable first and hook it up last!*
Caution: *On models equipped with the Theft-lock audio system, be sure the lockout feature is turned off before performing any procedure which requires disconnecting the battery (see the front of this manual):*
1   Battery maintenance is an important procedure that will help ensure you aren't stranded because of a dead battery. Several tools are required for this procedure **(see illustration)**.
2   A sealed battery is standard equipment on all vehicles covered by this manual. Although this type of battery has many advantages over the older, capped cell type, and never requires the addition of water, it should still be routinely maintained according to the procedures that follow.

Check

3   The battery is located in the left front

corner of the engine compartment next to the windshield washer fluid reservoir. The battery removal procedure is described in Chapter 5.
4   Check the tightness of the battery cable terminals and connections to ensure good electrical connections and check the entire length of each cable for cracks and frayed conductors **(see illustration)**.
5   If corrosion (visible as white, fluffy deposits) is evident, remove the cables from the terminals, clean them with a battery brush and reinstall the cables **(see illustrations)**. Corrosion can be kept to a minimum by using special treated fiber washers available at auto parts stores or by applying a layer of petroleum jelly to the terminals and cables after they are assembled.
6   Make sure that the battery tray is in good condition and the hold-down clamp

9.5a A tool like this one (available at auto parts stores) is used to clean the side terminal type battery contact area

9.5b Use the brush to finish the cleaning job

9.5c The result should be a clean, shiny terminal area

bolt is tight. If the battery is removed from the tray, make sure no parts remain in the bottom of the tray when the battery is reinstalled. When reinstalling the hold-down clamp bolt, do not overtighten it. **Note:** *Always reinstall the battery's protective insulating sleeve when putting the battery back in the vehicle.*

7   Information on removing and installing the battery can be found in Chapter 5. Information on jump-starting can be found at the front of this manual. For more detailed battery checking procedures, refer to the *Haynes Automotive Electrical Manual*.

Cleaning

8   Corrosion on the hold-down components, battery case and surrounding areas can be removed with a solution of water and baking soda. Thoroughly rinse all cleaned areas with plain water.

9   Any metal parts of the vehicle damaged by corrosion should be covered with a zinc-based primer, then painted.

Charging

Warning: *When batteries are being charged, hydrogen gas, which is very explosive and flammable, is produced. Do not smoke or allow open flames near a charging or a recently charged battery. Wear eye protection when near the battery during charging. Also, make sure the charger is unplugged before connecting or disconnecting the battery from the charger.*

10   Slow-rate charging is the best way to restore a battery that's discharged to the point where it will not start the engine. It's also a good way to maintain the battery charge in a vehicle that's only driven a few miles between starts. Maintaining the battery charge is particularly important in the winter when the battery must work harder to start the engine and electrical accessories that drain the battery are in greater use.

11   It's best to use a one or two-amp bat-

tery charger (sometimes called a "trickle" charger). They are the safest and put the least strain on the battery. They are also the least expensive. For a faster charge, you can use a higher amperage charger, but don't use one rated more than 1/10th the amp/hour rating of the battery. Rapid boost charges that claim to restore the power of the battery in one to two hours are hardest on the battery and can damage batteries that aren't in good condition. This type of charging should only be used in emergency situations.

12   The average time necessary to charge a battery should be listed in the instructions that come with the charger. As a general rule, a trickle charger will charge a battery in 12 to 16 hours.

13   Remove all of the cell caps (if equipped) and cover the holes with a clean cloth to prevent spattering electrolyte. Disconnect the negative battery cable and hook the battery charger leads to the battery posts (positive to positive, negative to negative), then plug in the charger. Make sure it is set at 12-volts if it has a selector switch.

14   If you're using a charger with a rate higher than two amps, check the battery regularly during charging to make sure it doesn't overheat. If you're using a trickle charger, you can safely let the battery charge overnight after you've checked it regularly for the first couple of hours.

15   If the battery has removable cell caps, measure the specific gravity with a hydrometer every hour during the last few hours of the charging cycle. Hydrometers are available inexpensively from auto parts stores - follow the instructions that come with the hydrometer. Consider the battery charged when there's no change in the specific gravity reading for two hours and the electrolyte in the cells is gassing (bubbling) freely. The specific gravity reading from each cell should be very close to the others. If not, the battery probably has a bad cell(s).

16   Your original sealed factory battery has a built-in hydrometer on the top that indicate the state of charge by the color displayed in

the hydrometer window. Normally, a bright-colored hydrometer indicates a full charge and a dark hydrometer indicates the battery still needs charging. Check the battery manufacturer's instructions to be sure you know what the colors mean.

17   If the battery has a sealed top and no built-in hydrometer, you can hook up a digital voltmeter across the battery terminals to check the charge. A fully charged battery should read 12.5-volts or higher.

10   Cooling system check (every 6000 miles or 6 months)

Refer to illustration 10.4
Caution: *Never mix green-colored ethylene glycol anti-freeze and orange-colored "DEX-COOL" silicate-free coolant because doing so will destroy the efficiency of the "DEX-COOL" coolant which is designed to last for 100,000 miles or five years.*

1   Many major engine failures can be attributed to a faulty engine cooling system. If the vehicle is equipped with an automatic transmission, the cooling system also cools the transmission fluid and plays an important role in prolonging transmission life.

2   The cooling system should be checked with the engine cold. Do this before the vehicle is driven for the day or after the engine has been shut off for at least three hours.

3   Remove the coolant pressure cap on the surge tank by turning it slowly counterclockwise. If you hear any hissing sounds (indicating there is still pressure in the system), wait until it stops. Now continue turning to the left until the cap can be removed. Thoroughly clean the cap, inside and out, with clean water. Also clean the filler neck on the surge tank. All traces of corrosion should be removed. The coolant inside the surge tank should be relatively transparent. If it is rust colored, the system should be drained and refilled (see Section 25). If the coolant level is

Check for a chafed area that could fail prematurely.

Check for a soft area indicating the hose has deteriorated inside.

Overtightening the clamp on a hardened hose will damage the hose and cause a leak.

Check each hose for swelling and oil-soaked ends. Cracks and breaks can be located by squeezing the hose

10.4 Hoses, like drivebelts, have a habit of failing at the worst possible time - to prevent the inconvenience of a blown radiator or heater hose, inspect them carefully as shown here

not up to the top, add additional anti-freeze/coolant mixture (see Section 4).

4   Carefully check the large upper and lower radiator hoses along with any smaller diameter heater hoses that run from the engine to the firewall. Inspect each hose along its entire length, replacing any hose that is cracked, swollen or shows signs of deterioration. Cracks may become more apparent if the hose is squeezed **(see illustration)**.

5   Make sure all hose connections are tight. A leak in the cooling system will usually show up as white or rust-colored deposits on the areas adjoining the leak. If wire-type clamps are used at the ends of the hoses, it

may be wise to replace them with more secure, screw-type clamps.

6   Use compressed air or a soft brush to remove bugs, leaves, etc. from the front of the radiator or air conditioning condenser. Be careful not to damage the delicate cooling fins or cut yourself on them.

7   Every other inspection, or at the first indication of cooling system problems, have the cap and system pressure tested. If you don't have a pressure tester, most gas stations and repair shops will do this for a minimal charge.

11   Underhood hose check and replacement (every 6000 miles or 6 months)

General

1   **Warning:** *Replacement of air conditioning hoses must be left to a dealer service department or air conditioning shop that has the equipment to depressurize the system safely and recover the refrigerant. Never remove air conditioning components or hoses until the system has been depressurized.*

2   High temperatures under the hood can cause the deterioration of the rubber and plastic hoses used for engine, accessory and emission systems operation. Periodic inspection should be made for cracks, loose clamps, material hardening and leaks. Information specific to the cooling system hoses can be found in Section 10.

3   Some, but not all, hoses are secured to the fittings with clamps. Where clamps are used, check to be sure they haven't lost their tension, allowing the hose to leak. If clamps aren't used, make sure the hose hasn't expanded and/or hardened where it slips over the fitting, allowing it to leak.

Vacuum hoses

4   It's quite common for vacuum hoses, especially those in the emissions system, to be color-coded or identified by colored stripes molded into each hose. Various systems require hoses with different wall thicknesses, collapse resistance and temperature resistance. When replacing hoses, be sure the new ones are made of the same material.

5   Often the only effective way to check a hose is to remove it completely from the vehicle. If more than one hose is removed, be sure to label the hoses and fittings to ensure correct installation.

6   When checking vacuum hoses, be sure to include any plastic T-fittings in the check. Inspect the fittings for cracks and the hose where it fits over the fitting for distortion, which could cause leakage.

7   A small piece of vacuum hose (1/4-inch inside diameter) can be used as a stethoscope to detect vacuum leaks. Hold one end of the hose to your ear and probe around

vacuum hoses and fittings, listening for the "hissing" sound characteristic of a vacuum leak. **Warning:** *When probing with the vacuum hose stethoscope, be careful not to allow your body or the hose to come into contact with moving engine components such as the drivebelt, cooling fan, etc.*

Fuel hose

Warning: *Gasoline is extremely flammable, so take extra precautions when you work on any part of the fuel system. Don't smoke or allow open flames or bare light bulbs near the work area, and don't work in a garage where a natural gas-type appliance (such as a water heater or clothes dryer) is present. Since gasoline is carcinogenic, wear latex gloves when there's a possibility of being exposed to fuel, and, if you spill any fuel on your skin, rinse it off immediately with soap and water. Mop up any spills immediately and do not store fuel-soaked rags where they could ignite. When you perform any kind of work on the fuel system, wear safety glasses and have a Class B type fire extinguisher on hand. The fuel system is under pressure, so if any lines must be disconnected, the pressure in the system must be relieved first (see Chapter 4 for more information).*

8   Check all rubber fuel lines for deterioration and chafing. Check especially for cracks in areas where the hose bends and just before fittings, such as where a hose attaches to the fuel filter and fuel injection unit.

9   High quality fuel line, specifically designed for high-pressure fuel injection applications, must be used for fuel line replacement. Never, under any circumstances, use regular fuel line, unreinforced vacuum line, clear plastic tubing or water hose for fuel lines.

10   Spring-type clamps are commonly used on fuel lines. These clamps often lose their tension over a period of time, and can be "sprung" during the removal process. As a result, spring-type clamps should be replaced with screw-type clamps whenever a hose is replaced.

Metal lines

11   Sections of steel tubing often used for fuel line between the fuel pump and fuel injection unit. Check carefully for cracks, kinks and flat spots in the line.

12   If a section of metal fuel line must be replaced, only seamless steel tubing should be used, since copper and aluminum tubing do not have the strength necessary to withstand normal engine vibration.

13   Check the metal brake lines where they enter the master cylinder and brake proportioning unit (if used) for cracks in the lines and loose fittings. Any sign of brake fluid leakage calls for an immediate, thorough inspection of the brake system.

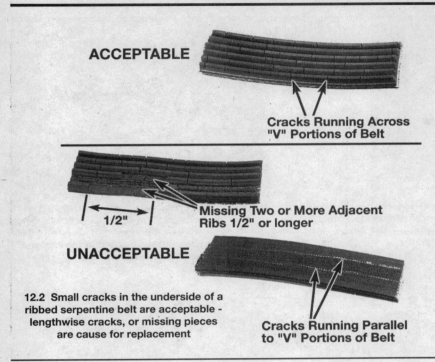

ACCEPTABLE

Cracks Running Across
"V" Portions of Belt

Missing Two or More Adjacent
Ribs 1/2" or longer

1/2"

UNACCEPTABLE

**12.2 Small cracks in the underside of a
ribbed serpentine belt are acceptable -
lengthwise cracks, or missing pieces
are cause for replacement**

Cracks Running Parallel
to "V" Portions of Belt

**12.4 When the mark on the tensioner body is
outside the margins of the marks on the front,
replace the belt**

1

12 Drivebelt and tensioner check and replacement (every 6000 miles or 6 months)

Drivebelt
Check
Refer to illustrations 12.2 and 12.4

1   A single serpentine drivebelt is located at the front of the engine and plays an important role in the overall operation of the engine and its components. Due to its function and material make up, the belt is prone to wear and should be periodically inspected. The serpentine belt drives the alternator, power steering pump, water pump and air conditioning compressor.

2   With the engine off, open the hood and use your fingers (and a flashlight, if necessary), to move along the belt checking for cracks and separation of the belt plies. Also check for fraying and glazing, which gives the belt a shiny appearance **(see illustration)**. Both sides of the belt should be inspected, which means you will have to twist the belt to check the underside.

3   Check the ribs on the underside of the belt. They should all be the same depth, with none of the surface uneven.

4   The tension of the belt is maintained by the tensioner assembly and isn't adjustable. The belt should be checked at the specified mileage; if the belt shows noticeable damage or wear during these checks it should be replaced **(see illustration)**.

Replacement
Refer to illustrations 12.5a, 12.5b and 12.8

5   To replace the belt on four-cylinder engines, rotate the tensioner counterclockwise to release belt tension **(see illustration)**. Note the routing of the belt before removing it. **Note:** *These models have a drivebelt routing decal on the engine to help during drivebelt installation* **(see illustration)**. *If the decal is missing, make a sketch.*

6   Remove the belt from the auxiliary components and slowly release the tensioner.

7   Route the new belt over the various pulleys, again rotating the tensioner to allow the belt to be installed, then release the belt tensioner.

8   To replace the belt on V6 engines, rotate the tensioner counterclockwise to release belt tension, then slip the belt from the alternator pulley **(see illustration)**.

9   To remove the lower end of the belt from the V6 engine, refer to Chapter 2B and support the engine, then remove the right engine mount. Lower the engine enough to remove

**12.5a On 2.4L four-cylinder engines, use a
wrench to rotate the tensioner by the bolt
(arrow) - a socket and breaker bar won't
fit here**

**12.5b A serpentine drivebelt routing
diagram is located on the engine
(3.1L V6 shown)**

**12.8 On V6 engines, rotate the drivebelt
tensioner (arrow) counterclockwise to
remove or install the belt, using a 3/8-
inch-drive breaker bar in the square hole**

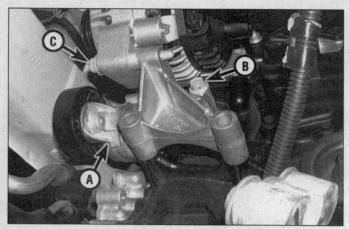

12.14 To remove the tensioner (A) on four-cylinder engines, remove the air conditioning compressor, the bracket bolt (B), then the alternator bolt (C)

12.17 Remove the retaining bolt (arrow) located at the center of the tensioner assembly on V6 engines

the old belt and install the new belt, then raise the engine and reinstall the engine mount (see Chapter 2B). **Caution:** *Never work underneath the engine while it is supported only by a jack.*

10   The remainder of 3.1L belt installation is the reverse of the removal procedure.

Tensioner replacement

Refer to illustrations 12.14 and 12.17

11   Remove the engine drivebelt as described previously.

Four-cylinder engine

12   Refer to Chapter 5 and remove the alternator.

13   Refer to Chapter 3 and remove the air conditioning compressor, without disconnecting the refrigerant lines.

14   Remove the bolts securing the tensioner assembly to the block and replace it **(see illustration)**.

15   The remainder of installation is the reverse of the removal procedure.

V6 engines

16   Support the vehicle and remove the right front wheel and the inner fender liner (see Chapter 11).

17   Remove the tensioner retaining bolt **(see illustration)** and detach the tensioner assembly from the front of the engine.

18   Installation is the reverse of the removal procedure.

---

13   Tire rotation (every 6000 miles or 6 months)

Refer to illustrations 13.2a and 13.2b

1   The tires should be rotated at the specified intervals and whenever uneven wear is noticed. Since the vehicle will be raised and the tires removed anyway, this is a good time to check the brakes (see Section 16).

2   Radial tires must be rotated in a specific

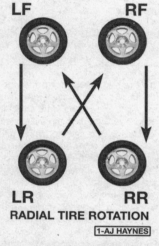

13.2a The recommended four-tire rotation pattern for non-directional radial tires

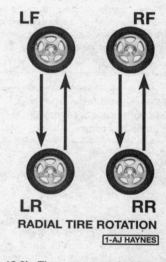

13.2b The recommended four-tire rotation pattern for directional radial tires

pattern **(see illustrations)**. Note: *Most vehicles are sold with non-directional radial tires, but some replacement performance tires are available that are directional, and have a different rotation pattern. Directional tires have an arrow on the sidewall indicating the direction they must turn when mounted on the vehicle.*

3   See the information in *Jacking and towing* at the front of this manual for the proper procedures to follow when raising the vehicle and changing a tire; however, if the brakes are to be checked, don't apply the parking brake as stated. Make sure the tires are blocked to prevent the vehicle from rolling.

4   Preferably, the entire vehicle should be raised at the same time. This can be done on a hoist or by jacking up each corner of the vehicle and lowering it onto jackstands. Always use four jackstands and make sure the vehicle is safely supported.

5   After the tire rotation, check and adjust

the tire pressures as necessary and be sure to check wheel lug nut tightness.

---

14   Seat belt check (every 6,000 miles or 6 months)

1   Check the seat belts, buckles, latch plates and guide loops for obvious damage and signs of wear.

2   See if the seat belt reminder light comes on when the key is turned to the Run or Start position. A chime should also sound.

3   The seat belts are designed to lock up during a sudden stop or impact, yet allow free movement during normal driving. Make sure the retractors return the belt against your chest while driving and rewind the belt fully when the buckle is unlatched.

4   If any of the above checks reveal problems with the seat belt system, replace parts as necessary.

15.6 Pull up the boot on front and rear struts to check for signs of fluid leakage at the point (arrow) where the shaft enters the cartridge (front strut shown, rear similar)

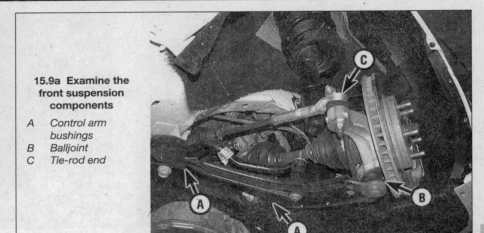

15.9a Examine the front suspension components

A   Control arm bushings
B   Balljoint
C   Tie-rod end

15   Steering, suspension and driveaxle boot check (every 15,000 miles or 12 months)

Note: *The steering linkage and suspension components should be checked periodically. Worn or damaged suspension and steering linkage components can result in excessive and abnormal tire wear, poor ride quality and vehicle handling, and reduced fuel economy. For detailed illustrations of the steering and suspension components, refer to Chapter 10.*

Strut check

Refer to illustration 15.6

1   Park the vehicle on level ground, turn the engine off and set the parking brake. Check the tire pressures.

2   Push down at one corner of the vehicle, then release it while noting the movement of the body. It should stop moving and come to rest in a level position within one or two bounces.

3   If the vehicle continues to move up-and-down or if it fails to return to its original position, a worn or weak strut assembly is probably the reason.

4   Repeat the above check at each of the three remaining corners of the vehicle.

5   Raise the vehicle and support it securely on jackstands.

6   Check the struts for evidence of fluid leakage (see illustration). A light film of fluid is no cause for concern. Make sure that any fluid noted is from the struts and not from some other source. If leakage is noted, replace the struts as a set.

7   Check the struts to be sure that they are securely mounted and undamaged. Check the upper mounts for damage and wear. If damage or wear is noted, replace the struts as a set (front or rear).

8   If the shocks must be replaced, refer to Chapter 10 for the procedure.

Steering and suspension check

Refer to illustrations 15.9a, 15.9b, 15.10 and 15.12

9   Visually inspect the steering and suspension components (front and rear) for damage and distortion. Look for damaged seals, boots and bushings and leaks of any kind At the front end, examine the bushings where the control arm meets the chassis, and at the rear where the trailing arm is bushed, either at the rear knuckle or the chassis end (see illustrations).

10   Inspect the rack-and-pinion steering gear boots for signs of cracking or lubricant leakage (see illustration). If the boots need replacing, refer to Chapter 10.

11   Clean the lower end of the steering knuckle. Have an assistant grasp the lower edge of the tire and move the wheel in-and-

15.9b Examine the rear suspension components

A   Lateral link bushings (there are four links)
B   Stabilizer bar bushings
C   Trailing arm bushings

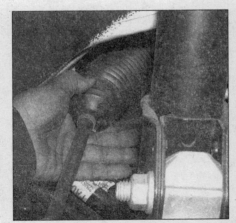

15.10 Flex the steering gear boots (arrow) to check for cracks or signs of leakage

15.12 Check for tie-rod end play by moving the wheel/tire front and rear, and check for balljoint play by moving the top and bottom of the tire

15.15 Inspect the inner and outer driveaxle boots for loose clamps, cracks of signs of leaking lubricant (outer boot shown)

out while you look for movement at the steering knuckle-to-control arm balljoint. If there is any movement, the suspension balljoint(s) must be replaced.

12   Grasp each front tire at the front and rear edges, push in at the front, pull out at the rear, then reverse the motions and feel for play in the steering system components. If any freeplay is noted, check the tie-rod ends for looseness **(see illustration)**.

13   Additional steering and suspension system information and illustrations can be found in Chapter 10.

Driveaxle boot check

Refer to illustration 15.15

14   The driveaxle boots are very important because they prevent dirt, water and foreign material from entering and damaging the constant velocity (CV) joints. Oil and grease can cause the boot material to deteriorate prematurely, so it's a good idea to wash the boots with soap and water. Because it constantly pivots back and forth following the steering action of the front hub, the outer CV boot wears out sooner and should be inspected regularly.

15   Inspect the boots for tears and cracks as well as loose clamps **(see illustration)**. If there is any evidence of cracks or leaking lubricant, they must be replaced as described in Chapter 8.

16   Brake check (every 15,000 miles or 12 months)

Warning: *The dust created by the brake system may contain asbestos, which is harmful to your health. Never blow it out with compressed air and don't inhale any of it. An approved filtering mask should be worn when working on the brakes. Do not, under any circumstances, use petroleum-based solvents to clean brake parts. Use brake system cleaner only! Try to use non-asbestos replacement parts whenever possible.*

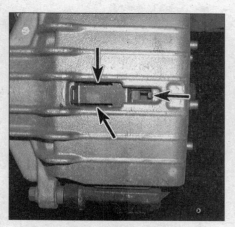

16.7a With the wheel off, check the thickness of the inner pad through the slits in the anti-rattle spring (arrows) - the outer pad thickness can be seen at right

Note: *For detailed photographs of the brake system, refer to Chapter 9.*

1   In addition to the specified intervals, the brakes should be inspected every time the wheels are removed or whenever a defect is suspected.

2   Any of the following symptoms could indicate a potential brake system defect: The vehicle pulls to one side when the brake pedal is depressed; the brakes make squealing or dragging noises when applied; brake pedal travel is excessive; the pedal pulsates; or brake fluid leaks, usually onto the inside of the tire or wheel.

3   Loosen the wheel lug nuts.

4   Raise the vehicle and place it securely on jackstands.

5   Remove the wheels (see *Jacking and towing* at the front of this book, or your owner's manual, if necessary).

Disc brakes

Refer to illustrations 16.7a, 16.7b, 16.9 and 16.11

6   There are two pads (an outer and an

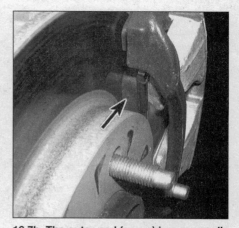

16.7b The outer pad (arrow) is more easily checked at the edge of the caliper

inner) in each caliper. The pads are visible with the wheels removed.

7   Check the pad thickness by looking at each end of the caliper and through the inspection window in the caliper body **(see illustrations)**. If the lining material is less than the thickness listed in this Chapter's Specifications, replace the pads. **Note:** *Keep in mind that the lining material is riveted or bonded to a metal backing plate and the metal portion is not included in this measurement.*

8   If it is difficult to determine the exact thickness of the remaining pad material by the above method, or if you are at all concerned about the condition of the pads, remove the caliper(s), then remove the pads from the calipers for further inspection (refer to Chapter 9).

9   Once the pads are removed from the calipers, clean them with brake cleaner and re-measure them with a ruler or a vernier caliper **(see illustration)**.

10   Measure the disc thickness with a micrometer to make sure that it still has service life remaining. If any disc is thinner than the specified minimum thickness, replace it

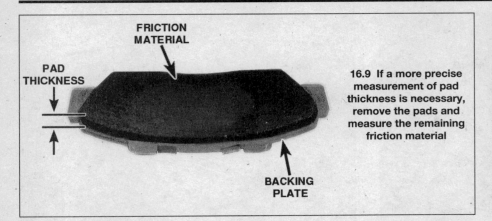

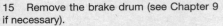

FRICTION
MATERIAL

PAD
THICKNESS

BACKING
PLATE

16.9 If a more precise measurement of pad thickness is necessary, remove the pads and measure the remaining friction material

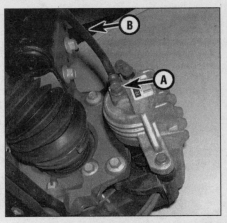

16.11 Check the fitting at the caliper (A), look along the brake hose (B) for signs of cracking or fluid leakage, and check the fitting where the flexible brake hose meets the steel line on the chassis

1

(refer to Chapter 9). Even if the disc has service life remaining, check its condition. Look for scoring, gouging and burned spots. If these conditions exist, remove the disc and have it resurfaced (see Chapter 9).

11   Before installing the wheels, check all brake lines and hoses for damage, wear, deformation, cracks, corrosion, leakage, bends and twists, particularly in the vicinity of the rubber hoses at the calipers **(see illustration)**. Check the clamps for tightness and the connections for leakage. Make sure that all hoses and lines are clear of sharp edges, moving parts and the exhaust system. If any of the above conditions are noted, repair, reroute or replace the lines and/or fittings as necessary (see Chapter 9).

Drum brakes

Refer to illustrations 16.17 and 16.19

12   Raise the vehicle and support it securely on jackstands. Block the front tires to prevent the vehicle from rolling; however, don't apply the parking brake or it will lock the drums in place.

13   Remove the wheels, referring to *Jacking and towing* at the front of this manual if necessary.

14   Mark the hub so it can be reinstalled in the same position. Use a scribe, chalk, etc. on the drum, hub and backing plate.

15   Remove the brake drum (see Chapter 9 if necessary).

16   With the drum removed, carefully clean the brake assembly with brake system cleaner. **Warning:** *Don't blow the dust out with compressed air and don't inhale any of it (it may contain asbestos, which is harmful to your health).*

17   Note the thickness of the lining material on both front and rear brake shoes. If the material has worn away to within 1/16-inch of the recessed rivets or metal backing, the shoes should be replaced **(see illustration)**. The shoes should also be replaced if they're cracked, glazed (shiny areas), or covered with brake fluid.

18   Make sure all the brake assembly springs are connected and in good condition.

19   Check the brake components for signs of fluid leakage. With your finger or a small screwdriver, carefully pry back the rubber boots on the wheel cylinder located at the top of the brake shoes **(see illustration)**. Any leakage here is an indication that the wheel cylinders should be overhauled immediately (see Chapter 9). Also, check all hoses and connections for signs of leakage.

20   Wipe the inside of the drum with a clean rag and denatured alcohol or brake cleaner. Again, be careful not to breathe the dangerous asbestos dust.

21   Check the inside of the drum for cracks, score marks, deep scratches and "hard spots" which will appear as small discolored areas. If imperfections cannot be removed with fine emery cloth, the drum must be taken to an automotive machine shop for resurfacing.

22   Repeat the procedure for the remaining wheel. If the inspection reveals that all parts are in good condition, reinstall the brake drums, install the wheels and lower the vehicle to the ground.

Brake booster check

23   Sit in the driver's seat and perform the following sequence of tests.

24   With the brake fully depressed, start the engine - the pedal should move down a little when the engine starts.

25   With the engine running, depress the brake pedal several times - the travel distance should not change.

26   Depress the brake, stop the engine and hold the pedal in for about 30 seconds - the

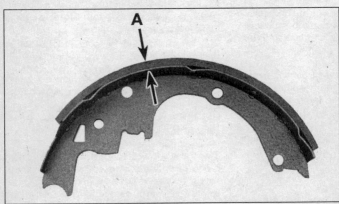

A

16.17 If the lining is bonded to the brake shoe, measure the lining thickness from the outer surface to the metal shoe, as shown here; if the lining is riveted to the shoe, measure from the lining outer surface to the rivet head

16.19 Check for fluid leakage at both ends of the wheel cylinder dust boots

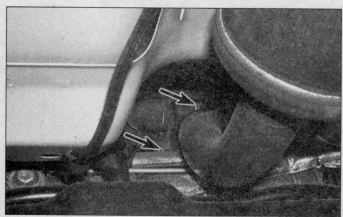

17.2a Check the flange connections for exhaust leaks - also check that the retaining nuts (arrows) are securely tightened

17.2b Check the exhaust system hangers (arrows) for damage and cracks

18.3 Gently pry off the trim cap and check the tightness of the wiper arm retaining nut

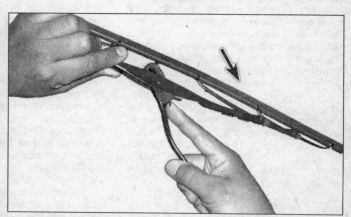

18.5 Press on the release tab (finger is on it here), then slide the blade assembly down and out of the hook in the arm

pedal should neither sink nor rise.

27   Restart the engine, run it for about a minute and turn it off. Then firmly depress the brake several times - the pedal travel should decrease with each application.

28   If your brakes do not operate as described, the brake booster has failed. Refer to Chapter 9 for the replacement procedure.

Parking brake

29   One method of checking the parking brake is to park the vehicle on a steep hill with the parking brake set and the transmission in Neutral. If the parking brake cannot prevent the vehicle from rolling, it's in need of adjustment (see Chapter 9).

17   Exhaust system check (every 15,000 miles or 12 months)

Refer to illustrations 17.2a and 17.2b

1   With the engine cold (at least three hours after the vehicle has been driven), check the complete exhaust system from the engine to the end of the tailpipe. Ideally, the inspection should be done with the vehicle on

a hoist to permit unrestricted access. If a hoist is not available, raise the vehicle and support it securely on jackstands.

2   Check the exhaust pipes and connections for evidence of leaks, severe corrosion and damage. Make sure that all brackets and hangers are in good condition and tight (see illustrations).

3   At the same time, inspect the underside of the body for holes, corrosion, open seams, etc. which may allow exhaust gases to enter the interior. Seal all body openings with silicone or body putty.

4   Rattles and other noises can often be traced to the exhaust system, especially the mounts and hangers. Try to move the pipes, muffler and catalytic converter. If the components can come in contact with the body or suspension parts, secure the exhaust system with new mounts.

5   This is also an ideal time to check the running condition of the engine by inspecting the very end of the tailpipe. The exhaust deposits here are an indication of engine state-of-tune. If the pipe is black and sooty or coated with white deposits, the engine may be in need of a tune-up (including a thorough fuel injection system inspection).

18   Wiper blade inspection and replacement (every 15,000 miles or 12 months)

Refer to illustrations 18.3, 18.5 and 18.6

1   The windshield wiper and blade assemblies should be inspected periodically for damage, loose components and cracked or worn blade elements.

2   Road film can build up on the wiper blades and affect their efficiency, so they should be washed regularly with a mild detergent solution.

3   The action of the wiping mechanism can loosen the bolts, nuts and fasteners, so they should be checked and tightened, as necessary, at the same time the wiper blades are checked (see illustration).

4   If the wiper blade elements (sometimes called inserts) are cracked, worn or warped, they should be replaced with new ones.

5   Lift the arm assembly away from the glass for clearance, press on the release lever, then slide the wiper blade assembly out of the hook in the end of the arm (see illustration).

6   Use needle-nose pliers to compress the

18.6 Use needle-nose pliers to compress the prongs on the rubber element, then slide the element out - slide the new element in and lock the blade assembly fingers into the prongs of the wiper element - on models with rubber prongs, the element can be pulled out by hand without pliers

19.1a Remove the screws and separate the cover from the air cleaner housing

1

blade element clips, then slide the element out of the frame and discard it **(see illustration)**.

7   Compare the new element with the old for length, design, etc. Some replacement elements come in a three-piece design (two metal strips, one on either side of the rubber) that is held together by several small plastic sleeves. Keep the sleeves in place on this design until you start sliding the element into the frame. Remove each of the plastic sleeves as needed when they reach the frame.

8   Slide the new element into the frame, notched end last and secure the clips into the notches of the frame.

9   Reinstall the blade assembly on the arm, wet the windshield and test for proper operation.

19   Air filter replacement (every 30,000 miles or 24 months)

Refer to illustration 19.1a and 19.1b

1   The air filter is located inside the air cleaner housing at the left (driver's) side of the engine compartment. To remove the air filter, release the screws **(see illustration)** that secure the two halves of the air cleaner housing together, then separate the cover halves and remove the air filter element **(see illustration)**.

2   Inspect the outer surface of the filter element. If it is dirty, replace it. If it is only moderately dusty, it can be reused by blowing it clean from the back to the front surface with compressed air. Because it is a pleated paper type filter, it cannot be washed or oiled. If it cannot be cleaned satisfactorily with compressed air, discard and replace it. While the

19.1b Lift the cover up and slide the element out of the housing

cover is off, be careful not to drop anything down into the housing. **Caution:** *Never drive the vehicle with the air cleaner removed. Excessive engine wear could result and backfiring could even cause a fire under the hood.*

3   Wipe out the inside of the air cleaner housing.

4   Place the new filter into the air cleaner housing, making sure it seats properly.

5   Installation of the housing is the reverse of removal.

20   Positive Crankcase Ventilation (PCV) valve check and replacement (V6 engines only) (every 30,000 miles or 24 months)

Check

Refer to illustration 20.2

1   On V6 engines the PCV valve is located in the forward valve cover.

2   With the engine idling at normal operating temperature, pull the valve (with hose attached) out of the rubber grommet in the

20.2 The PCV valve (arrow) on V6 engines is located in the front valve cover - pull it out and check for vacuum with your finger with the engine idling

valve cover **(see illustration)**.

3   Place your finger over the end of the valve. If there is no vacuum at the valve, check for a plugged hose, manifold port, or the valve itself. Replace any plugged or deteriorated hoses.

4   Turn off the engine and shake the PCV valve, listening for a rattle. If the valve doesn't rattle, replace it with a new one.

Replacement

5   To replace the valve, pull it out of the end of the hose, noting its installed position and direction.

6   When purchasing a replacement PCV valve, make sure it's for your particular vehicle, model year and engine size. Compare the old valve with the new one to make sure they are the same.

7   Push the valve into the end of the hose until it's seated.

8   Inspect the rubber grommet for damage and replace it with a new one if necessary.

9   Push the PCV valve and hose securely into position.

21 Spark plug wire check and replacement (V6 engines only) (every 30,000 miles or 24 months)

Refer to illustration 21.4

1   The spark plug wires should be checked and, if necessary, replaced at the same time new spark plugs are installed.

2   The easiest way to identify bad wires is to make a visual check while the engine is running. In a dark, well-ventilated garage, start the engine and look at each plug wire. Be careful not to come into contact with any moving engine parts. If there is a break in the wire, you will see arcing or a small spark at the damaged area. If arcing is noticed, make a note to obtain new wires.

3   The spark plug wires should be inspected one at a time, beginning with the spark plug for the number one cylinder (see the wire diagram at the beginning of this Chapter) to prevent confusion. Clearly label each plug wire with a piece of tape marked with the correct number. The plug wires must be reinstalled in the correct order to ensure proper engine operation.

4   Disconnect the plug wire from the first spark plug. A removal tool can be used, or you can grab the wire boot, twist it a half-turn and pull the boot free. Do not pull on the wire itself, only on the rubber boot **(see illustration)**.

5   Push the wire and boot back onto the end of the spark plug. It should fit snugly. If it doesn't, detach the wire and boot once more and use a pair of pliers to carefully crimp the metal connector inside the wire boot until it does.

6   Using a clean rag, wipe the entire length of the wire to remove built-up dirt and grease.

7   Once the wire is clean, check for burns, cracks and other damage. Do not bend the wire sharply or you might break the conductor.

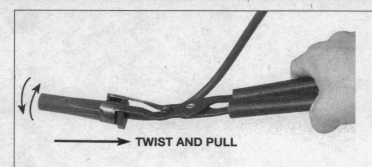

21.4 Using a spark plug boot puller tool like this one will make the job of removing the spark plug boots much easier

TWIST AND PULL

8   Disconnect the wire from the coil pack. Pull only on the rubber boot. Check for corrosion on the coil pack towers and the plug wire ends, and that they fit tight. Reinstall the wire.

9   Inspect each of the remaining spark plug wires, making sure that each one is securely fastened on each end.

10   If new spark plug wires are required, purchase a set for your specific engine model. Pre-cut wire sets with the boots already installed are available. Remove and replace the wires one at a time to avoid mix-ups in the firing order. Should a mix up occur, refer to the Specifications at the beginning this Chapter.

22 Spark plug check and replacement (see maintenance schedule for service intervals)

Note: *The engines covered by this manual are equipped with either a distributorless Direct Ignition System (DIS) (V6 engines) or an Integrated Direct Ignition system (IDI) (2.4L four-cylinder engine). On the DIS system, the coil pack and ignition module are mounted at the top-rear of the engine, with spark plug wires connecting the spark plugs to the coil packs. The IDI system on 2.4L engines incorporates the ignition module, coils and spark plug boots as one complete unit mounted directly on top of the cylinder head, without spark plug wires.*

Caution: *The engine should be cool when the spark plugs are removed.*

Four-cylinder engine

Refer to illustrations 22.2 and 22.3

1   Disconnect the negative battery cable (see Chapter 1). **Caution:** *On models equipped with the Theftlock audio system, be sure the lockout feature is turned off before performing any procedure which requires disconnecting the battery (see the front of this manual).*

2   Disconnect the electrical connector at the IDI cover, then remove the four cover mounting bolts **(see illustration)**.

3   Pull the IDI cover straight up until the boots come free of the spark plugs, and set the cover aside **(see illustration)**. **Note:** *Some of the boots may stay stick on the spark plugs and not come up with the cover. Pull the boots from the plugs and reattach them to the coil towers on the bottom of the IDI cover.*

22.2 On 2.4L engines, remove the four bolts (arrow) on the ignition coil cover

22.3 Pull the cover straight up - some of the plug boots (arrows) may stick to the plugs, but can be pulled from the plugs with a boot puller

V6 engines

4   A special plug wire removal tool is available for separating the wire boots from the spark plugs, and is a good idea on these models because the boots fit very tightly **(see illustration 21.4)**.

5   Twist the boot one-half turn and remove it from the plug.

All models

Refer to illustrations 22.6, 22.9a, 22.9b, 22.12, 22.13 and 22.14

6   In most cases, the tools necessary for spark plug replacement include a spark plug socket which fits onto a ratchet (spark plug sockets are padded inside to prevent damage to the porcelain insulators on the new plugs), various extensions and a gap gauge to check and adjust the gaps on the new plugs **(see illustration)**. A torque wrench should be used to tighten the new plugs. It is a good idea to allow the engine to cool before removing or installing the spark plugs.

7   The best approach when replacing the spark plugs is to purchase the new ones in advance, adjust them to the proper gap and replace the plugs one at a time. When buying the new spark plugs, be sure to obtain the correct plug type for your particular engine. The plug type can be found in the Specifications at the front of this Chapter.

8   Allow the engine to cool completely before attempting to remove any of the plugs. While you are waiting for the engine to cool, check the new plugs for defects and adjust the gaps.

9   Check the gap by inserting the proper thickness gauge between the electrodes at the tip of the plug **(see illustration)**. The gap between the electrodes should be the same as the one specified on the Emissions Control Information label or as listed in this Chapter's Specifications. The wire should slide between the electrodes with a slight amount of drag. If the gap is incorrect, use the

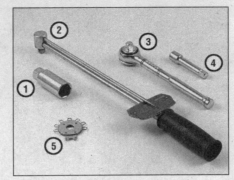

22.6 Tools required for changing spark plugs

1   **Spark plug socket** - This will have special padding inside to protect the spark plug's porcelain insulator
2   **Torque wrench** - Although not mandatory, using this tool is the best way to ensure the plugs are tightened properly
3   **Ratchet** - Standard hand tool to fit the spark plug socket
4   **Extension** - Depending on model and accessories, you may need special extensions and universal joints to reach one or more of the plugs
5   **Spark plug gap gauge** - This gauge for checking the gap comes in a variety of styles. Make sure the gap for your engine is included

adjuster on the gauge body to bend the curved side electrode slightly until the proper gap is obtained **(see illustration)**. If the side electrode is not exactly over the center electrode, bend it with the adjuster until it is. Check for cracks in the porcelain insulator (if any are found, the plug should not be used). **Note:** *Manufacturers recommend using a tapered thickness gauge when checking platinum-type spark plugs. Other types of gauges may scrape the thin platinum coating from the*

22.9a Spark plug manufacturers recommend using a tapered-thickness gauge when checking the gap - slide the thin side into the gap and turn until the gauge just fills the gap, then read the thickness on the gauge - do not force the tool into the gap or use the tapered portion to widen a gap

electrodes, thus dramatically shortening the life of the plugs.

10   If compressed air is available, use it to blow any dirt or foreign material away from the spark plug hole. The idea here is to eliminate the possibility of debris falling into the cylinder as the spark plug is removed.

11   The spark plugs on V6 models are, for the most part, difficult to reach so a spark plug socket incorporating a universal joint will be necessary.

12   Place the spark plug socket over the plug and remove it from the engine by turning it in a counterclockwise direction **(see illustration)**.

13   Compare the spark plug with the chart shown on the inside back cover of this manual to get an indication of the general running condition of the engine. Before installing the new plugs, it is a good idea to apply a thin coat of anti-seize compound to the threads **(see illustration)**.

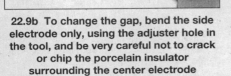

22.9b To change the gap, bend the side electrode only, using the adjuster hole in the tool, and be very careful not to crack or chip the porcelain insulator surrounding the center electrode

22.12 On four-cylinder engines, use a straight extension and spark plug socket to remove the plugs from between the cam housings

22.13 Apply a thin coat of anti-seize compound to the spark plug threads, being careful not to get any near the lower threads (arrows)

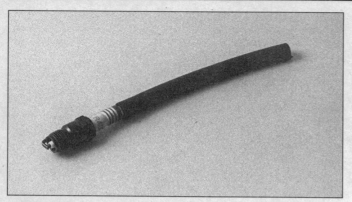

22.14 A length of snug-fitting rubber hose will save time and prevent damaged threads when installing the spark plugs

23.3 The fuel filter (arrow) is located underneath the vehicle in front of the gas tank

14   Thread one of the new plugs into the hole until you can no longer turn it with your fingers, then tighten it with a torque wrench (if available) or the ratchet. It's a good idea to slip a short length of rubber hose over the end of the plug to use as a tool to thread it into place **(see illustration)**. The hose will grip the plug well enough to turn it, but will start to slip if the plug begins to cross-thread in the hole - this will prevent damaged threads and the accompanying repair costs.

15   On V6 engines, before pushing the spark plug wire onto the end of the plug, inspect it following the procedures outlined in Section 21. Attach the plug wire to the new spark plug, again using a twisting motion on the boot until it's seated on the spark plug.

16   Repeat the procedure for the remaining spark plugs. On V6 engines, replace them one at a time to prevent mixing up the spark plug wires.

17   After replacing all four plugs on 2.4L engines, push the IDI assembly straight down, making sure the boots align with the spark plugs, then tighten the four mounting bolts and reconnect the electrical connector, making sure it is properly oriented to the socket.

23   Fuel filter replacement (every 30,000 miles or 24 months)

Refer to illustrations 23.3 and 23.5
Warning: *Gasoline is extremely flammable, so take extra precautions when you work on any part of the fuel system. Don't smoke or allow open flames or bare light bulbs near the work area, and don't work in a garage where a natural gas-type appliance (such as a water heater or clothes dryer) is present. Since gasoline is carcinogenic, wear latex gloves when there's a possibility of being exposed to fuel, and, if you spill any fuel on your skin, rinse it off immediately with soap and water. Mop up any spills immediately and do not store fuel-soaked rags where they could ignite. The fuel system is under constant pressure, so, if any fuel lines are to be disconnected, the fuel pressure in the system must be relieved first (see Chapter 4 for more infor-*

23.5 Squeeze the white plastic quick-disconnect tabs together and pull the inlet line (A) away from the filter, then detach the outlet line (B) from the fuel filter - unbolt the filter mounting bracket to remove the fuel filter

mation). When you perform any kind of work on the fuel system, wear safety glasses and have a Class B type fire extinguisher on hand.

1   Relieve the fuel system pressure (see Chapter 4). Disconnect the cable from the negative terminal of the battery. **Caution:** *On models equipped with the Theftlock audio system, be sure the lockout feature is turned off before performing any procedure which requires disconnecting the battery (see the front of this manual).*

2   Raise the vehicle and support it securely on jackstands.

3   The fuel filter is mounted to the floor pan just in front of the fuel tank **(see illustration)**.

4   Use compressed air or carburetor cleaner to clean any dirt surrounding the fuel inlet and outlet line fittings.

5   Once all the dirt has been removed, twist the two lines a quarter-turn in opposite directions to loosen the fitting. Depress the white plastic quick-disconnect tabs and detach the inlet line from the fuel filter **(see illustration)**. **Note:** *Have spare rags or a small container to catch or wipe up extra gasoline that will spill from the filter assembly.*

6   Use an open end wrench to steady the outlet side of the filter and a flare nut wrench to unscrew the fuel line nut, then separate the outlet line from the filter while noting the installed position of the O-ring. Using a flare nut wrench will help to avoid rounding the corners off of the fuel line nut.

7   Detach the fuel filter mounting bracket bolt and remove the fuel filter.

8   Installation is the reverse of removal. **Note:** *Be sure to install a new O-ring on the outlet line before reassembling.*

24   Fuel system check (every 30,000 miles or 24 months)

Refer to illustration 24.4
Warning: *Gasoline is extremely flammable, so take extra precautions when you work on any part of the fuel system. Don't smoke or allow open flames or bare light bulbs near the work area, and don't work in a garage where a natural gas-type appliance (such as a water heater or clothes dryer) is present. Since gasoline is carcinogenic, wear latex gloves when there's a possibility of being exposed to fuel, and, if you spill any fuel on your skin, rinse it off immediately with soap and water. Mop up any spills immediately and do not store fuel-soaked rags where they could ignite. The fuel system is under constant pressure, so, if any fuel lines are to be disconnected, the fuel pressure in the system must be relieved first (see Chapter 4 for more information). When you perform any kind of work on the fuel system, wear safety glasses and have a Class B type fire extinguisher on hand.*

1   The fuel system is most easily checked with the vehicle raised on a hoist so the components underneath the vehicle are readily visible and accessible.

2   If the smell of gasoline is noticed while

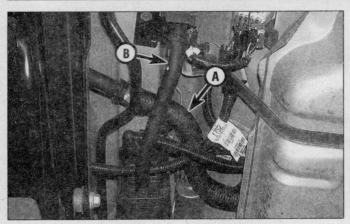

24.4 At the fuel tank, check the fuel filler hose (A) and the hose to the EVAP canister (B) for cracks and deterioration and the hose clamps for tightness

25.3 Use the square end of a 1/4-inch drive extension to twist the drain plug (arrow) - the coolant will drain out of the hole in the radiator's lower left rubber mount (not seen here)

driving or after the vehicle has been in the sun, the system should be thoroughly inspected immediately.

3   Remove the fuel filler cap and check for damage, corrosion and an unbroken sealing imprint on the gasket. Replace the cap with a new one if necessary.

4   With the vehicle raised, inspect the fuel tank and filler neck for cracks and other damage **(see illustration)**. The connection between the filler neck and tank is especially critical. Sometimes a filler neck will leak due to cracks, problems a home mechanic can't repair. **Warning:** *Do not, under any circumstances, try to repair a fuel tank yourself (except rubber components). A welding torch or any open flame can easily cause the fuel vapors to explode if the proper precautions are not taken.*

5   Carefully check all rubber hoses and metal lines leading away from the fuel tank. Check for loose connections, deteriorated hoses, crimped lines and other damage. Follow the lines to the front of the vehicle, carefully inspecting them all the way. Repair or replace damaged sections as necessary.

25 Cooling system servicing (draining, flushing and refilling) (see maintenance schedule for service intervals)

Warning: *Make sure the engine is completely cool before performing this procedure. Do not allow antifreeze to come in contact with your skin or painted surfaces of the vehicle. Flush contacted areas immediately with plenty of water. Do not store new coolant or leave old coolant lying around where it is easily accessible to children and pets, because they are attracted by its sweet taste. Ingestion of even a small amount can be fatal. Wipe up the garage floor and drip pan coolant spills immediately. Keep antifreeze containers covered and repair leaks in your cooling system immediately. Antifreeze is flammable - be*

sure to read the precautions on the container. **Note:** *Non-toxic coolant is available at local auto parts stores. Although the coolant is non-toxic when fresh, proper disposal is still required.*
Caution: *Never mix green-colored ethylene glycol anti-freeze and orange-colored "DEX-COOL" silicate-free coolant because doing so will destroy the efficiency of the "DEX-COOL" coolant which is designed to last for 100,000 miles or five years.*

Draining

Refer to illustrations 25.3 and 25.4

1   Periodically, the cooling system should be drained, flushed and refilled to replenish the antifreeze mixture and prevent formation of rust and corrosion, which can impair the performance of the cooling system and cause engine damage. When the cooling system is serviced, all hoses and the radiator cap should be checked and replaced if necessary.

2   Apply the parking brake and block the wheels. **Warning:** *If the vehicle has just been driven, wait several hours to allow the engine to cool down before beginning this procedure.*

3   Move a large container under the radiator drain to catch the coolant. The drain plug

is located on the lower right side of the radiator **(see illustration)**. Attach a 3/8-inch diameter hose to the drain fitting (if possible) to direct the coolant into the container, then open the drain fitting (a pair of pliers may be required to turn it). Remove the coolant reservoir cap.

4   After coolant stops flowing out of the radiator, move the container under the engine block drain plugs - there's one on each side of the block **(see illustration)**. Remove the plugs and allow the coolant in the block to drain. **Note:** *Frequently, the coolant will not drain from the block after the plug is removed. This is due to a rust layer that has built up behind the plug. Insert a Phillips screwdriver into the hole to break the rust barrier.*

5   While the coolant is draining, check the condition of the radiator hoses, heater hoses and clamps (refer to Section 10 if necessary).

6   Replace any damaged clamps or hoses.

Flushing

Refer to illustration 25.9

7   Once the system is completely drained, remove the thermostat from the engine (see Chapter 3). Then reinstall the thermostat housing without the thermostat. This will allow the system to be thoroughly flushed.

25.4 Location of the block drain plugs (arrow indicates one of two, four-cylinder engine shown)

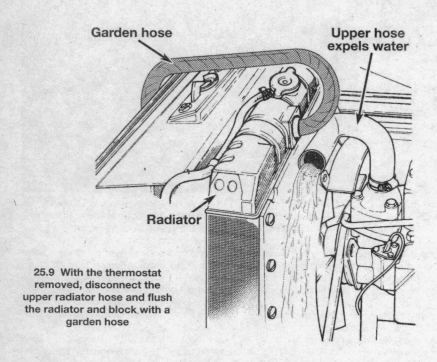

Garden hose

Upper hose expels water

Radiator

25.9 With the thermostat removed, disconnect the upper radiator hose and flush the radiator and block with a garden hose

8　Reinstall the lower radiator hose and twist the radiator drain valve shut. Turn your heating system controls to Hot, so that the heater core will be flushed at the same time as the rest of the cooling system.

9　Disconnect the upper radiator hose, then place a garden hose in the upper radiator inlet and flush the system until the water runs clear at the upper radiator hose **(see illustration)**.

10　In severe cases of contamination or clogging of the radiator, remove the radiator (see Chapter 3) and have a radiator repair facility clean and repair it if necessary.

11　Many deposits can be removed by the chemical action of a cleaner available at auto parts stores. Follow the procedure outlined in the manufacturer's instructions. **Note:** *When the coolant is regularly drained and the system refilled with the correct antifreeze/water mixture, there should be no need to use chemical cleaners or descalers.*

Refilling

Refer to illustration 25.13

12　Install the coolant reservoir, reconnect the hoses, close the drain fitting hand tight and install the block drain plugs, using Permatex #2 sealant on the threads of the plugs.

13　On V6 engines, connect one end of a four foot long piece of 1/4-inch diameter clear plastic tubing to the bleed screw and run the other end into an empty coolant container **(see illustration)**. **Caution:** *Keep the bleeder hose away from hot or moving com-*

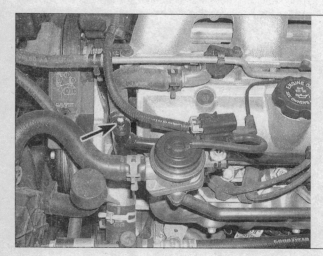

25.13 Use a small wrench to open the bleeder screw (arrow) two or three turns on V6 engines

ponents on the engine.

14　Open the bleed screw and slowly add coolant to the surge tank until the coolant stream running from the hose is bubble-free. Squeeze the upper radiator hose gently to make sure all remaining air is expelled, then close the bleed valve screw and remove the hose. Fill the coolant surge tank until the level is above the Full Cold mark, then remove the clamp from the connecting hose so that a little coolant goes into the secondary chamber.

15　Run the engine until normal operating temperature is reached and, with the engine idling, add coolant up the correct level. Install the cap on the surge tank.

16　Always refill the system with a mixture of antifreeze and water in the proportion called for on the antifreeze container; in this Chapter's Specifications or in your owner's manual. Chapter 3 also contains information on antifreeze mixtures.

17　Keep a close watch on the coolant level and the various cooling system hoses during the first few miles of driving. Tighten the hose clamps and add more coolant mixture as necessary.

26 Brake fluid change (every 30,000 miles or 24 months)

Warning: *Brake fluid can harm your eyes and damage painted surfaces, so use extreme caution when handling or pouring it. Do not use brake fluid that has been standing open or is more than one year old. Brake fluid absorbs moisture from the air. Excess moisture can cause a dangerous loss of braking effectiveness.*

1　At the specified intervals, the brake fluid should be drained and replaced. Since the brake fluid may drip or splash when pouring it, place plenty of rags around the master cylinder to protect any surrounding painted surfaces.

2　Before beginning work, purchase the specified brake fluid (see *Recommended lubricants and fluids* at the beginning of this Chapter).

3　Remove the cap from the master cylinder reservoir.

4　Using a hand suction pump or similar device, withdraw the fluid from the master cylinder reservoir.

5　Add new fluid to the master cylinder until it rises to the base of the filler neck.

6　Bleed the brake system as described in Chapter 9 at all four brakes until new and uncontaminated fluid is expelled from the bleeder screw. Be sure to maintain the fluid level in the master cylinder as you perform the bleeding process. If you allow the master cylinder to run dry, air will enter the system.

7　Refill the master cylinder with fluid and check the operation of the brakes. The pedal should feel solid when depressed, with no sponginess. **Warning:** *Do not operate the vehicle if you are in doubt about the effectiveness of the brake system.*

27.7 After removing the front and side pan bolts, loosen the rear bolts and allow the fluid to drain, then remove the bolts and lower the pan from the vehicle

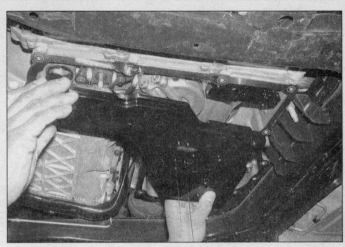

27.10a Pull the transaxle filter straight down and out of the transaxle - there are no fasteners

27.10b Pry out the old seal, being careful not to damage the aluminum housing

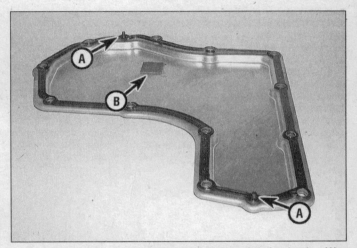

27.12 Place the gasket on the pan, aligning the plastic pins (A) with the holes in the pan for alignment (they also help align the pan to the transaxle) - make sure the magnet (B) is clean and in place before installing the pan

27 Automatic transaxle fluid and filter change (every 60,000 miles or 48 months)

Refer to illustrations 27.7, 27.10a, 27.10b 27.12, 27.17 and 27.19

1   At the specified time intervals, the transaxle fluid should be drained and replaced. Since the fluid will remain hot long after driving, perform this procedure only after everything has cooled down completely.

2   Before beginning work, purchase the specified transaxle fluid (see *Recommended lubricants and fluids* at the front of this Chapter) and a new filter.

3   Other tools necessary for this job include jackstands to support the vehicle in a raised position, a drain pan capable of holding several quarts, newspapers and clean rags.

4   Raise and support the vehicle on jackstands.

5   With a drain pan in place, remove the front and side transaxle pan mounting bolts.

6   Loosen the rear pan bolts one turn.

7   Carefully pry the transaxle pan loose with a screwdriver, allowing the fluid to drain **(see illustration)**.

8   Remove the remaining bolts, pan and gasket. Carefully clean the gasket surface of the transaxle to remove all traces of the old gasket and sealant.

9   Drain the fluid from the transaxle pan, clean the pan with solvent and dry it with compressed air. Be careful not to lose the magnet.

10   Remove the filter and pry out the seal **(see illustrations)**.

11   Push a new filter seal fully into its bore, then install the new filter.

12   Make sure the gasket surface on the transaxle pan is clean, then install the new gasket **(see illustration)**. Put the pan in place against the transaxle and install the bolts. Working around the pan, tighten each bolt a little at a time until the final torque figure is reached. **Note:** *The manufacturer recommends using new bolts, and coating the threads with a thin film of RTV sealant.*

13   Lower the vehicle and add the specified amount of automatic transmission fluid through the vent/fill cap and check the fluid level (see below).

14   Check under the vehicle for leaks during the first few trips.

Fluid level check

15   The automatic transaxle fluid level should be carefully maintained. Low fluid level can lead to slipping or loss of drive, while overfilling can cause foaming and loss of fluid. **Warning:** *This procedure is potentially dangerous and is best left to a profes-*

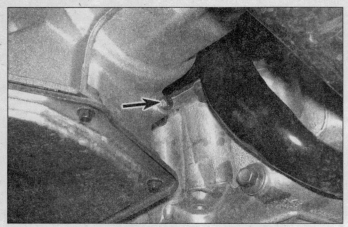

27.17 Location of the check plug (arrow) on the automatic transaxle

27.19 Location of the vent/fill cap (arrow) on the automatic transaxle

sional shop with a safe lifting apparatus. The vehicle must be kept level while being safely raised high enough for access to the check plug on the transaxle.

16   With the vehicle raised and safely supported, start the engine, then move the shift lever through all the gear ranges, ending in Park. **Note:** *Incorrect fluid level readings will result if the vehicle has just been driven at high speeds for an extended period, in hot weather in city traffic, or if it has been pulling a trailer. If any of these conditions apply, wait until the fluid has cooled (about 30 minutes).*

17   With the engine running and the transaxle at normal operating temperature (having idled for 3 to 5 minutes), locate the check plug on the transaxle. The check plug is located near the pan, adjacent to the engine oil drain plug **(see illustration)**.

18   Place an oil container under the check plug and remove it. Observe the fluid as it drips into the pan, indicating correct fluid level.

19   The fluid level should be at the bottom of the check hole. If fluid pours out excessively, the transaxle may have been over-filled. Double-check to make sure the vehicle is level. If no fluid drips from the check hole, add small amounts of fluid through the vent/fill cap at the top of the transaxle until the level is at the bottom of the check hole **(see illustration)**. A long-necked funnel will be necessary to add fluid.

20   The condition of the fluid should also be checked along with the level. If the fluid in the drain pan is a dark reddish-brown color, or if the fluid has a burned smell, the fluid should be changed (see above). If you're in doubt about the condition of the fluid, purchase some new fluid and compare the two for color and smell.

21   Be sure to install the check plug and tighten it securely when you're done.

Chapter 2 Part A
Four-cylinder engine

Contents

Specifications

General

Displacement	
2.4 liter	146 cubic inches
Bore	3.54 inches
Stroke	3.70 inches
Cylinder numbers (drivebelt end-to-transaxle end)	1-2-3-4
Firing order	1-3-4-2
Oil pressure (minimum)	
At 900 rpm	10 psi
At 3000 rpm	30 psi

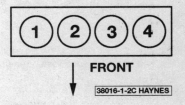

**Cylinder identification diagram -
four-cylinder engine**

Camshafts and housings

Lobe lift (intake and exhaust)	
Intake	0.3540 inch
Exhaust	0.3460 inch
Lobe taper limit	0.0018 to 0.0033 inch
Endplay	0.0009 to 0.0088 inch
Journal diameter	
No. 1	1.5720 to 1.5728 inches
All others	1.3751 to 1.3760 inches
Bearing oil clearance	0.0019 to 0.0043 inch
Lifters	
Bore diameter	1.3381 to 1.3393 inches
Outside diameter	1.3369 to 1.3375 inches
Lifter-to-bore clearance	0.0006 to 0.0023 inch
Camshaft housing warpage limit	0.001 inch per 4.0 inches

Balance shafts and housing

Chain slack (with 3 lb. applied to chain guide)	0.040 inch
End play	0.0073 to 0.0179 inch
Gear backlash	0.0029 to 0.0057 inch
Housing warpage maximum	0.003 inch
Journal diameter	1.1791 to 1.1801 inches
Bearing oil clearance	0.0017 to
Thrust plate thickness	0.1159 to 0.1199 inch
Balance shaft chain guide wear limit	0.100 inch

Oil pump

Gerotor pocket depth	0.6023 to 0.6043 inch
Outer rotor thickness	0.5994 to 0.6003 inch
Outer rotor diameter	1.773 to 1.775 inches
Outer rotor-to-housing clearance	0.0019 to 0.0059 inch
Outer rotor-to-inner rotor tip clearance (maximum)	0.0059 inch
Oil pump drive to driven gear backlash	0.0029 to 0.0055 inch

Torque specifications

Ft-lbs (unless otherwise indicated)

Balance shaft housing bolts (see illustration 15.19)	
Step 1	
Bolts 1, 2, 4, 5, 6 and 7	89 in-lbs
Bolts 3 and 8	132 in-lbs
Step 2	Tighten all bolts an additional 40-degrees
Balance shaft housing assembly-to-block bolts (see illustration 15.19)	
Step 1	
Bolts 9, 10 and 12	18
Bolt 11	30
Bolt 13	39
Step 2	
Bolts 9, 10 and 12	Tighten an additional 70-degrees
Bolt 11	Tighten an additional 60-degrees
Balance shaft sprocket bolt (left hand thread)	
Step 1	30
Step 2	Tighten an additional 45-degrees
Balance shaft chain tensioner bolt	115 in-lbs
Balance shaft chain cover nut and bolt	115 in-lbs
Balance shaft thrust plate bolts	115 in-lbs
Camshaft housing-to-cylinder head bolts	
Step 1	16
Step 2	Tighten an additional 90-degrees
Camshaft cover-to-camshaft housing bolts	
(rear two on the intake camshaft housing)	
Step 1	16
Step 2	Tighten an additional 30-degrees
Camshaft sprocket-to-camshaft bolt	52
Crankshaft pulley bolt	
Step 1	129
Step 2	Tighten an additional 90-degrees
Crankshaft rear main oil seal housing bolts	106 in-lbs
Cylinder head bolts (in sequence - **see illustration 10.15)**	
Step 1	
Bolts 1 thru 8	40
Bolts 9 and 10	30
Step 2	Tighten all bolts an additional 90-degrees
Engine mount-to-bracket bolts	96
Engine mount-to-body nuts	49
Engine mount bracket-to-engine bolts *	
Step 1	44
Step 2	Tighten an additional 60-degrees
Engine mount strut bolts	
Step 1	74
Step 2	Tighten an additional 90-degrees
Exhaust manifold nuts-to-cylinder head	132 in-lbs
Exhaust manifold brace	
Bolts	41
Nuts	19
Flywheel-to-crankshaft bolts	
Step 1	22
Step 2	Tighten an additional 45-degrees

Torque specifications

	Ft-lbs (unless otherwise indicated)
Intake manifold-to-cylinder head nuts/bolts...	19
Oil pan bolts	
M6	106 in-lbs
M8	18
Oil pump-to-balance shaft housing bolts	
Short	89 in-lbs
Long	106 in-lbs
Timing chain cover-to-housing bolts...	106 in-lbs
Timing chain housing-to-camshaft housing bolts	19
Timing chain housing-to-block bolts	
8 mm	21
10 mm	37
Timing chain tensioner..	89 in-lbs

* Replace with new bolts anytime they are removed.

1   General information

Caution: *On models equipped with the Theft-lock audio system, be sure the lockout feature is turned off before performing any procedure which requires disconnecting the battery (see the front of this manual).*

This Part of Chapter 2 is devoted to in-vehicle repair procedures for the 2.4L four-cylinder (Twin Cam) engine. All information concerning engine removal and installation and engine block and cylinder head overhaul can be found in Part C of this Chapter. The following repair procedures are based on the assumption the engine is installed in the vehicle. If the engine has been removed from the vehicle and mounted on a stand, many of the steps outlined in this Part of Chapter 2 will not apply. The Specifications included in this Part of Chapter 2 apply only to the procedures contained in this Part. Part C of Chapter 2 contains the Specifications necessary for cylinder head and engine block rebuilding.

These engines utilize a number of advanced design features to increase power output and improve durability. The aluminum cylinder head contains four valves per cylinder. A double-row timing chain drives two overhead camshafts - one for intake and one for exhaust. Lightweight bucket-type hydraulic lifters actuate the valves. Rotators are used on all valves for extended service life.

A balance shaft assembly has been added to smooth power pulsations. This assembly is bolted to the bottom of the main bearing webs and is chain driven from the rear of the crank. The gerotor oil pump is driven by the trailing balance shaft and is mounted at the rear of the balance shaft housing.

2   Repair operations possible with the engine in the vehicle

Many major repair operations can be accomplished without removing the engine from the vehicle. Clean the engine compartment and the exterior of the engine with some type of degreaser before any work is done. It'll make the job easier and help keep dirt out of the internal areas of the engine.

Depending on the components involved, it may be helpful to remove the hood to improve access to the engine as repairs are performed (refer to Chapter 11 if necessary). Cover the fenders to prevent damage to the paint. Special pads are available, but an old bedspread or blanket will also work.

If vacuum, exhaust, oil or coolant leaks develop, indicating a need for gasket or seal replacement, the repairs can generally be made with the engine in the vehicle. The intake and exhaust manifold gaskets, timing chain housing gasket, oil pan gasket, crankshaft oil seals and cylinder head gasket are all accessible with the engine in place.

Exterior engine components, such as the intake and exhaust manifolds, the oil pan (and the oil pump), the water pump, the starter motor, the alternator and the fuel system components can be removed for repair with the engine in place.

Since the cylinder head can be removed without pulling the engine, camshaft and valve component servicing can also be accomplished with the engine in the vehicle. Replacement of the timing chain and sprockets is also possible with the engine in the vehicle.

In extreme cases caused by a lack of necessary equipment, repair or replacement of piston rings, pistons, connecting rods and rod bearings is possible with the engine in the vehicle. However, this practice is not recommended because of the cleaning and preparation work that must be done to the components involved.

3   Top Dead Center (TDC) for number one piston - locating

Refer to illustration 3.7

1   Top Dead Center (TDC) is the highest point in the cylinder that each piston reaches as it travels up the cylinder bore. Each piston reaches TDC on the compression stroke and again on the exhaust stroke, but TDC generally refers to piston position on the compression stroke.

2   Positioning the piston(s) at TDC is an essential part of certain procedures such as camshaft removal and timing chain/sprocket removal.

3   Before beginning this procedure, be sure to place the transaxle in Neutral (or Park on automatic transaxle models), apply the parking brake and block the rear wheels. Disconnect the cable from the negative terminal of the battery. **Caution:** *On models equipped with the Theftlock audio system, be sure the lockout feature is turned off before performing any procedure which requires disconnecting the battery (see the front of this manual).*

4   Remove the spark plugs (see Chapter 1).

5   When looking at the drivebelt end of the engine, normal crankshaft rotation is clockwise. In order to bring any piston to TDC, the crankshaft must be turned with a socket and ratchet attached to the bolt threaded into the center of the lower drivebelt pulley on the crankshaft.

6   Have an assistant turn the crankshaft with a socket and ratchet as described above while you hold a finger over the number one spark plug hole. **Note:** *See the Specifications for the number one cylinder location.*

7   When the piston approaches TDC, pressure will be felt at the spark plug hole. Have your assistant stop turning the crankshaft when the timing marks are aligned **(see illustration).**

3.7 Align the notch on the crankshaft pulley with the "0" or "TDC" mark on the timing cover

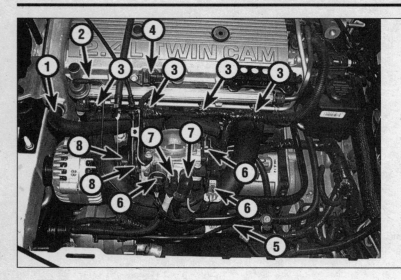

4.4 Label and disconnect the following components required for intake manifold removal

1   Oil/air separator hose
2   Fuel pressure regulator vacuum hose
3   Fuel injector electrical connectors
4   Camshaft position sensor electrical connector
5   Electrical conduit (at bottom of intake manifold)
6   IAC, TPS and MAP sensor electrical connectors
7   Vacuum hoses
8   Accelerator and cruise control cables

8   If the timing marks are bypassed, turn the crankshaft two complete revolutions clockwise until the timing marks are properly aligned.

9   After the number one piston has been positioned at TDC on the compression stroke, TDC for any of the remaining pistons can be located by turning the crankshaft one-half turn (180-degrees) to get to TDC for the next cylinder in the firing order.

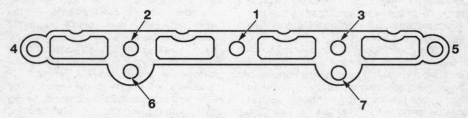

4.14 Intake manifold fastener tightening sequence

4   Intake manifold - removal and installation

Removal

Refer to illustration 4.4

1   Refer to Chapter 4 and relieve the fuel system pressure, then remove the air intake resonator and the air intake duct from the engine compartment.

2   Detach the cable from the negative terminal of the battery. **Caution:** *On models equipped with the Theftlock audio system, be sure the lockout feature is turned off before performing any procedure which requires disconnecting the battery (see the front of this manual).*

3   Remove the accelerator cable and the cruise control cable (if, equipped) (see Chapter 4).

4   Label and disconnect the vacuum hoses, breather hoses and electrical wires from the various sensors and position the wiring harness aside **(see illustration)**.

5   Remove the alternator mounting bolt closest to the engine block.

6   Remove the EGR pipe from the EGR adapter, if equipped (see Chapter 6).

7   Loosen the manifold mounting nuts/bolts in 1/4-turn increments until they can be removed by hand.

8   The manifold will probably be stuck to the cylinder head and force may be required to break the gasket seal. If necessary, dislodge the manifold with a soft-face hammer.

Caution: *Don't pry between the cylinder head and manifold or damage to the gasket sealing surfaces will result and vacuum leaks could develop.*

Installation

Refer to illustration 4.14

9   Remove all traces of sealant and old gasket material from the cylinder head and the intake manifold. Thoroughly clean the gasket mating surfaces. If there's old gasket material or oil on the mating surfaces when the manifold is reinstalled, vacuum leaks may develop.

10   Make sure the intake manifold bolt threads and the bolt holes/stud threads in the cylinder head are clean and free of debris.

11   Position the gasket on the cylinder head with the markings or numbers facing out towards the intake manifold. Make sure all intake port openings, coolant passage holes and bolt holes are aligned correctly.

12   Install the manifold, taking care to avoid damaging the gasket.

13   Thread the nuts/bolts into place by hand.

14   Tighten the nuts/bolts to the torque listed in this Chapter's Specifications following the recommended sequence **(see illustration)**. Work up to the final torque in three steps.

15   The remaining installation steps are the reverse of removal. Start the engine and check carefully for leaks at the intake manifold joints.

5   Exhaust manifold - removal and installation

Removal

Refer to illustrations 5.3 and 5.4
Warning: *Allow the engine to cool completely before performing this procedure.*

1   Detach the cable from the negative terminal of the battery. **Caution:** *On models equipped with the Theftlock audio system, be sure the lockout feature is turned off before performing any procedure which requires disconnecting the battery (see the front of this manual).*

5.3 Exhaust manifold heat shield retaining bolts (arrows)

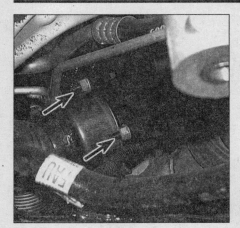

5.4 Exhaust pipe flange-to-exhaust manifold mounting nuts (two nuts visible in this photo)

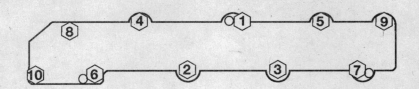

5.10 Exhaust manifold fastener tightening sequence

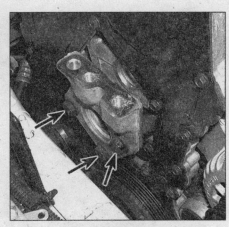

6.3 Remove the three bolts and the engine mount bracket

6.5 Timing chain cover fastener locations (arrows)

2   Unplug the oxygen sensor (see Chapter 6).

3   Remove the manifold heat shield **(see illustration)**.

4   Raise the vehicle and support it securely on jackstands, then remove the exhaust pipe-to-manifold nuts **(see illustration)**. The nuts are usually rusted in place, so penetrating oil should be applied to the stud threads before attempting to remove them. Loosen them a little at a time, working from side-to-side to prevent the flange from jamming.

5   Separate the exhaust pipe flange from the manifold studs, then pull the pipe down slightly to break the seal at the manifold joint. **Caution:** *DO NOT bend the exhaust flex coupler more than 3 degrees in any direction or damage to the flex coupler may occur.*

6   Remove the exhaust manifold brace from the bottom of the exhaust manifold (if equipped).

7   Working from above in the engine compartment remove the oil dipstick tube. Loosen the exhaust manifold mounting nuts 1/4-turn at a time each, working from the inside out, until they can be removed by hand. **Note:** *It may be helpful to also detach the brake booster vacuum tube from the top of the rear cam cover and from the side of the engine block to help facilitate removal of the exhaust manifold.*

8   Separate the manifold from the cylinder head and remove it.

Installation

Refer to illustration 5.10

9   The manifold and cylinder head mating surfaces must be clean when the manifold is reinstalled. Use a gasket scraper to remove all traces of old gasket material and carbon deposits.

10   Using a new gasket, install the manifold and hand tighten the fasteners. Following the recommended sequence **(see illustration)**, tighten the bolts/nuts to the torque listed in this Chapter's Specifications.

11   The remaining installation steps are the reverse of removal.

6   Timing chain and sprockets - removal, inspection and installation

Note: *Special tools are required for this procedure. Read through the entire procedure and acquire the necessary tools and equipment before beginning work.*

Removal

Refer to illustrations 6.3, 6.5, 6.8, 6.10, 6.11, 6.12, 6.13, 6.15, 6.17 and 6.18

1   Detach the cable from the negative terminal of the battery. **Caution:** *On models equipped with the Theftlock audio system, be sure the lockout feature is turned off before performing any procedure which requires disconnecting the battery (see the front of this manual).*

2   Remove the coolant reservoir (see Chapter 3). Remove the drivebelt (see Chapter 1).

3   Support the engine from above, using an engine support fixture (available at rental yards), or from below using a floor jack. Use a wood block between the floor jack and the engine to prevent damage. Remove the front engine mount (see Section 18). Remove the engine mount bracket and discard the bolts **(see illustration)**. **Caution:** *Replace the engine support bracket bolts with the manufacturers original type bolt anytime they are removed.*

4   Remove the crankshaft pulley (see Section 11).

5   Working from above, remove the upper

timing chain cover fasteners **(see illustration)**.

6   Working from below, remove the lower timing chain cover fasteners.

7   Detach the timing chain cover and gaskets from the housing.

8   Slide the oil slinger off the crankshaft if equipped **(see illustration)**.

9   Temporarily reinstall the crankshaft pulley bolt to use when turning the crankshaft.

10   Turn the crankshaft clockwise until the camshaft sprocket's timing pin holes align with the holes in the timing chain housing.

6.8 Note how it's installed, then remove the oil slinger (arrow) from the crankshaft (if equipped)

6.10 Insert two 8 mm bolts (arrows) through the holes in the camshaft sprockets and into the holes in the timing chain housing - this locks the camshafts in the "timed" position

6.11 The mark on the crankshaft sprocket must align with the mark on the block (arrows) when the engine is locked into the "timed" position

Insert 8 mm pins or bolts into the holes to maintain alignment (see illustration).

11 The mark on the crankshaft sprocket should align with the mark on the engine block (see illustration). The crankshaft sprocket keyway should point up and align with the centerline of the cylinder bores.

12 Remove the three timing chain guides (see illustration).

13 Make sure all the slack in the timing chain is above the tensioner assembly, then remove the chain tensioner mounting bolts (see illustration).

14 Early models are equipped with a one piece tensioner and shoe assembly that must be disengaged from the wear grooves in the tensioner shoe in order to remove the tensioner assembly. On these models slide a screwdriver blade under the timing chain while pulling the tensioner shoe out. Note: If difficulty is encountered when removing the chain tensioner, proceed as follows:

a) Hold the intake camshaft sprocket with an appropriate tool and remove the sprocket bolt and washer.

b) Remove the washer from the bolt and thread the bolt back into the camshaft by hand.

c) Remove the intake camshaft sprocket, using a three-jaw puller in the three relief holes in the sprocket, if necessary. Caution: Don't try to pry the sprocket off the camshaft or damage to the sprocket could occur.

d) Disengage the chain from the shoe, then remove the tensioner and shoe assembly from the engine.

15 Later models are equipped with a two piece tensioner and shoe assembly that must be removed in several steps. On these models, first remove the timing chain tensioner as shown in illustration 6.13, then remove the tensioner shoe. Use snap-ring pliers to disengage the plastic tab securing the tensioner shoe to the pivot pin (see illustration).

16 If you intend to reuse the timing chain, use white paint or chalk to make match marks to align the camshaft sprockets with the chain.

6.12 The three timing chain guides are wedged into the housing at four points (arrows) - just pull them out to remove them

6.13 Timing chain tensioner mounting bolts (arrows) (late models shown, early models similar)

Note: If a used timing chain is reinstalled with the wear pattern in the opposite direction noise and increased wear may occur.

17 Slip the timing chain off the sprockets (see illustration).

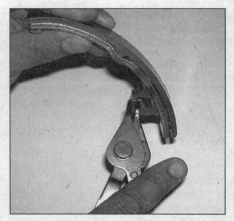

6.15 On later models it will be necessary to use snap-ring pliers to disengage the locking tab on the tensioner shoe from the pivot pin

18 To remove the camshaft sprockets, loosen the bolts while holding the sprockets with a screwdriver or punch inserted through one of the holes. Mark the sprockets for iden-

6.17 Begin removing the chain at the exhaust camshaft sprocket

6.18 Mark the sprockets *exhaust* and *intake* (arrows), then remove the bolts and pull the sprockets off

2A

6.21 The camshaft sprocket dowel pin(s) should be near the top prior to sprocket installation

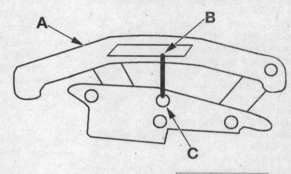

38016-2B-7.15 HAYNES

6.26a After retracting the tensioner shoe on early models, insert a piece of wire (B) bent into a "U" shape between the tensioner shoe (A) and reset hole (C)

tification **(see illustration)**, remove the bolts, then pull on the sprockets by hand until they slip off the dowels. If necessary, use a small puller, with the legs inserted in the relief holes, to pull the sprockets off.

19   The crankshaft sprocket should slip off the crankshaft by hand. If not, use a puller.

Inspection

20   Visually inspect all parts for wear and damage. Check the timing chain for loose pins, cracks, worn rollers and side plates. Check the sprockets for hook-shaped, chipped and broken teeth. **Note:** *Some scoring of the timing chain shoe and guides is normal. Replace the timing chain, sprockets, chain shoe and guides as a set if the engine has high mileage or fails the visual parts inspection. Replace the tensioner shoe and chain guides if scoring or wear exceeds 0.045 inch.*

Installation

Refer to illustrations 6.21, 6.26a, 6.26b and 6.29

21   Make sure the camshafts are positioned with the dowel pins at the top **(see illustration)**. Install both camshaft sprockets (if removed). Apply thread locking compound to the camshaft sprocket bolt threads and make

sure the washers are in place. Hold the camshaft from turning as described in Step 18 and tighten the bolts to the torque listed in this Chapter's Specifications.

22   Recheck the positions of the camshaft and crankshaft sprockets for correct valve timing **(see illustrations 6.10 and 6.11)**. Install the alignment pins. **Note:** *If the camshafts are out of position and must be rotated more than 1/8-turn in order to install the alignment pins:*

a) *The crankshaft must be rotated 90-degrees clockwise past Top Dead Center to give the valves adequate clearance to open.*

b) *Once the camshafts are in position and the alignment pins installed, rotate the crankshaft counterclockwise back to Top Dead Center.* **Caution:** *Do not rotate the crankshaft clockwise to TDC (valve or piston damage could occur).*

23   Slip the timing chain over the exhaust camshaft sprocket, then around the water pump sprocket and the crankshaft sprocket.

24   Remove the alignment pin from the intake camshaft. Using an appropriate tool, rotate the intake camshaft sprocket counterclockwise just enough to slide the timing chain over the intake camshaft sprocket, then

release the tool. The length of chain between the two camshaft sprockets should now tighten when the tool is released. If the valve timing is correct, the intake camshaft alignment pin should slide in easily. If it doesn't index, the camshafts aren't timed correctly; repeat the procedure.

25   Leave the alignment pins installed for now and check the timing marks. With slack removed from the timing chain between the intake camshaft sprocket and the crankshaft sprocket, the timing marks on the crankshaft sprocket and the engine block should be aligned. If the marks aren't aligned, move the chain one tooth forward or backward, remove the slack and recheck the marks.

26   Reload the timing chain tensioner assembly to its "reset" position as follows:

Early models

a) *Bend a piece of heavy wire into a "U" shape*

b) *Apply light force on the tensioner shoe to compress the plunger.*

c) *Insert a small screwdriver into the reset access hole and pry the ratchet pawl away from the ratchet teeth.*

d) *Install the locking wire or keeper into the access hole and the shoe* **(see illustration).**

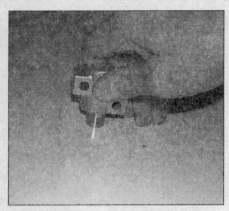

6.26b On later models depress the plunger until it locks into the reset position

Late models

e) *Turn the tensioner up side down and compress the tensioner plunger on a flat surface until all the oil has been purged from the tensioner and the plunger locks into place in the tensioner body (see illustration).*

27   On later models install the tensioner shoe on the engine block pivot pin. When fully seated the locking tab on the tensioner shoe should snap into the groove on the pivot pin.

28   On all models install the tensioner assembly in the chain housing. Tighten the bolts to the torque listed in this Chapter's Specifications.

29   On early models remove the bent piece of wire, squeeze the plunger into the tensioner body, then let go to unload the plunger assembly. On later models unload the tensioner plunger by inserting a pry bar or hooked tool between the back of the tensioner shoe and the tip of the tensioner plunger, then press inward on plunger until it releases **(see illustration)**.

30   Remove the camshaft sprocket alignment pins.

31   Slowly rotate the crankshaft clockwise two full turns (720-degrees). Do not force it; if

resistance is felt, recheck the installation procedure. Align the crankshaft timing mark with the mark on the engine block and temporarily reinstall the 8 mm alignment pins. The pins should slide in easily if the valve timing is correct. **Caution:** *If the valve timing is incorrect, severe engine damage could occur.*

32   Install the remaining components in the reverse order of removal, noting the following:

a) *Replace the engine mount bracket bolts with new bolts and tighten them to the torque listed in this Chapter's Specifications.*

b) *Check fluid levels, start the engine and check for proper operation and coolant/oil leaks.*

7   Timing chain housing - removal and installation

Refer to illustration 7.11

Warning: *Wait until the engine is completely cool before beginning this procedure.*

1   Detach the cable from the negative terminal of the battery. **Caution:** *On models equipped with the Theftlock audio system, be sure the lockout feature is turned off before performing any procedure which requires disconnecting the battery (see the front of this manual).*

2   Drain the cooling system (see Chapter 1), then remove the heater hose from the thermostat housing to drain the coolant from the engine block.

3   Remove the timing chain and sprockets (see Section 6).

4   Remove the water pump (see Chapter 3).

5   Remove the four oil pan-to-timing chain housing bolts.

6   Remove the timing chain housing-to-block lower fasteners.

7   Remove the lowest cover retaining stud from the timing chain housing.

8   Remove the eight chain housing-to-camshaft housing bolts **(see illustration 6.21)**.

9   Using an engine support fixture or floor jack, raise the engine slightly for additional clearance.

10   Remove the timing chain housing and gaskets. Thoroughly clean the mating surfaces to remove any traces of old sealant or gasket material.

11   Install the timing chain housing with new gaskets **(see illustration)**. Tighten the bolts to the torque listed in this Chapter's Specifications.

12   The remaining steps are the reverse of removal.

8   Camshafts, lifters and housings - removal, inspection and installation

Note: *Special tools are required for this procedure. Read through the entire procedure and acquire the necessary tools and equipment before beginning work.*

Removal

Refer to illustrations 8.9, 8.10, 8.11, 8.12 and 8.13

1   Relieve the fuel system pressure (see Chapter 4).

2   Detach the cable from the negative terminal of the battery. **Caution:** *On models equipped with the Theftlock audio system, be sure the lockout feature is turned off before performing any procedure which requires disconnecting the battery (see the front of this manual).*

3   Position the engine at TDC on the compression stroke (see Section 3). Remove the timing chain and camshaft sprockets (see Section 6).

4   Remove the timing chain housing-to-camshaft housing bolts **(see illustration 6.21)**.

5   Remove the ignition coil and module assembly (see Chapter 5).

Intake (front) camshaft housing

6   Remove the power steering pump without disconnecting the hoses and secure it

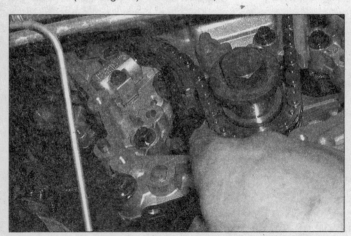

6.29 After the tensioner is installed on later models, press inward on the tensioner plunger to unload the tensioner

7.11 Position the lower gasket over the dowel pins (arrows)

8.9 When removing the intake camshaft housing, detach the camshaft position sensor (A), the air intake resonator bracket (B) and the two cover-to-housing nuts (C)

8.10 When removing the exhaust camshaft housing, detach the bolts (arrows) securing the engine lifting bracket to the rear of the camshaft housing

8.11 Gently lift the camshaft housing off the cylinder head and turn it over so the lifters don't fall out

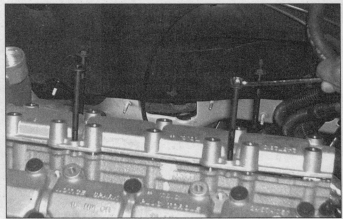

8.12 Push the housing cover off the camshaft housing by threading four of the housing-to-cylinder head bolts into the tapped holes in the cover - tighten the bolts evenly to prevent

aside (see Chapter 10).

7   Remove the fuel rail and injectors from the cylinder head (see Chapter 4), then disconnect the electrical connector from the oil pressure sending unit.

8   Remove the camshaft position sensor (see Chapter 6) and the air intake resonator bracket from the intake camshaft housing cover.

9   Loosen and remove the cover-to-housing nuts located at the left side (power steering pump end) of the housing (see illustration).

Exhaust (rear) camshaft housing

10   Unbolt the oil level dipstick tube and the engine lifting bracket from the exhaust camshaft housing (see illustration).

Intake and exhaust camshaft housing

11   Loosen the camshaft housing-to-cylinder head bolts in 1/4-turn increments, in the reverse of the tightening sequence (see illustration 8.29). Then lift the camshaft housing

off the cylinder head (see illustration).

12   Working on bench, push the camshaft housing cover off the housing by threading four of the housing-to-cylinder head bolts into the tapped holes in the cover (see illustration). Carefully lift the camshaft out of the camshaft housing.

13   Remove the oil seal from the intake camshaft (see illustration) and discard it.

14   Remove the lifters from the housing and store them in order so they can be reinstalled in their original locations. To minimize lifter bleed-down, store the lifters valve-side up, submerged in clean engine oil.

15   Remove all traces of old gasket material from the mating surfaces and clean them with lacquer thinner or acetone to remove any traces of oil.

8.13 Remove the oil seal (arrow) from the intake camshaft

8.17a Check the camshaft lobe surfaces and the bore surfaces of the lifters for wear (arrows)

8.17b Check the valve-side of the lifters too, especially the valve stem contact area (arrow)

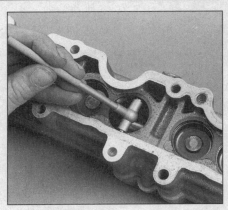

8.18 Use a telescoping gauge and micrometer to measure the lifter bores . . .

Inspection

Refer to illustrations 8.17a, 8.17b, 8.18 and 8.19

16   Refer to Chapter 2, Part C, for camshaft inspection procedures, but use this Chapter's Specifications. If the camshaft is damaged or worn beyond the specifications, replace the camshaft, do not attempt to salvage worn camshafts. Whenever a camshaft is replaced, replace all the lifters actuated by the camshaft as well.

17   Visually inspect the lifters for wear, galling, score marks and discoloration from overheating **(see illustrations)**.

18   Measure each lifter bore inside diameter and record the results **(see illustration)**.

19   Measure each lifter outside diameter and record the results **(see illustration)**.

20   Subtract the lifter outside diameter from the corresponding bore inside diameter to determine the clearance. Compare the results to this Chapter's specifications and replace parts as necessary.

Installation

Refer to illustrations 8.25 and 8.29

21   Inspect the gasket(s) that seal the rear of

the timing chain housing to camshaft housing for tears and cuts. If the gasket(s) are damaged replace them. **Note:** *If new gaskets are used it will be necessary to apply a small amount of sealer to hold the gasket in place while installing the camshaft housing.*

22   Install a new camshaft housing gasket over the dowels pins on the cylinder head, then position the camshaft housing on the cylinder head and temporarily hold it in place with two bolts.

23   Coat the lifters with camshaft assembly lube and install them in their original locations in the camshaft housing.

24   On the intake camshaft only, lubricate the lip of the oil seal with clean engine oil, then position the seal over the (left) end of the camshaft journal with the spring side facing in.

25   Coat the camshaft journals and lobes with camshaft assembly lube and install the camshaft(s) in the housing with the sprocket dowel pin UP (12 o'clock position) **(see illustration)**. **Note:** *When lowering the intake camshaft into place it will be necessary to properly align the oil seal with the bore in the housing.*

26   Install new camshaft cover seals into the

camshaft cover(s). **Note:** *Each camshaft cover seal is different shape and color. The intake cover seals are green with the inner seal configured differently than the outer seal. The exhaust cover seals are orange and they are also configured differently.*

27   Remove the two bolts temporarily installed to hold the housing to the cylinder head and position the camshaft cover on the housing over the dowel pins.

28   Apply thread sealant to the threads of the camshaft housing and cover bolts.

29   Tighten the bolts in the sequence shown **(see illustration)** to the torque and angle of rotation listed in this Chapter's Specifications. Be sure to tighten the camshaft cover-to-housing bolts/nuts (rear two on the intake camshaft housing) at the prescribed lower torque setting.

30   Install the power steering pump.

31   Install the remaining parts in the reverse order of removal.

32   Change the oil and filter (see Chapter 1), then start the engine and check for normal operation. **Note:** *If new lifters have been installed or the lifters bled down while the engine was disassembled, excessive lifter noise may be experienced after startup - this is normal. Use the following procedure to*

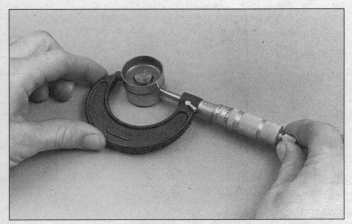

8.19 . . . then measure the lifters with a micrometer - subtract each lift diameter from the corresponding bore diameter to obtain the lifter-to-bore clearance

8.25 Install the camshafts with the dowel pins (arrows) at the top (12 o'clock position)

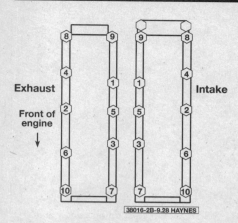

8.29 Camshaft housing-to-cylinder head bolt tightening sequence

9.5 You'll need an air hose adapter this long to reach down into the spark plug wells on the cylinder head - they're commonly available in auto parts stores

purge the lifters of air:

a) Start the engine and allow it to warm up at idle for five minutes.
b) Increase engine speed to 2,000 rpm until the lifter noise is gone.
c) Return the engine to idle for an additional five minutes.

33 Road test the vehicle and check for oil and coolant leaks.

9 Valve springs, retainers and seals - replacement

Refer to illustrations 9.5 and 9.17

Note: Broken valve springs and defective valve stem seals can be replaced without removing the cylinder head. Two special tools and a compressed air source are normally required to perform this operation, so read through this Section carefully and rent or buy the tools before beginning the job.

1 Remove the spark plug from the cylinder which has the defective part. Due to the design of this engine, the intake and exhaust camshaft housings can be removed separately to service their respective components. If all of the valve stem seals are being replaced, all of the spark plugs and both camshaft housings should be removed.

2 Refer to Chapter 5 and remove the ignition coil assembly.

3 Remove the camshaft housing(s) as described in Section 8. Note: Removal of the camshaft covers, camshafts and lifters is not necessary unless camshaft and lifter inspection is required.

4 Turn the crankshaft until the piston in the affected cylinder is at top dead center on the compression stroke (refer to Chapter 2, Part C, for instructions). If you're replacing all of the valve stem seals, begin with cylinder number one and work on the valves for one cylinder at a time. Move from cylinder-to-cylinder following the firing order sequence (see this Chapter's Specifications).

5 Thread an adapter into the spark plug hole (see illustration) and connect an air

hose from a compressed air source to it. Most auto parts stores can supply the air hose adapter. Note: Many cylinder compression gauges utilize a screw-in fitting that may work with your air hose quick-disconnect fitting.

6 Apply compressed air to the cylinder. Warning: The piston may be forced down by compressed air, causing the crankshaft to turn suddenly. If the wrench used when positioning the number one piston at TDC is still attached to the bolt in the crankshaft nose, it could cause damage or injury when the crankshaft moves.

7 The valves should be held in place by the air pressure.

8 Stuff shop rags into the cylinder head holes adjacent to the valves to prevent parts and tools from falling into the engine, then use a valve spring compressor to compress the spring. Remove the keepers with small needle-nose pliers or a magnet.

9 Remove the retainer and valve spring, then remove the valve guide seal and the rotator (intake) or the valve spring seat (exhaust). Note: If air pressure fails to hold the valve in the closed position during this operation, the valve face or seat is probably damaged. If so, the cylinder head will have to be removed for additional repair operations.

10 Wrap a rubber band or tape around the top of the valve stem so the valve won't fall into the combustion chamber, then release the air pressure.

11 Inspect the valve stem for damage. Rotate the valve in the guide and check the end for eccentric movement, which would indicate the valve stem is bent.

12 Move the valve up-and-down in the guide and make sure it does not bind. If the valve stem binds, either the valve is bent or the guide is damaged. In either case, the cylinder head will have to be removed for repair.

13 Reapply air pressure to the cylinder to retain the valve in the closed position, then remove the tape or rubber band from the valve stem.

14 Reinstall the valve rotator (intake) or the valve spring seat (exhaust).

15 Lubricate the valve stem with engine oil and install a new guide seal.

16 Install the spring in position over the valve.

17 Install the valve spring retainer. Compress the valve spring and carefully install the keepers in the groove. Apply a small dab of grease to the inside of each keeper to hold it

in place if necessary (see illustration). Remove the pressure from the spring tool and make sure the keepers are seated.

18 Disconnect the air hose and remove the adapter from the spark plug hole.

19 Refer to Section 8 and install the camshaft(s), lifters and housing(s).

20 Install the spark plug(s) and the coil assembly.

21 Start and run the engine, then check for oil leaks and unusual sounds coming from the camshaft housings.

10 Cylinder head - removal and installation

Warning: Wait until the engine is completely cool before beginning this procedure.

Removal

Refer to illustration 10.5

1 Detach the cable from the negative terminal of the battery. Caution: On models equipped with the Theftlock audio system, be sure the lockout feature is turned off before performing any procedure which requires disconnecting the battery (see the front of this manual).

2 Drain the cooling system (see Chapter 1), then remove the engine block drain plug to drain the coolant from the engine block. Also remove the upper radiator hose and the coolant temperature sensor electrical connector from the coolant outlet housing on the left (driver side) of the cylinder head.

3 Refer to Section 4 and remove the

9.17 Apply a small dab of grease to each keeper before installation to hold it in place on the valve stem until the spring is released

10.5 To avoid mixing up the cylinder head bolts, use a new gasket to transfer the bolt hole pattern to a piece of cardboard, then punch holes to accept the bolts

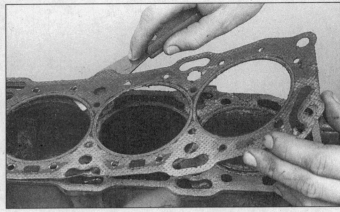

10.10 Remove the old gasket and clean the cylinder head thoroughly

intake manifold. Refer to Section 5 and detach the exhaust manifold.

4   Remove the timing chain, the timing chain housing and the camshaft housings as described in Sections 6, 7 and 8. **Note:** *Removal of the camshaft housing covers, camshafts and lifters is not necessary unless camshaft and lifter inspection is required.*

5   Using the new cylinder head gasket, outline the cylinders and bolt pattern on a piece of cardboard to make a holder for the cylinder head bolts. Be sure to indicate the front of the engine for reference. Punch holes at the bolt locations **(see illustration)**.

6   Loosen the cylinder head bolts in 1/4-turn increments until they can be removed by hand. Work from bolt-to-bolt in a pattern that's the reverse of the tightening sequence **(see illustration 10.15)**. Store the bolts in the cardboard holder as they're removed - this will ensure they are reinstalled in their original locations.

7   Lift the cylinder head off the engine. If resistance is felt, don't pry between the cylinder head and block as damage to the mating surfaces will result. To dislodge the cylinder head, place a wood block against the end of it and strike the wood block with a hammer. Store the cylinder head on wood blocks to prevent damage to the gasket sealing surfaces.

8   Cylinder head disassembly and inspection procedures are covered in detail in Chapter 2, Part C.

Installation

Refer to illustrations 10.10 and 10.15

9   The mating surfaces of the cylinder head and block must be perfectly clean when the cylinder head is installed.

10   Use a gasket scraper to remove all traces of carbon and old gasket material **(see illustration)**, then clean the mating surfaces with lacquer thinner or acetone. If there's oil on the mating surfaces when the cylinder head is installed, the gasket may not seal correctly and leaks could develop. **Note:** *Since the cylinder head is made of aluminum, aggressive scraping can cause damage. Be*

10.15 Cylinder head bolt TIGHTENING sequence

extra careful not to nick or gouge the mating surface with the scraper. Use a vacuum cleaner to remove debris that falls into the cylinders. **Caution:** *Do not use a wire brush or an abrasive pad to clean the cylinder head mating surface.*

11   Check the block and cylinder head mating surfaces for nicks, deep scratches and other damage. If damage is slight, it can be removed with a flat mill file; if it's excessive, machining may be the only alternative.

12   Use a nylon bristle brush to clean the threads of the cylinder head bolt holes in the block. Mount each bolt in a vise and run a die down the threads to remove corrosion and restore the threads. Dirt, corrosion, sealant and damaged threads will affect torque readings.

13   Position the new gasket over the dowel pins in the block.

14   Carefully position the cylinder head on the block without disturbing the gasket.

15   Install the bolts in their original locations and tighten them finger tight. Following the recommended sequence **(see illustration)**, tighten the bolts in several steps to the torque and angle of rotation listed in this Chapter's Specifications.

16   The remaining installation steps are the reverse of removal.

17   Refill the cooling system and change the oil and filter (see Chapter 1, if necessary).

18   Run the engine and check for leaks and proper operation.

11   Crankshaft pulley – removal and installation

Refer to illustrations 11.4, 11.5 and 11.6

1   Detach the cable from the negative terminal of the battery. **Caution:** *On models equipped with the Theftlock audio system, be sure the lockout feature is turned off before performing any procedure which requires disconnecting the battery (see the front of this manual).*

2   With the parking brake applied and the shifter in Park (automatic) or in gear (manual), loosen the lugnuts from the right front wheel, then raise the front of the vehicle and support it securely on jackstands.

3   Remove the right front wheel and the right splash shield from the wheelwell. Remove the drivebelt (see Chapter 1).

4   Remove the bolt from the front of the crankshaft. A breaker bar will probably be necessary, since the bolt is very tight. Insert a bar through a hole in the pulley to prevent the crankshaft from turning **(see illustration)**.

5   Using a puller that bolts to the crankshaft hub, remove the crankshaft pulley from the crankshaft **(see illustration)**. **Note:** *Depending on the type of puller you have it may be necessary to support the engine from above, remove the right side engine mount and lower to engine to gain sufficient clearance to use the puller.*

11.4 The crankshaft pulley can be held with a bar while the bolt is loosened or tightened

11.5 Use a puller that applies force to the crankshaft pulley hub - don't use a jaw-type puller that applies force to the outer edge or damage to the crankshaft pulley will occur

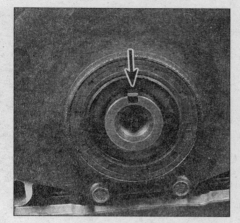

11.6 Align the keyway in the crankshaft pulley hub with the Woodruff key in the crankshaft (arrow)

2A

12.2 Pry the old seal out with a seal removal tool (shown here) or a screwdriver

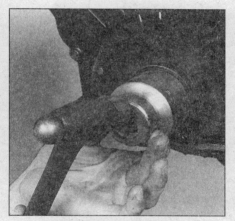

12.5 Install the new seal with a large socket or section of pipe

12.6 If the sealing surface on the pulley hub has a wear groove from contact with the seal, repair sleeves are available at most auto parts stores

6   To install the crankshaft pulley slide the pulley onto the crankshaft as far as it will slide on, then use a vibration damper installation tool to press the pulley onto the crankshaft. Note that the slot (keyway) in the hub must be aligned with the Woodruff key in the end of the crankshaft **(see illustration)** and that the crankshaft bolt can also be used to press the crankshaft pulley into position.
7   Tighten the crankshaft bolt to the torque and angle of rotation listed in this Chapter's Specifications.
8   The remaining installation steps are the reverse of removal.

12 Crankshaft front oil seal - replacement

Refer to illustrations 12.2, 12.5 and 12.6

1   Remove the crankshaft pulley (see Section 11).
2   Pry the old oil seal out with a seal removal tool **(see illustration)** or a screw-

driver. Be very careful not to nick or otherwise damage the crankshaft in the process and don't distort the timing chain cover.
3   Apply a thin coat of RTV-type sealant to the outer edge of the new seal. Lubricate the seal lip with multi-purpose grease or clean engine oil.
4   Place the seal squarely in position in the bore with the spring side facing in.
5   Carefully tap the seal into place with a large socket or section of pipe and a hammer **(see illustration)**. The outer diameter of the socket or pipe should be the same size as the seal outer diameter.
6   Check the surface of the pulley hub that the oil seal rides on. If the surface has been grooved from long-time contact with the seal, a press-on sleeve may be available to renew the sealing surface **(see illustration)**. This sleeve is pressed into place with a hammer and a block of wood and is commonly available from auto parts stores.
7   Apply a thin layer of clean multi-purpose grease or clean engine oil to the seal contact surface of the crankshaft pulley hub.
8   Reinstall the crankshaft pulley as

described in Section 11.
9   The remaining installation steps are the reverse of removal.
10   Start the engine and check for oil leaks at the seal.

13 Oil pan - removal and installation

Warning: *Wait until the engine is completely cool before beginning this procedure.*
Note: *The following procedure is based on the assumption the engine is installed in the vehicle. If it has been removed, simply unbolt the oil pan and detach it from the block.*

Removal

Refer to illustration 13.12

1   Detach the cable from the negative terminal of the battery. **Caution:** *On models equipped with the Theftlock audio system, be sure the lockout feature is turned off before performing any procedure which requires dis-*

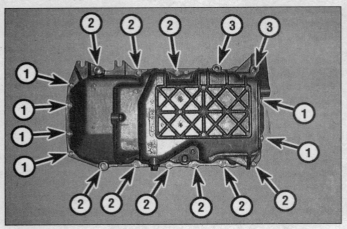

13.12 Oil pan bolt locations

| 1 | *M6 x 1.00 x 25 bolts* | 2 | *M8 x 1.25 x 80 bolts* |
| 2 | *M8 x 1.25 x 22 bolts* | | |

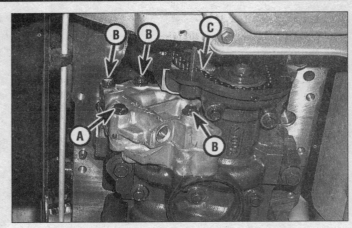

14.3 Remove the oil pump cover bolt (A) and the oil pump-to-balance shaft housing bolts (B) – note that a fourth oil pump housing bolt (C) is accessed through a hole in the balance shaft housing

connecting the battery (see the front of this manual).

2 Remove the right front wheel and the lower splash shield from the fenderwell.

3 Drain the engine oil and the coolant (see Chapter 1).

4 Remove the drivebelt (see Chapter 1).

5 Remove the air conditioning compressor from the bracket and set it aside without disconnecting the hoses (see Chapter 3).

6 On some models it may be necessary to raise the engine and remove the engine mount strut and bracket from the front of the oil pan (see Section 18).

7 Detach the radiator outlet pipe and the thermostat from the engine (see Chapter 3).

8 Unbolt the exhaust manifold brace (if equipped).

9 Disconnect and remove the oil level sensor from the oil pan (if equipped).

10 Remove the plastic flywheel/driveplate inspection cover.

11 Remove the oil pan-to-transaxle bolt/nuts, then remove the transaxle-to-oil pan stud and spacer (see Chapter 7).

12 Remove the oil pan mounting bolts **(see illustration)**.

13 Unclip the right front brake line from the frame to allow removal of the oil pan, then carefully separate the pan from the block.

14 Don't pry between the block and pan or damage to the sealing surfaces may result and oil leaks may develop. **Note:** *The crankshaft may have to be rotated to gain clearance for oil pan removal.*

Installation

15 Clean the sealing surfaces with lacquer thinner or acetone. Make sure the bolt holes in the block are clean.

16 The gasket should be checked carefully and replaced with a new one if damage is noted. Minor imperfections can be repaired with RTV sealant. **Caution:** *Use only enough sealant to restore the gasket to its original size and shape. Excess sealant may cause part misalignment and oil leaks.*

17 Carefully install the pan gasket and hold the pan against the block and install the bolts finger tight.

18 Tighten the bolts in three steps to the torque listed in this Chapter's Specifications **(see illustration 13.12)**. Start at the center of the pan and work out toward the ends in a

criss-cross pattern. Note that the bolts are not all tightened to the same torque figure.

19 The remaining steps are the reverse of removal. **Caution:** *Don't forget to refill the engine with oil and coolant before starting it* (see Chapter 1).

20 Start the engine and check carefully for oil leaks at the oil pan.

14 Oil pump - removal, inspection and installation

Removal

Refer to illustrations 14.3, 14.4a and 14.4b

1 Remove the oil pan as described in Section 13.

2 Remove the balance shaft chain cover and tensioner (see Section 15).

3 Remove the oil pump mounting bolts **(see illustration)**.

4 Separate the oil pump cover from the oil pump housing, then remove the oil pump housing and gears from the balance shaft assembly **(see illustrations)**.

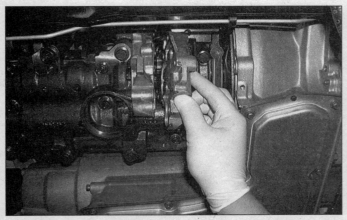

14.4a Separate the pump cover from the pump housing . . .

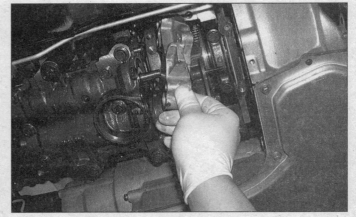

14.4b . . . then remove the oil pump housing and gears from the balance shaft

14.5 Remove the oil pressure relief valve for cleaning and inspection

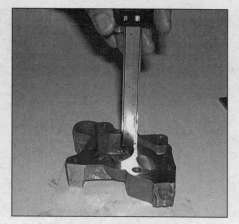

14.9 Use a dial caliper to measure the depth of the gerotor pocket

14.10 Use a micrometer or dial calipers to check the thickness and the diameter of the outer rotor

2A

Inspection

Refer to illustrations 14.5, 14.9, 14.10, 14.11 and 14.12

5   Remove the pressure relief valve and spring by removing the roll pin **(see illustration)**. Use an 1/8-inch punch to drive the roll pin out of the cover, allowing inspection and cleaning.

6   Remove the inner and outer rotors from the oil pump housing, noting their installed direction for reassembly.

7   Clean all parts thoroughly in solvent and carefully inspect the rotors, pump cover and oil pump housing for nicks, scratches or burrs. Replace the assembly if it is damaged.

8   Clean the relief valve plunger and inspect it for wear. Small burrs can be removed with 400-grit wet sandpaper and oil. After cleaning and inspection reinstall it back into the pump cover.

9   Use a dial caliper or a depth micrometer to measure the depth of the gerotor pocket in the oil pump housing **(see illustration)**. If the depth more than the limit listed in this Chapter's Specifications, the pump should be replaced.

10   Measure the thickness and the diameter of the outer rotor. If either measurement is less than the value listed in this Chapter's Specifications, the pump should be replaced **(see illustration)**.

11   Insert the outer rotor into the oil pump housing and, while holding the rotor against one side of the housing with your finger, measure the clearance at the opposite side between the rotor and housing **(see illustration)**. If the measurement is more than the maximum allowable clearance listed in this Chapter's Specifications, the pump should be replaced.

12   Install the inner rotor in the oil pump assembly and measure the clearance between the lobes on the inner and outer rotors **(see illustration)**. If the clearance is more than the value listed in this Chapter's Specifications, the pump should be replaced.

Installation

13   Lubricate the oil pump gears with clean engine oil, then pack the cavities between the oil pump gears with petroleum jelly to ensure proper lubrication during start up.

14   Position the oil pump housing onto the balance shaft assembly.

15   Install the pump cover onto the pump housing and tighten the bolts to the torque listed in this Chapter's Specifications.

16   Install the balance shaft chain tensioner and cover (see Section 15).

17   Install the oil pan.

18   Add oil and run the engine. Check for oil pressure and leaks.

15   Balance shaft assembly - removal, inspection and installation

Note: *Special tools are normally required to perform this operation. Read through the entire Section carefully and acquire the necessary tools before beginning this procedure.*

Removal

Refer to illustrations 15.1, 15.2, 15.5a, 15.5b and 15.7

1   Remove the oil pan (see Section 13). Remove the balance shaft chain cover **(see illustration)**.

14.11 Check the outer rotor-to-housing clearance

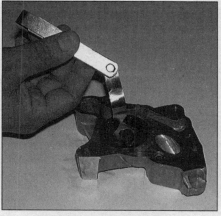

14.12 Check the clearance between the lobes of the inner and outer rotors

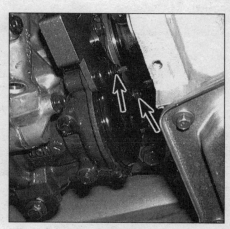

15.1 Balance shaft chain cover nut and bolt (arrow)

15.2 Loosen, but don't remove, the balance shaft chain guide

15.5a Mark the face of the balance shaft driven sprocket (arrow) so it can be installed the same way it came off

15.5b It will be necessary to purchase a balance shaft alignment tool (arrow) to hold the shafts as the driven sprocket is removed and to time the shafts during installation

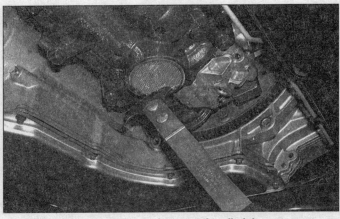

15.7 Using a suitable tool pry out the oil pick-up screen

2   Loosen, but don't remove, the balance shaft chain guide **(see illustration)**.

3   Remove the oil pump (see Section 14).

4   Place the number 1 piston at TDC (Top Dead Center), see Section 3.

5   Make a mark on the face of the driven sprocket so it can be installed the same way it came off **(see illustration)**. Install the balance shaft alignment tool to keep the balance shafts from rotating as the drive sprocket is removed **(see illustration)**.

6   Remove the balance shaft driven sprocket. **Caution:** *The bolt is a left handed thread and must be loosened in a clockwise direction.*

7   Pry out the oil pump pick-up screen **(see illustration)**. **Note:** *It will be necessary to replace the screen if damaged to the screen has occurred during removal.*

8   Break loose the bolts in the middle of the housing (bolts 1 through 8), holding the upper and lower housing halves together **(see illustration 15.19)**. DO NOT loosen bolts 1 through 8 more than one turn or remove them at this time.

9   Remove the bolts from the outer edges of the housing (bolts 9 through 13), securing the balance shaft assembly-to-block **(see illustration 15.19)**. **Warning:** *Support the*

assembly securely before removal of the bolts. Remove the balance shaft assembly and place it on a workbench for disassembly and inspection.

Inspection

Refer to illustrations 15.11 and 15.12

10   Check the balance shaft endplay. Set up a dial indicator at the rear of the shaft(s), then move the shafts foreword and rearward in the housing and note the endplay. If the endplay

is more than the value listed in this Chapter's Specifications, remove the thrust plate and inspect the thrust plate for wear. Measure the thrust plate thickness and compare it to the value listed in this Chapter's Specifications. **Note:** *Excessive balance shaft endplay can only be remedied by replacing a worn thrust plate or by replacing the entire balance shaft assembly.*

11   Remove the bolts and separate the upper and lower housing **(see illustration)**.

15.11 Remove the bolts and separate the upper and lower housing (arrows indicate the thrust plate retaining bolts)

15.12 Hold one shaft in place and rotate the other shaft inward (toward the shaft being held) to get an accurate backlash reading

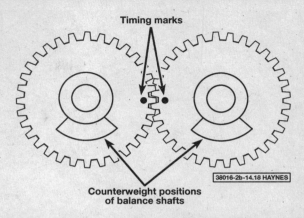

15.18 Be sure the timing marks are set correctly on the balance shaft gears

12   Check the balance shaft gear backlash **(see illustration)**. If the clearance is more than the value listed in this Chapter's Specifications, replace the balance shaft assembly.

13   Remove each balance shaft and gear assembly from the housing. Inspect the shafts for damage such as nicks, cracks, scored bearing journals, broken or cracked teeth, etc. and replace any necessary parts. **Caution:** *Balance shafts must be replaced together. Any time balance shafts are replaced, the bearings must also be replaced.*

14   Remove the bearings from the upper and lower housings and inspect the bearings for scoring, overheating, etc. Also check the housing for cracks, scored bearing bores and damaged threaded holes. Replace all parts as necessary. **Caution:** *If the housing is damaged in any way, replace the entire assembly.*

15   Inspect the chain for damaged links. **Caution:** *DO NOT replace individual links in the chain. The entire chain must be replaced if any damage is found. If the chain is to be replaced, the sprockets must also be replaced. Chain replacement requires removal of the crankshaft, which in turn will require removal of the engine.*

Installation

Refer to illustrations 15.18 and 15.19

16   Assemble the thrust plate and tighten the bolts to the torque listed in this Chapter's Specifications.

17   Install the bearing halves in the upper and lower housings and lubricate the bearing faces with engine assembly lube.

18   Align the timing marks and install the balance shafts in the housings **(see illustration)**. **Caution:** *The engine will make noise or vibrate if the marks are not properly aligned.*

19   Assemble the upper and lower housings and tighten bolts 1 through 8 to 44 inch-lbs following the correct sequence **(see illustration)**. Final tightening will be done after the balance shaft assembly is installed on the block.

20   Bolt the balance shaft assembly to the engine block. **Note:** *Use Locktite 242, or*

equivalent, thread locking compound on the housing-to-block bolts.

21   Tighten the housing-to-block bolts (bolts 9 through 13), in sequence, to 44 inch-lbs **(see illustration 15.19)**. Make sure the balance shafts spin freely.

22   Tighten all bolts, in sequence **(see illustration 15.19)** to the torque listed in this Chapter's Specifications.

23   Install the oil pick-up screen. **Caution:** *The screen must not be installed until all bolts have been tightened to the final specification.*

24   Reconfirm that the number 1 piston is still at TDC (Top Dead Center). If the crankshaft moved during this procedure it will be necessary to realign the timing marks with the number 1 piston at TDC (see Section 3).

25   Rotate the crankshaft, clockwise, 90-degrees from TDC and install the balance shaft alignment tool as described in Step 5. Note that the "flats" on the rear of the balance shafts should be pointing downward with the tool installed **(see illustration 15.5b)**. **Caution:** *This is a very important Step, if it is not followed exactly, the balance shafts will not be properly timed with the crankshaft.*

26   Insert the driven sprocket teeth into the balance shaft drive chain and bolt the sprocket to the balance shaft. **Caution:** *If reusing the old sprocket, be sure the mark, made on disassembly, shows. The manufacturer recommends replacing the balance shaft driven gear bolt after every time it is removed.*

27   Tighten the driven sprocket bolt to torque listed in this Chapter's Specifications. The balance shafts must not turn while the driven sprocket is being tightened. Remember, the balance shaft sprocket bolt is reverse threaded, so turn it counter-clockwise to tighten.

28   Loosely install the chain tensioner and bolts.

29   Adjust the chain tension by inserting a 0.040-inch brass feeler gauge between the chain and chain guide. Apply light pressure (about 3 lbs) to the chain guide and tighten the chain guide bolt to the torque listed in this Chapter's Specifications. **Caution:** *A brass*

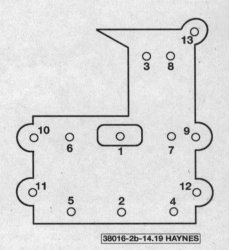

15.19 Balance shaft assembly tightening sequence

feeler gauge is necessary to measure chain clearance. A steel gauge will not bend and will give a incorrect chain-to-guide clearance.

30   The remainder of the installation is the reverse of the removal procedure.

31   Add oil and a new filter, run the engine and check for leaks.

16   Flywheel/driveplate - removal and installation

This procedure is essentially the same for all engines. Refer to Chapter 2B and follow the procedure outlined there. Be sure to use the bolt torque listed in this Chapter's Specifications.

17   Rear main oil seal - replacement

Refer to illustrations 17.5, 17.6, 17.7 and 17.8

1   Remove the transaxle (see Chapter 7). Support the engine from above using an

17.5 Remove the seal housing bolts (arrows)

17.6 After removing the housing from the engine, support it on wood blocks and drive out the old seal with a punch and hammer

17.7 Drive the new seal into the housing with a wood block - be careful not to cock the seal in the housing bore

17.8 Position a new gasket over the dowel pins (arrows)

engine support fixture (available at rental yards). If the special support fixture is unavailable, position a jack under the engine oil pan. Place a large wood block between the jack head and the oil pan, then carefully raise the engine just enough to support the weight. **Warning:** *DO NOT place any part of your body under the engine when it's supported only by a jack!*

2   If equipped with a manual transaxle, remove the pressure plate and clutch disc (see Chapter 8).

3   Remove the flywheel or driveplate (see Section 16).

4   Remove the seal housing-to-oil pan bolts.

5   Remove the seal housing-to-engine block bolts **(see illustration)**. Detach the seal housing and remove the old gasket material.

6   Support the seal housing between two wood blocks on a workbench and drive the old seal out from the back side with a punch and hammer **(see illustration)**.

7   Drive the new seal into the housing with a wood block **(see illustration)**.

8   Lubricate the crankshaft seal journal and the lip of the new seal with multi-purpose grease or clean engine oil. Position a new

gasket on the engine block **(see illustration)**.

9   Inspect the oil pan gasket. The gasket should be checked carefully and replaced with a new one if damage is noted. Minor imperfections can be repaired with RTV sealant. **Caution:** *Use only enough sealant to restore the gasket to its original size and shape. Excess sealant may cause part mis-alignment and oil leaks.*

10   Slowly and carefully push the new seal onto the crankshaft. The seal lip is stiff, so work it onto the crankshaft with a smooth object such as the end of an extension as you push the housing against the block.

11   Install and tighten the housing bolts and the oil pan bolts to the torque listed in this Chapter's specifications.

12   Install the flywheel and clutch components.

13   Reinstall the transaxle.

---

18 Powertrain mounts - check and replacement

---

1   Powertrain mounts seldom require attention, but broken or deteriorated mounts

should be replaced immediately or the added strain placed on the driveline components may cause damage or wear. The four-cylinder engine is equipped with four powertrain mounts, the front mount, the rear mount and the driver's side mount attach to the transaxle and the frame while the passenger side mount attaches to the front of the timing chain cover and the frame. **Note:** *Some earlier models may also be equipped with an additional torque strut that mounts to a bracket at the front of the oil pan and the frame.*

Check

2   During the check, the engine must be raised slightly to remove the weight from the mounts.

3   Raise the vehicle and support it securely on jackstands. Support the engine from above using an engine support fixture (available at rental yards). If the special support fixture is unavailable, position a jack under the engine oil pan. Place a large wood block between the jack head and the oil pan, then carefully raise the engine just enough to take the weight off the mounts. **Warning:** *DO NOT place any part of your body under the engine when it's supported only by a jack!*

18.9a Front powertrain mount installation details

A	Powertrain mount	C	Mount through-bolt
B	Mount-to-transaxle bolts (two)	D	Mount-to-chassis bracket
		E	Chassis bracket mounting bolts

4   Check the mounts to see if the rubber is cracked, hardened or separated from the metal plates. Sometimes the rubber will split right down the center.

5   Check for relative movement between the mount plates and the engine or frame (use a large screwdriver or pry bar to attempt to move the mounts). If movement is noted, lower the engine and tighten the mount fasteners.

Replacement

6   Detach the cable from the negative terminal of the battery. **Caution:** *On models equipped with the Theftlock audio system, be sure the lockout feature is turned off before performing any procedure which requires disconnecting the battery (see the front of this manual).*

7   Raise the engine slightly to take the weight off the mount.

Front, rear and driver's side mounts

Refer to illustrations 18.9a, 18.9b and 18.9c

8   Remove the mount through-bolt(s).

9   Remove the mount to transaxle bolts and remove the engine mount **(see illustrations)**.

10   Place the new mount in position and install the bolts and nuts. Gently lower the engine and tighten the nuts to the torque

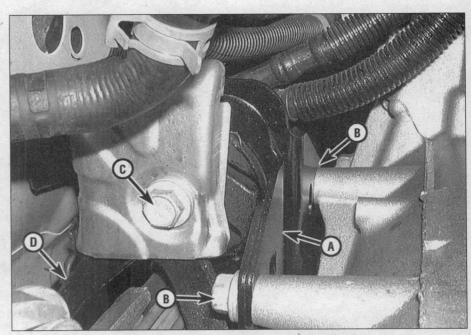

18.9b Rear powertrain mount installation details

A	Powertrain mount	C	Mount through-bolt
B	Mount-to-transaxle bolts (three)	D	Mount-to-chassis bracket

18.9c Driver's side powertrain mount installation details

A	Powertrain mount	C	Mount through-bolt
B	Mount-to-transaxle bolts (four)	D	Mount-to-chassis bracket

listed in this Chapter's Specifications. **Note:** *Rubber preservative should be applied to the mounts to slow deterioration.*

Passenger side mount

Refer to illustration 18.13

11 Remove the coolant reservoir (see Chapter 3).

12 Raise the engine slightly to take the weight off the mount.

13 Remove the engine mount bracket, then remove the engine mount from the chassis **(see illustration)**.

14 Place the new mount in position and install the nuts. Install the engine mount bracket, then lower the engine and tighten the nuts to the torque listed in this Chapter's Specifications. **Note:** *Rubber preservative should be applied to the mounts to slow deterioration.*

Engine mount torque strut and bracket

15 Raise the vehicle and support it securely on jackstands. Remove the right lower splash shield.

16 Working under the strut, remove the through-bolts securing the strut to the chassis and to the engine.

17 Raise the engine slightly to take the weight off the strut and separate the strut from the bracket.

18.13 Passenger side powertrain mount installation details

A	Powertrain mount	D	Mount-to-engine bracket
B	Engine mount bracket		retaining nuts
C	Engine mount bracket bolts	E	Mount-to-chassis retaining nuts

18 Remove the engine mount strut bracket bolts from the oil pan and the engine block and remove the bracket.

19 Installation is the reverse of removal. Tighten the bolts to the torque listed in this Chapter's Specifications.

Chapter 2 Part B
V6 engines

Contents

Specifications

General

Displacement
3.1L V6	191 cubic inches
3.4L V6	204 cubic inches

Bore and stroke
3.1L V6	3.50 x 3.31 inches
3.4L V6	3.62 x 3.31 inches

Cylinder numbers (drivebelt end-to-transaxle end)
Front bank (radiator side)	2-4-6
Rear bank	1-3-5

Firing order 1-2-3-4-5-6

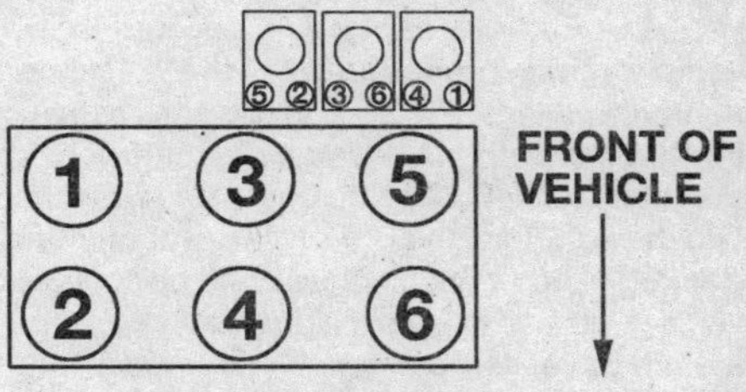

Cylinder location and coil terminal
identification diagram - V6 engines

24048-1-B HAYNES

FRONT OF VEHICLE

Torque specifications

Ft-lbs
(unless otherwise indicated)

Camshaft sprocket bolt
1997	81
1998 and later	103

Cylinder head bolts
Step 1
1997	33
1998 and later	37

Step 2 .... Rotate an additional 90-degrees (1/4-turn)

Torque specifications

	Ft-lbs (unless otherwise indicated)
Exhaust manifold retaining nuts	
1997	89 in-lbs
1998 and later	144 in-lbs
Exhaust heat shield bolts	89 in-lbs
Exhaust crossover pipe nuts	18
Flywheel/Driveplate-to-crankshaft bolts	
1997	61
1998 and later	52
Intake manifold bolts (lower)	115 in-lbs
Intake manifold bolts/studs (upper)	18
Oil pan bolts/nuts	
To block	18
Side bolts	37
Oil pump mounting bolt	30
Rocker arm bolts	
Step 1	
1997	89 in-lbs
1998 and later	168 in-lbs
Step 2	Rotate an additional 30-degrees
Timing chain cover bolts	
1997	
Small	15
Large	35
1998 and later	
Small	15
Medium	35
Large	41
Timing chain damper bolts	15
Valve cover-to-cylinder head bolts	89 in-lbs
Vibration damper bolt	76
Front and rear engine mount-through bolt	89
Front and rear engine mount-to-transaxle bolts	99
Front engine mount-bracket to cradle bolts	89
Rear engine mount-bracket to cradle bolts	49
Driver's side engine mount-through bolt	89
Driver's side engine mount-to-transaxle bolts	99
Passenger side upper engine mount-to-frame nuts	49
Passenger side upper engine mount-to-engine support bracket bolts	96
Passenger side lower engine mount nuts	
3.4L engine	35
Passenger side lower engine mount bracket bolts	
3.4L engine	43

1   General information

This Part of Chapter 2 is devoted to in-vehicle repair procedures for the 3.1L and 3.4L V6 engines. These engines utilize cast-iron blocks with six cylinders arranged in a "V" shape at a 60-degree angle between the two banks. The overhead valve aluminum cylinder heads are equipped with replaceable valve guides and seats. Hydraulic lifters actuate the valves through tubular pushrods.

The engines are easily identified by looking for the designations printed directly on top of the upper intake plenum.

All information concerning engine removal and installation and engine block and cylinder head overhaul can be found in Part C of this Chapter. The following repair procedures are based on the assumption that the engine is installed in the vehicle. If the engine has been removed from the vehicle and mounted on a stand, many of the steps outlined in this Part of Chapter 2 will not apply.

The Specifications included in this Part of Chapter 2 apply only to the procedures contained in this Part. Part C of Chapter 2 contains the Specifications necessary for cylinder head and engine block rebuilding.

2   Repair operations possible with the engine in the vehicle

Many major repair operations can be accomplished without removing the engine from the vehicle.

Clean the engine compartment and the exterior of the engine with some type of degreaser before any work is done. It'll make the job easier and help keep dirt out of the internal areas of the engine.

Depending on the components involved, it may be helpful to remove the hood to improve access to the engine as repairs are performed (refer to Chapter 11 if necessary). Cover the fenders to prevent damage to the paint. Special pads are available, but an old bedspread or blanket will also work.

If vacuum, exhaust, oil or coolant leaks develop, indicating a need for gasket or seal replacement, the repairs can generally be done with the engine in the vehicle. The intake and exhaust manifold gaskets, timing chain cover gasket, oil pan gasket, crankshaft oil seals and cylinder head gaskets are all accessible with the engine in place.

Exterior engine components, such as the intake and exhaust manifolds, the oil pan (and the oil pump), the water pump, the starter motor, the alternator and the fuel system components can be removed for repair with the engine in place.

Since the cylinder heads can be removed without pulling the engine, valve component servicing can also be accomplished with the engine in the vehicle. Replacement of the timing chain and sprockets is also possible with the engine in the vehicle, although camshaft removal can not be performed with the engine in the chassis (see Part C of this Chapter).

In extreme cases caused by a lack of

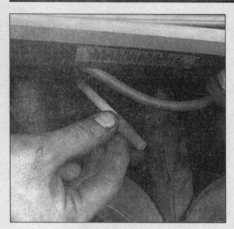

3.8 A plastic object inserted in the number one spark plug hole can be used to determine the highest point reached by that piston

necessary equipment, repair or replacement of piston rings, pistons, connecting rods and rod bearings is possible with the engine in the vehicle. However, this practice is not recommended because of the cleaning and preparation work that must be done to the components involved.

3 Top Dead Center (TDC) - locating

Refer to illustration 3.8

1 Top Dead Center (TDC) is the highest point in the cylinder each piston reaches as it travels up-and-down when the crankshaft turns. Each piston reaches TDC on the compression stroke and again on the exhaust stroke, but TDC generally refers to piston position on the compression stroke.
2 Positioning the piston(s) at TDC is an essential part of certain procedures such as timing chain/sprocket removal and camshaft removal.
3 Before beginning this procedure, be sure to place the transaxle in Park, apply the

parking brake and block the rear wheels. Raise the front of the vehicle and support it securely on jackstands.
4 Remove the spark plugs (see Chapter 1).
5 When looking at the drivebelt end of the engine, normal crankshaft rotation is clockwise. In order to bring any piston to TDC, the crankshaft must be turned with a socket and ratchet attached to the bolt threaded into the center of the vibration damper on the crankshaft.
6 Have an assistant turn the crankshaft with a socket and ratchet as described above while you hold a finger over the number one spark plug hole. **Note:** *See the cylinder numbering diagram in the specifications for this Chapter.*
7 When the piston approaches TDC, air pressure will be felt at the spark plug hole. Instruct your assistant to turn the crankshaft slowly.
8 Insert a long blunt object into the spark plug hole **(see illustration)**. As the piston rises the object will be pushed out. Note the point where the object stops moving out - this is TDC. **Note:** *It is preferred, that a long plastic object be used during this procedure to ensure that object won't fall into the cylinder and will not scratch the cylinder walls. Always hold the object upright while the engine is being rotated so that the object will not get wedged as the piston travels upward.*
9 After the number one piston has been positioned at TDC on the compression stroke, TDC for any of the remaining pistons can be located by repeating the procedure described above and following the firing order.

4 Valve covers - removal and installation

Removal

1 Disconnect the cable from the negative terminal of the battery. **Caution:** *On models*

equipped with the Theftlock audio system, be sure the lockout feature is turned off before performing any procedure which requires disconnecting the battery (see the front of this manual).

Front cover

Refer to illustration 4.7

2 Remove the air cleaner assembly (see Chapter 4).
3 Remove the spark plug wires from the spark plugs (see Chapter 1). Be sure each wire is labeled before removal to ensure correct reinstallation.
4 Detach the spark plug wire harness clamps from the coolant tube.
5 Remove the PCV tube from the valve cover.
6 Drain the coolant (see Chapter 1), and disconnect the coolant bypass pipe.
7 Loosen the valve cover mounting bolts **(see illustration)**. **Note:** *Some models are equipped with Torx type bolts. Remove them with a Torx driver.*
8 Detach the valve cover. **Note:** *If the cover sticks to the cylinder head, use a block of wood and a hammer to dislodge it. If the cover still won't come loose, pry on it carefully, but don't distort the sealing flange.*

Rear cover

Refer to illustration 4.13

9 Remove the spark plug wires from the spark plugs (see Chapter 1). Be sure each wire is labeled before removal to ensure correct reinstallation.
10 Remove the ignition coil assembly (see Chapter 5), which also includes the solenoids for the vacuum canister and purge control (if equipped). Tag all disconnected wires and hoses.
11 Detach the brake booster vacuum hose from the upper intake manifold.
12 Remove the serpentine drivebelt (see Chapter 1).
13 Remove the alternator (see Chapter 5) and the alternator support bracket **(see illustration)**.

2B

4.7 Loosen the valve cover mounting bolts (arrows indicate three) - the bolts will stay with the cover

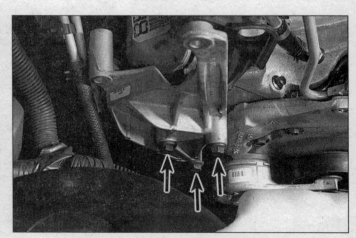

4.13 Remove the bolts (arrows) and detach the alternator support bracket

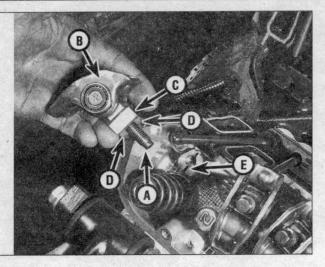

5.3 Rocker arm installation details - the rocker arms are kept as an assembly by a small sleeve between the bolt and the pedestal - note the projections on the pedestal; they fit into grooves in the head

A   Rocker arm bolt
B   Rocker arm
C   Rocker arm pedestal
D   Pedestal projections
E   Grooves in the head

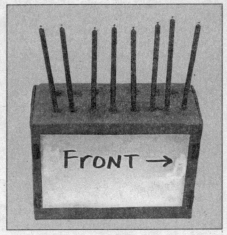

5.4 A perforated cardboard box can be used to store the pushrods to ensure they are reinstalled in their original locations - note the label indicating the front end of the engine

14   Loosen the valve cover mounting bolts. **Note:** *Some models are equipped with Torx-head type bolts. Remove them with a Torx driver.*
15   Detach the valve cover. **Note:** *If the cover sticks to the cylinder head, use a block of wood and a hammer to dislodge it. If the cover still won't come loose, pry on it carefully, but don't distort the sealing flange.*

Installation

16   The mating surfaces of each cylinder head and valve cover must be perfectly clean when the covers are installed. Use a gasket scraper to remove all traces of sealant or old gasket material, then clean the mating surfaces with lacquer thinner or acetone (if there's sealant or oil on the mating surfaces when the cover is installed, oil leaks may develop). The valve covers are made of aluminum, so be extra careful not to nick or gouge the mating surfaces with the scraper.
17   Clean the mounting bolt threads with a die if necessary to remove any corrosion and restore damaged threads. Use a tap to clean the threaded holes in the heads.
18   Apply a dab of RTV sealant to the two joints where the intake manifold and cylinder head meet.
19   Place the valve cover and new gasket in position, then install the bolts. Tighten the bolts in several steps to the torque listed in this Chapter's Specifications.
20   Complete the installation by reversing the removal procedure. Start the engine and check carefully for oil leaks at the valve cover-to-head joints.

5   Rocker arms and pushrods - removal, inspection and installation

Refer to illustrations 5.3 and 5.4

Removal

1   Disconnect the cable from the negative terminal of the battery. **Caution:** *On models*

equipped with the Theftlock audio system, be sure the lockout feature is turned off before performing any procedure which requires disconnecting the battery (see the front of this manual).
2   Remove the valve cover(s) (see Section 4).
3   Beginning at the drivebelt end of one cylinder head, remove the rocker arm mounting bolts one at a time and detach the rocker arms, pivot balls and pedestals **(see illustration)**. Store each set of rocker arm components separately in a marked plastic bag to ensure they're reinstalled in their original locations. **Note:** *The rocker arms have the pedestal mount "captured" on the rocker arm bolt by a metal sleeve inside. The components can be separated if necessary by tapping the bolt out of the pedestal, but normally all components for a particular valve will stay as an assembly.*
4   Remove the pushrods and store them separately to make sure they don't get mixed up during installation **(see illustration)**. **Note:** *Intake and exhaust pushrods are different lengths. Intake pushrods are approximately 5 3/4 inches long, while exhausts are 6.0 inches long. They may also have color codes to easily tell them apart.*

Inspection

5   Inspect each rocker arm for wear, cracks and other damage, especially where the pushrods and valve stems make contact.
6   Make sure the rollers operate freely as well.
7   Make sure the hole at the pushrod end of each rocker arm is open.
8   Inspect the pushrods for cracks and excessive wear at the ends. Roll each pushrod across a piece of plate glass to see if it's bent (if it wobbles, it's bent).

Installation

9   Lubricate the lower end of each pushrod with clean engine oil or moly-base grease and install them in their original locations. Make sure each pushrod seats completely in

the lifter socket.
10   Apply moly-base grease to the ends of the valve stems and the upper ends of the pushrods.
11   Apply clean engine oil to the pivot balls and to the bearing surfaces of each rocker arm to prevent damage to the mating surfaces before engine oil pressure builds up. Install the rocker arms, pivot balls, pedestals and bolts and tighten them to the torque listed in this Chapter's Specifications. As the bolts are tightened, make sure the pushrods engage properly in the rocker arms and that the projections on the bottom of the pedestals fit into the grooves on the head before tightening the bolts **(see illustration 5.3)**.
12   Install the valve covers. Start and run the engine, then check for oil leaks and unusual sounds coming from the valve cover area.

6   Valve springs, retainers and seals - replacement

Refer to illustrations 6.4, 6.6, 6.12 and 6.14
Note: *Broken valve springs and defective valve stem seals can be replaced without removing the cylinder head. Two special tools and a compressed air source are normally required to perform this operation, so read through this Section carefully and rent or buy the tools before beginning the job.*
1   Remove the valve cover (see Section 4).
2   Remove the spark plugs from the cylinders which have the defective components. If all of the valve stem seals are being replaced, all of the spark plugs should be removed.
3   Turn the crankshaft until the piston in the affected cylinder is at top dead center (see Section 3). If you're replacing all of the valve stem seals, begin with cylinder number one and work on the valves for one cylinder

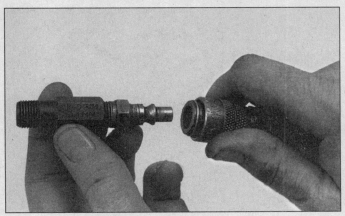

6.4 This is what the typical air hose adapter that threads into the spark plug hole looks like - they're commonly available at auto parts stores

6.6 While the valve spring tool is compressing the spring, remove the keepers with a small magnet or pliers

2B

at a time. Move from cylinder-to-cylinder following the firing order sequence (see this Chapter's Specifications).

4   Thread an adapter into the spark plug hole **(see illustration)** and connect an air hose from a compressed air source to it. Most auto parts stores can supply the air hose adapter. **Note:** *Many cylinder compression gauges utilize a screw-in fitting that may work with your air hose quick-disconnect fitting.*

5   Apply compressed air to the cylinder. **Warning:** *The piston may be forced down by compressed air, causing the crankshaft to turn suddenly. If the wrench used when positioning the number one piston at TDC is still attached to the bolt in the crankshaft nose, it could cause damage or injury when the crankshaft moves. The valves should be held in place by the air pressure. If the valve faces or seats are in poor condition, leaks may prevent air pressure from retaining the valves - a "valve job" is necessary to correct this problem.*

6   Stuff shop rags into the cylinder head holes above and below the valves to prevent parts and tools from falling into the engine, then use a valve spring compressor to compress the spring. Remove the keepers **(see illustration)** with small needle-nose pliers or a magnet.

7   Remove the spring retainer and valve spring, then remove the valve guide seal.

8   Wrap a rubber band or tape around the top of the valve stem so the valve won't fall into the combustion chamber, then release the air pressure.

9   Inspect the valve stem for damage. Rotate the valve in the guide and check the end for eccentric movement, which would indicate the valve stem is bent.

10   Move the valve up-and-down in the guide and make sure it doesn't bind. If the valve stem binds, either the valve is bent or the guide is damaged. In either case, the head will have to be removed for repair.

11   Reapply air pressure to the cylinder to retain the valve in the closed position, then

6.12 Tap the new seal in place on the guide with a socket

remove the tape or rubber band from the valve stem.

12   Lubricate the valve stem with engine oil and install a new valve guide seal. An appropriate-size socket can be used to install the new seal, just don't force it once it bottoms **(see illustration)**.

13   Install the spring in position over the valve. **Note:** *The large end of the spring goes toward the cylinder head.*

14   Install the valve spring retainer. Compress the valve spring and carefully install the keepers in the groove. Apply a small dab of grease to the inside of each keeper to hold it in place if necessary **(see illustration)**. Remove the pressure from the spring tool and make sure the keepers are seated.

15   Disconnect the air hose and remove the adapter from the spark plug hole.

16   Install the spark plug(s) and hook up the wire(s).

17   Install the valve cover(s).

18   Start and run the engine, then check for oil leaks and unusual sounds coming from the valve cover area.

6.14 Keepers don't always stay in place, so apply a small dab of grease to each one as shown here before installation - the grease will hold the keepers in place on the valve stem

7   Intake manifold - removal and installation

Upper intake manifold (plenum)

Refer to illustration 7.8

1   Disconnect the cable from the negative terminal of the battery. **Caution:** *On models equipped with the Theftlock audio system, be sure the lockout feature is turned off before performing any procedure which requires disconnecting the battery (see the front of this manual).*

2   Refer to Chapter 4 and relieve the fuel system pressure, then remove the air intake duct and detach the throttle cable and the cruise control cable from the throttle body.

3   Label and disconnect the hoses and electrical connectors attached to the plenum and throttle body.

4   Refer to Chapter 6 and remove the EGR valve-to-plenum bolts. On some models it may also be necessary to remove the alternator (see Chapter 5).

5   Remove the spark plug wires and the

7.8 Upper intake manifold TIGHTENING sequence

7.13a Loosen the heater pipe fitting from the thermostat housing/lower intake manifold . . .

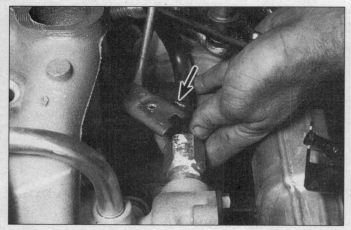

7.13b . . . then remove bolt securing the heater pipe bracket to the cylinder head and pull it out the (arrow) from the lower intake manifold - use a screwdriver under the bracket to twist the pipe until it comes out of the fitting

7.14 Disconnect the thermostat bypass hose (arrow) from the pipe

ignition coil and module assembly (see Chapter 5).

6   Loosen the upper intake manifold bolts in the reverse order of the tightening sequence **(see illustration 7.8)** and remove the upper plenum with the throttle body attached.

7   To install the upper manifold, clean the mounting surfaces of the lower intake manifold and the upper plenum with lacquer thinner and remove all traces of the old gasket material or sealant.

8   Install the new gasket over the lower intake manifold, then install the upper plenum onto the lower intake manifold and tighten the bolts in the recommended tightening sequence **(see illustration)** to the torque listed in this Chapter's Specifications. The remainder of the installation is the reverse of removal.

Lower intake manifold

Refer to illustrations 7.13a, 7.13b, 7.14, 7.16, 7.20, 7.21 and 7.23

9   Remove the upper intake manifold (see Steps 1 through 6). Drain the cooling system

(see Chapter 1).

10   Label and disconnect any remaining wires, fuel and vacuum lines from the lower intake manifold.

11   Refer to Chapter 4 and remove the fuel rail and injectors from the lower intake manifold.

12   Remove the valve covers (see Section 4). Remove the power steering pump without disconnecting the hoses and set it aside (see Chapter 10).

13   Remove the heater pipe from the transaxle end of the lower intake manifold **(see illustrations)**.

14   Disconnect the thermostat bypass hose from the bypass pipe **(see illustration)**.

15   Loosen the manifold mounting bolts/nuts in 1/4-turn increments until they can be removed by hand **(see illustration)**.

16   The manifold will probably be stuck to the cylinder heads and force may be required to break the gasket seal **(see illustration)**. **Caution:** *Don't pry between the manifold and the heads or damage to the gasket sealing surfaces may occur, leading to vacuum leaks.*

17   Loosen the rocker arm bolts, rotate the rocker arms out of the way and remove the pushrods that go through the manifold gaskets (see Section 5).

18   Lift the old gaskets off. Use a gasket scraper to remove all traces of sealant and old gasket material, then clean the mating surfaces with lacquer thinner or acetone. **Note:** *The mating surfaces of the cylinder heads, block and manifold must be perfectly clean when the manifold is installed. Gasket removal solvents are available at most auto parts stores and may be helpful when removing old gasket material that's stuck to the heads and manifold (since the manifold is made of aluminum, aggressive scraping can cause damage). Be sure to follow the directions printed on the container. If there's old sealant or oil on the mating surfaces when the manifold is installed, oil or vacuum leaks may develop. Use a vacuum cleaner to remove any gasket material that falls into the intake ports or the lifter valley.*

19   Use a tap of the correct size to chase the threads in the bolt holes, if necessary,

7.16 Pry the manifold loose at a casting boss (arrow) - don't pry between the gasket surfaces!

7.20 Install the intake gaskets (arrows) against each cylinder head

7.21 Apply a bead of sealant to the end ridges between the heads (arrow indicates ridge at transaxle end)

2B

7.23 Intake manifold TIGHTENING sequence - make sure the bolts in the center (1 through 4) are completely tightened before tightening the end bolts (5 through 8)

then use compressed air (if available) to remove the debris from the holes. **Warning:** *Wear safety glasses or a face shield to protect your eyes when using compressed air!*

20   Place the intake manifold gaskets in position on the heads **(see illustration)**. Then install the pushrods and rocker arms (see Section 4).

21   Apply a 3/16-inch (5 mm) bead of RTV sealant to the front and rear ridges of the engine block between the heads **(see illustration)**.

22   Carefully lower the manifold into place and install the mounting bolts/nuts finger tight. **Note:** *Coat the bolt threads with pipe sealant before installing them.*

23   Tighten the four vertical bolts (1 through 4) at the center of the manifold in the recommended tightening sequence **(see illustration)** to the torque listed in this Chapter's Specifications.

24   Tighten the four angled bolts (5 through 8) at the ends of the manifold in the recommended tightening sequence to the torque listed in this Chapter's Specifications. **Caution:** *To prevent oil leaks, tighten the vertical bolts first to ensure that the lower manifold stays centered on the gaskets, then tighten the angled bolts.*

25   Install the remaining components in the reverse order of removal.

26   Change the oil and filter and refill the cooling system (see Chapter 1). Start the engine and check for leaks.

8   Exhaust manifolds - removal and installation

Removal

1   Disconnect the cable from the negative

terminal of the battery. **Caution:** *On models equipped with the Theftlock audio system, be sure the lockout feature is turned off before performing any procedure which requires disconnecting the battery (see the front of this manual).*

2   Remove the air cleaner assembly and duct (see Chapter 4).

Front manifold

Refer to illustrations 8.7, 8.8 and 8.9

3   Allow the engine to cool completely, then drain the coolant (see Chapter 1).

4   Remove the upper radiator hose from the thermostat housing.

5   Remove the spark plug wires from the front spark plugs.

6   Disconnect the coolant by-pass pipe from the front exhaust manifold and the water pump.

7   Remove the crossover pipe heat shield

8.7 Remove the screws (A) and the exhaust manifold heat shield, then remove the exhaust crossover heat shield (B indicates the two front bolts, others are at the rear manifold)

8.8 Unbolt the crossover pipe where it joins the front manifold (arrows)

8.9 Remove the six nuts (arrows indicate the upper three) from the exhaust manifold studs

8.13 Remove the nuts (arrows) holding the exhaust pipe to the rear manifold

and the manifold heat shield **(see illustration)**.

8   Unbolt the crossover pipe where it joins the front manifold **(see illustration)**. You should first apply penetrating oil to the fastener threads - they're usually rusted. **Note:** *After removing the three nuts, pull the crossover pipe back (the flexible joint toward the rear will allow some movement) and remove the three studs. There are hex portions on the studs. If the studs are not removed from the manifold, it is difficult to pull the manifold from the cylinder head because the manifold is also mounted on studs.*

9   Remove the mounting nuts and detach the manifold from the cylinder head **(see illustration)**.

Rear manifold

Refer to illustration 8.13

10   Remove the crossover pipe heat shield and the manifold heat shield.

11   Unbolt the crossover pipe where it joins the rear manifold. You should first apply pen-

etrating oil to the fastener threads - they're usually rusted.

12   Disconnect the electrical connector from oxygen sensor(s) and remove the EGR tube from the rear manifold (see Chapter 6). **Note:** *To help prevent possible damage to the oxygen sensor(s) it is recommended that the sensor(s) be removed from the exhaust manifold before the manifold is removed from the engine (see Chapter 6).*

13   Working under the vehicle, remove the exhaust pipe-to-manifold bolts and position the front exhaust pipe aside **(see illustration)**. **Note:** *You may have to apply penetrating oil to the fastener threads - they're usually corroded.*

14   On early models remove the bracket bolt on the automatic transaxle dipstick tube and move the tube aside.

15   Unbolt and remove the rear exhaust manifold.

Installation (front or rear)

Refer to illustration 8.17

16   Clean the mating surfaces to remove all

traces of old gasket material, then inspect the manifold for distortion and cracks. Warpage can be checked with a precision straightedge held against the mating flange. If a feeler gauge thicker than 0.030-inch can be inserted between the straightedge and flange surface, take the manifold to an automotive machine shop for resurfacing.

17   Remove the exhaust manifold inner heat shield (the gasket material is part of the inner heat shield) and examine the gasket areas for signs of corrosion or leakage **(see illustration)**. If the shield/gasket seems reusable, reinstall it, place the manifold in position and install the mounting bolts finger tight.

18   Starting in the middle and working out toward the ends, tighten the mounting bolts a little at a time until all of them are at the torque listed in this Chapter's Specifications.

19   Install the remaining components in the reverse order of removal.

20   Start the engine and check for exhaust leaks between the manifold and cylinder head and between the manifold and exhaust pipe.

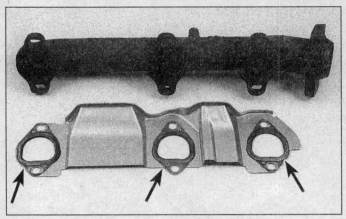

8.17 Examine the gasket areas (arrows) of both sides of the inner heat shield - if bad, the whole shield must be replaced

9.3 Remove the bolt (arrow) holding the oil dipstick tube to the front cylinder head

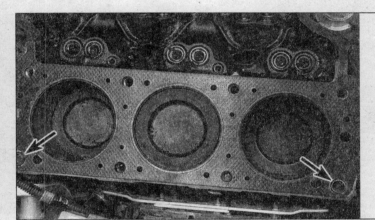

2B

9.14a Position the new gasket over the dowel pins (arrows) . . .

9.11 Remove the old gasket and carefully scrape off all old gasket material and sealant

9 Cylinder heads - removal and installation

Refer to illustrations 9.3, 9.11, 9.14a, 9.14b and 9.17

Removal

1 Disconnect the cable from the negative terminal of the battery. **Caution:** *On models equipped with the Theftlock audio system, be sure the lockout feature is turned off before performing any procedure which requires disconnecting the battery (see the front of this manual).*

2 Remove the air cleaner assembly (see Chapter 4) and then remove the upper and lower intake manifold as described in Section 7.

3 If you're removing the front cylinder head, remove the oil dipstick tube mounting bolt **(see illustration)**.

4 Disconnect all wires and vacuum hoses from the cylinder head(s). Be sure to label them to simplify reinstallation.

5 Detach the exhaust manifold from the cylinder head being removed (see Section 8).

6 Remove the rocker arms and pushrods (see Section 5).

7 Using the new head gasket, outline the cylinders and bolt pattern on a piece of cardboard. Be sure to indicate the front (drivebelt end) of the engine for reference. Punch holes at the bolt locations. Loosen each of the cylinder head mounting bolts 1/4-turn at a time until they can be removed by hand - work from bolt-to-bolt in a pattern that's the *reverse* of the tightening sequence **(see illustration 9.17)**. **Caution:** *The engine must be completely cool before loosening the cylinder head bolts*. Store the bolts in the cardboard holder as they're removed - this will ensure they are reinstalled in their original locations, which is absolutely essential. Note which ones are studs and their location.

8 Lift the head(s) off the engine. If resistance is felt, don't pry between the head and block as damage to the mating surfaces will result. Recheck for head bolts that may have been overlooked, then use a hammer and block of wood to tap up on the head and break the gasket seal. Be careful because there are locating dowels in the block which position each head. As a last resort, pry each head up at the rear corner only and be careful not to damage anything. After removal, place the head on blocks of wood to prevent damage to the gasket surfaces.

9 Refer to Chapter 2, Part C, for cylinder head disassembly, inspection and valve service procedures.

Installation

10 The mating surfaces of each cylinder head and block must be perfectly clean when the head is installed.

11 Use a gasket scraper to remove all traces of carbon and old gasket material **(see illustration)**, then clean the mating surfaces with lacquer thinner or acetone. If there's oil on the mating surfaces when the head is installed, the gasket may not seal correctly and leaks may develop. When working on the block, it's a good idea to cover the lifter valley with shop rags to keep debris out of the engine. Use a shop rag or vacuum cleaner to remove any debris that falls into the cylinders.

12 Check the block and head mating surfaces for nicks, deep scratches and other damage. If damage is slight, it can be removed with a file; if it's excessive, machining may be the only alternative.

13 Use a tap of the correct size to chase the threads in the head bolt holes. Dirt, corrosion, sealant and damaged threads will affect torque readings.

14 Position the new gasket over the dowel pins in the block. Some gaskets are marked TOP or THIS SIDE UP to ensure correct

9.14b ... with the correct side facing up

9.17 Cylinder head bolt TIGHTENING sequence

10.5 Remove the crankshaft bolt (arrow) - it's very tight, so use a six-point socket and a breaker bar

10.6 Use a puller that bolts to the crankshaft pulley hub; jaw-type pullers will damage the crankshaft pulley

installation **(see illustrations)**.

15 Carefully position the head on the block without disturbing the gasket.

16 Clean the bolt threads and install the bolts in the correct locations - two different lengths are used. Here's where the cardboard holder comes in handy.

17 Tighten the bolts, using the recommended sequence **(see illustration)**, to the torque listed in this Chapter's Specifications. Then, using the same sequence, turn each bolt the amount of angle listed in this Chapter's Specifications.

18 The remaining installation steps are the reverse of removal.

19 Change the engine oil and filter (see Chapter 1).

10 Crankshaft pulley – removal and installation

Refer to illustrations 10.5, 10.6, and 10.7

1 Disconnect the cable from the negative terminal of the battery. **Caution:** *On models equipped with the Theftlock audio system, be sure the lockout feature is turned off before performing any procedure which requires disconnecting the battery (see the front of this manual).*

2 With the parking brake applied and the shifter in Park (automatic) or in gear (manual), loosen the lugnuts from the right front wheel, then raise the front of the vehicle and support it securely on jackstands.

3 Remove the right front wheel and the right splash shield from the wheelwell.

4 Remove the drivebelt (see Chapter 1).

5 Remove the bolt from the front of the crankshaft **(see illustration)**. The bolt is normally very tight, so use a large breaker bar and a six-point socket to remove it. **Note 1:** *On automatic transaxle equipped models, remove the driveplate cover and position a large screwdriver in the ring gear teeth to keep the crankshaft from turning while an assistant removes the crankshaft pulley bolt.* **Note 2:** *On some models it will be necessary to remove passenger side upper engine mount, then support the right side of the engine crossmember with a floor jack. Loosen the right side crossmember bolts and lower the crossmember frame to allow clearance for the removal of the crankshaft pulley and the pulley bolt.*

6 Using a puller that bolts to the crankshaft hub, remove the crankshaft pulley/balancer from the crankshaft **(see illustration)**. **Caution:** *On these engines a rubber sleeve connects the inertia weight to the balancer hub. Take care when working on the*

crankshaft pulley/balancer that you do not accidentally shift the inertia weight's position relative to the sleeve or balancer hub, as this will upset the tuning of the balancer.

7 Position the crankshaft pulley/balancer on the crankshaft and slide it on as far as it will go. Note that the slot (keyway) in the hub must be aligned with the Woodruff key in the end of the crankshaft **(see illustration)**.

10.7 The pulley keyway must be aligned with the Woodruff key (arrow) in the crankshaft nose

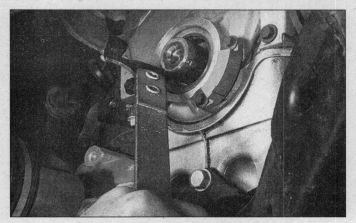

11.2 Carefully pry the old seal out of the timing chain cover - don't damage the crankshaft in the process

11.3 Drive the new seal into place with a large socket and hammer

8   Using a crankshaft balancer installation tool, press the crankshaft pulley/balancer onto the crankshaft. Note that the crankshaft bolt can also be used to press the crankshaft balancer into position, but when doing so, use a liberal amount of clean engine oil on the bolt threads to prevent galling.

9   Tighten the crankshaft bolt to the torque listed in this Chapter's Specifications.

10   The remaining installation steps are the reverse of removal.

11 Crankshaft front oil seal - removal and installation

Refer to illustrations 11.2, 11.3 and 11.4

1   Remove the crankshaft pulley (see Section 10).

2   Note how the seal is installed - the new one must be installed to the same depth and facing the same way. Carefully pry the oil seal out of the cover with a seal puller or a large screwdriver **(see illustration)**. Be very careful not to distort the cover or scratch the crankshaft! Wrap electrician's tape around the tip of the screwdriver to avoid damage to

the crankshaft.

3   Apply clean engine oil or multi-purpose grease to the outer edge of the new seal, then install it in the cover with the lip (spring side) facing IN. Drive the seal into place **(see illustration)** with a large socket and a hammer (if a large socket isn't available, a piece of pipe will also work). Make sure the seal enters the bore squarely and stop when the front face is at the proper depth.

4   Check the surface on the pulley hub that the oil seal rides on. if the surface has been grooved from long time contact with the seal, a press on sleeve may be available to renew the sealing surface **(see illustration)**. This sleeve is pressed into place with a hammer and a block of wood and is commonly available at auto parts stores for various applications.

5   Lubricate the pulley hub with clean engine oil and reinstall the crankshaft pulley. Use a vibration damper installation tool to press the pulley onto the crankshaft.

6   Install the crankshaft pulley retaining bolt and tighten it to the torque listed in this Chapter's Specifications.

7   The remainder of installation is the reverse of the removal.

12 Timing chain and sprockets - removal, inspection and installation

Removal

Refer to illustrations 12.5, 12.11, 12.15a, 12.15b, 12.18a, 12.18b and 12.20

1   Disconnect the cable from the negative terminal of the battery. **Caution:** *On models equipped with the Theftlock audio system, be sure the lockout feature is turned off before performing any procedure which requires disconnecting the battery (see the front of this manual).*

2   Loosen, but do not remove, the water pump pulley bolts, then remove the serpentine drivebelt (see Chapter 1).

3   Remove the water pump pulley (see Chapter 3).

4   Remove the crankshaft pulley (see Section 10).

5   Unbolt the drivebelt tensioner **(see illustration)** and idler, if equipped.

6   Drain the coolant and engine oil (see Chapter 1).

7   Remove the alternator and loosen the

11.4 If the sealing surface of the pulley hub has a wear groove from contact with the seal, repair sleeves are available at most auto parts stores

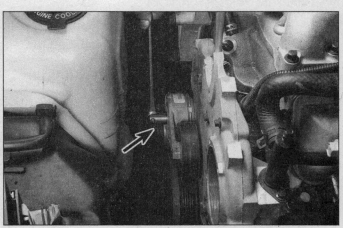

12.5 The drivebelt tensioner is secured to the timing chain cover by a bolt (arrow)

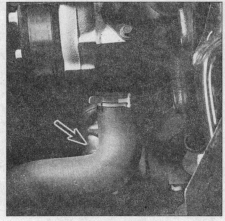

12.11 Support the engine with an engine support fixture and chains to the front and rear engine lifting eyes (arrows)

12.15a Disconnect the radiator hose (arrow) from the water pump housing

12.15b Disconnect the hose at the intake manifold pipe (A), then remove the bolt (B) at the bypass and pull the bypass from the front cover

12.18a Timing chain cover bolt locations (arrows), upper . . .

mounting bracket (see Chapter 5).

8   Unbolt the power steering pump (if equipped) and tie it aside (see Chapter 10). Leave the hoses connected.

9   Unbolt the flywheel/driveplate cover below the transaxle.

10   Remove the starter (see Chapter 5).

11   Remove the passenger side upper engine mount and the mount support bracket from the front of the timing chain cover (see Section 18). **Note:** *This procedure requires an engine support fixture* **(see illustration)** *to secure the engine from above when the mounts are removed. Make sure you have these tools or rent them before beginning the procedure.* Refer to Chapter 3 and remove the engine cooling fan assembly, unbolt the air-conditioning compressor and set it aside without disconnecting the refrigerant lines, then remove the compressor mounting bracket.

12   Remove the front exhaust manifold (see Section 8).

13   Remove the hood (see Chapter 11).

14   Remove the oil pan (see Section 14).

Note: *The front cover can be removed with the oil pan in place, but the pan must be removed for a good installation to seal against the bottom of the front cover.*

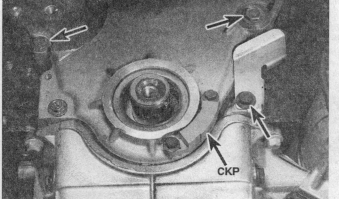

12.18b . . . and lower (arrows) - the crankshaft position sensor (CKP) should be unbolted and laid aside

15   Disconnect the coolant hoses from the bypass pipe and water pump, and remove the coolant bypass pipe from the front cover **(see illustrations)**.

16   Disconnect the electrical connector at the crankshaft position sensor (see Chapter 6).

17   Remove the right side ball joint and the lower control arm, then remove the suspension support (see Chapter 10).

18   Remove the timing chain cover-to-engine block bolts **(see illustrations)**.

19   Separate the cover from the engine. If it's stuck, tap it with a soft-face hammer, but don't try to pry it off.

20   Temporarily install the crankshaft pulley bolt and turn the crankshaft with the bolt to align the timing marks on the crankshaft and camshaft sprockets. When aligned at TDC for number 1 piston, the crankshaft sprocket timing mark should align with the mark on the bottom of the chain tensioner plate, and the

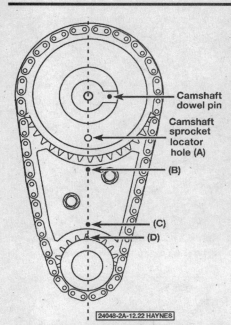

12.20 The timing marks on the sprockets should align as shown - a straight line should pass through the center of the camshaft, camshaft sprocket timing hole (A), the upper mark on the tensioner (B), the lower mark on the tensioner (C), the crankshaft sprocket timing mark (D) and the center of the crankshaft

small hole in the camshaft sprocket should be at the 6 o'clock position, aligned with the timing mark in the top of the chain tensioner plate **(see illustration)**.

21   Remove the camshaft sprocket bolt. Do not turn the camshaft in the process (if you do, realign the timing marks before the bolt is removed).

22   Use two large screwdrivers to carefully pry the camshaft sprocket off the camshaft dowel pin. Slip the timing chain and camshaft sprocket off the engine.

Inspection

Refer to illustration 12.24

23   The timing chain should be replaced with a new one if the engine has high mileage, the chain has visible damage, or total freeplay midway between the sprockets exceeds one-inch. Failure to replace a worn timing chain may result in erratic engine performance, loss of power and decreased fuel mileage. Loose chains can "jump" timing. In the worst case, chain "jumping" or breakage will result in severe engine damage. Always replace the timing chain and sprockets in sets. If you intend to install a new timing chain, remove the crankshaft sprocket with a puller and install a new one. Be sure to align the key in the crankshaft with the keyway in the sprocket during installation.

24   Inspect the timing chain damper (guide) for cracks and wear and replace it if necessary. The damper is held to the engine block by two bolts **(see illustration)**. The damper

12.24 The timing chain damper (guide) is retained by two bolts (arrows)

should be reinstalled before installing the new timing chain and sprockets.

25   Clean the timing chain and sprockets with solvent and dry them with compressed air (if available). **Warning:** *Wear eye protection when using compressed air.*

26   Inspect the components for wear and damage. Look for teeth that are deformed, chipped, pitted and cracked.

Installation

27   If the camshaft has turned at all since removal of the sprocket, turn the camshaft to position the dowel pin at 3 o'clock. Mesh the timing chain with the camshaft sprocket, then engage it with the crankshaft sprocket. The timing marks should be aligned as shown in **illustration 12.20. Note:** *If the crankshaft has been disturbed, turn it until the "O" stamped on the crankshaft sprocket is exactly at the top.*

28   Install the camshaft sprocket bolt (make sure the dowel hole in the sprocket is aligned with the dowel pin in the camshaft) and tighten to the torque listed in this Chapter's Specifications.

29   Lubricate the chain and sprocket with clean engine oil.

30   Use a gasket scraper to remove all traces of old gasket material and sealant from the cover and engine block. The cover is

made of aluminum, so be careful not to nick or gouge it. Clean the gasket sealing surfaces with lacquer thinner or acetone.

31   Apply a thin layer of anaerobic sealant to both sides of the new gasket, then position the gasket on the engine block (the dowel pins should keep it in place). Apply sealant to the bottom of the gasket, where it meet the oil pan.

32   Attach the cover to the engine and install the bolts. Follow a criss-cross pattern when tightening the fasteners and work up to the torque listed in this Chapter's Specifications in three steps.

33   The remainder of installation is the reverse of removal. Refer to Section 14 for oil pan installation.

34   Add oil and coolant, start the engine and check for leaks.

13   Valve lifters - removal, inspection and installation

1   A noisy valve lifter can be isolated when the engine is idling. Hold a mechanic's stethoscope or a length of hose near the location of each valve while listening at the other end. Another method is to remove the valve cover and, with the engine idling, touch each of the valve spring retainers, one at a time. If a valve lifter is defective, it'll be evident from the shock felt at the retainer each time the valve seats.

2   The most likely causes of noisy valve lifters are dirt trapped inside the lifter and lack of oil flow, viscosity or pressure. Before condemning the lifters, check the oil for fuel contamination, correct level, cleanliness and correct viscosity.

Removal

Refer to illustrations 13.5, 13.6a, 13.6b and 13.7

3   Remove the valve cover(s) and intake manifold as described in Sections 4 and 7.

4   Remove the rocker arms and pushrods (see Section 5).

5   Remove the bolts holding the roller lifter guide to the block, and remove the two roller lifter guides **(see illustration)**. Mark the guides as to which side they came from.

2B

13.5 Remove the bolts (A) and pull up the roller lifter guides (B)

13.6a A magnetic pick-up tool . . .

13.6b . . . or a scribe can be used to remove the lifters

13.7 Store the lifters in order to ensure installation in their original locations

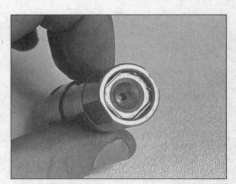

13.10a Check the pushrod seat in the top of each lifter for wear

13.10b The roller on the roller lifters must turn freely - check for wear and excessive play as well

6   There are several ways to extract the lifters from the bores. A special tool designed to grip and remove lifters is manufactured by many tool companies and is widely available, but it may not be required in every case. On newer engines without a lot of varnish buildup, the lifters can often be removed with a small magnet or even with your fingers. A machinist's scribe with a bent end can be used to pull the lifters out by positioning the point under the retainer ring in the top of each lifter (see illustrations). Caution: Don't use pliers to remove the lifters unless you intend to replace them with new ones (along with the camshaft). The pliers may damage the precision machined and hardened lifters, rendering them useless.

7   Before removing the lifters, arrange to store them in a clearly labeled box to ensure they're reinstalled in their original locations. Remove the lifters and store them where they won't get dirty (see illustration).

Inspection and installation

Refer to illustrations 13.10a and 13.10b

8   Parts for valve lifters are not available separately. The work required to remove them from the engine again if cleaning is unsuccessful outweighs any potential savings from repairing them.

9   Clean the lifters thoroughly with solvent

and dry them thoroughly, without mixing them up.

10   Check each lifter wall and plunger seat for scuffing, score marks or uneven wear (see illustration). Check the rollers carefully for wear or damage and make sure they turn freely without excessive play (see illustration). If the lifters walls are worn (not very likely), inspect the lifter bores in the block. If the pushrod seats are worn, inspect the pushrods also.

11   When reinstalling used lifters, make sure they're replaced in their original bores. Soak new lifters in oil to remove trapped air. Coat all lifters with moly-base grease or engine assembly lube prior to installation.

12   Install the push rods and the rocker arms (see Section 5).

13   The remaining installation steps are the reverse of removal.

14   Run the engine and check for oil leaks.

14   Oil pan - removal and installation

Removal

Refer to illustrations 14.3, 14.15, 14.16a, 14.16b and 14.17

1   Disconnect the cable from the negative terminal of the battery. Caution: On models

equipped with the Theftlock audio system, be sure the lockout feature is turned off before performing any procedure which requires disconnecting the battery (see the front of this manual).

2   Remove the serpentine drivebelt (see Chapter 1) and the belt tensioner (see illustration 12.5).

3   Raise the front of the vehicle and place it securely on jackstands. Apply the parking brake and block the rear wheels to keep it

14.3 Disconnect the oil level sensor connector (arrow) if equipped

14.15 Remove these bolts (arrows) at the pan and transmission, then remove the transmission-to-engine brace if equipped

2B

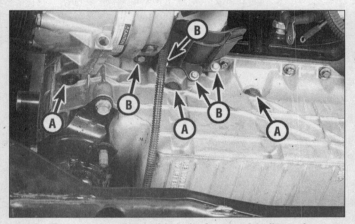

14.16a Remove the oil pan side bolts (arrows A indicate three on the radiator side) - also remove the oil filter shield bolts (B)

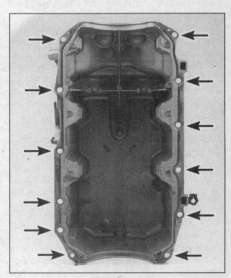

14.16b The side bolts on the rear side of the oil pan are more difficult to remove, but a box wrench with an offset bend in it can remove them

from rolling off the stands. Remove the lower splash pan and drain the engine oil (refer to Chapter 1 if necessary). Disconnect the oil-level sensor connector from the sensor **(see illustration)**.

4   Refer to Chapter 3 and unbolt the air conditioning compressor and set it aside without disconnecting the refrigerant lines. On some models it may be necessary to remove the A/C line at the accumulator. **Warning:** *The air conditioning system is under high pressure. Do not loosen any hose fittings or remove any components until after the system has been discharged by a dealer service department or service station. Always wear eye protection when disconnecting air conditioning system fittings.*

5   Remove the right side ball joint and the lower control arm, then remove the suspension support (see Chapter 10).

6   Remove the front exhaust pipe.

7   Remove the steering gear pinch bolt (see Chapter 10). **Caution:** *Be sure to separate the steering gear from the rack and pinion stub shaft to avoid damage to the steering gear and intermediate shaft.*

8   Remove the flywheel/driveplate lower cover.

9   Remove the starter (see Chapter 5).

10   Remove the hood (see Chapter 11) and support the engine with a support fixture **(see illustration 12.11)**.

11   Refer to Section 18 and remove the passenger side upper engine mount, then remove the passenger side lower mount and bracket if equipped. Remove the through bolt holding the driver's side transaxle mount to the frame, then remove the retaining bolts securing the mount to transaxle.

12   Place a floor jack under the frame front center crossmember.

13   Loosen the left side frame bolts - DO NOT REMOVE THEM!

14   Remove the right side frame bolts and lower the right side of the frame. Remove the crankshaft pulley.

15   At the rear side of the oil pan, remove the bolts and the brace (if equipped) from the transaxle-to-engine **(see illustration)**.

16   Remove the three side bolts (connecting the sides of the cast oil pan to the main cap supports) on each side of the oil pan **(see illustrations)**. Also remove the oil filter shield bolted to the top front of the pan.

17   Remove the remaining 12 oil pan-to-block bolts, then carefully separate the oil pan

from the block **(see illustration)**. Don't pry between the block and the pan or damage to the sealing surfaces could occur and oil leaks

14.17 Remove the 12 oil pan-to-block bolts (arrows) - pan removed for clarity

14.19 Apply a bead of RTV sealant on either side of the rear main cap, where the pan gasket will meet it (arrow)

15.2 Oil pump mounting bolt location (arrow)

16.2a Most flywheels/driveplates have locating dowels (arrow) - if the one you're working on doesn't have one, make some marks to ensure proper alignment on reassembly

may develop. Instead, tap the pan with a soft-face hammer to break the gasket seal.

Installation

Refer to illustration 14.19

18   Clean the pan with solvent and remove all old sealant and gasket material from the block and pan mating surfaces. Clean the mating surfaces with lacquer thinner or ace-tone and make sure the bolt holes in the block are clear.

19   Apply a bead of RTV sealant to the front of the gasket, where it contacts the front cover, and a short bead (9/32-inch wide) to either side of the rear main cap where it meets the block, then install the new one-piece oil pan gasket **(see illustration)**.

20   Place the oil pan in position on the block and install the nuts/bolts.

21   After the pan-to-block fasteners are installed, tighten them to the torque listed in this Chapter's Specifications. Starting at the center, follow a criss-cross pattern and work up to the final torque in three steps.

22   After all the pan-to-block bolts have been torqued, install the oil pan side bolts and tighten them to Specifications.

23   The remaining steps are the reverse of the removal procedure.

24   Refill the engine with oil, run it until normal operating temperature is reached and check for leaks.

15 Oil pump - removal and installation

Refer to illustration 15.2

1   Remove the oil pan (see Section 14).

2   Unbolt the oil pump and lower it from the engine **(see illustration)**. **Note:** *The oil pump driveshaft will come out with the pump as you lower it. It's a rod with a flat-sided por-tion at each end.*

3   If the pump is defective, replace it with a new one - don't reuse the original or attempt

to rebuild it. Inspect the ends of the oil pump driveshaft and the plastic collar that retains the driveshaft to the oil pump. If there are signs of wear on the shaft or if the plastic col-lar is cracked or missing, replace the shaft with a new one. **Note:** *The plastic collar cen-ters the oil pump driveshaft over the oil pump shaft. If the collar is not used or is missing, damage to the oil pump driveshaft and the oil pump will occur. A new plastic collar is usu-ally included with a new oil pump or drive-shaft.*

4   Prime the pump by pouring clean engine oil into the pick-up screen while turning the pump driveshaft.

5   To install the pump, turn the flat on the driveshaft so it mates with the slot in the oil pump shaft. Make sure the plastic collar is fit-ted over the oil pump-to-oil pump driveshaft joint, then install the oil pump and driveshaft assembly into the block while engaging the upper end of the oil pump driveshaft into the oil pump drive.

6   Install the pump mounting bolt and tighten it to the torque listed in this Chapter's Specifications.

7   The remainder of assembly is the reverse of the removal procedure.

16 Flywheel/driveplate - removal and installation

Removal

Refer to illustrations 16.2a and 16.2b

1   Raise the vehicle and support it securely on jackstands, then refer to Chapter 7 and remove the transaxle. If the vehicle is equipped with a manual transaxle, remove the clutch components (see Chapter 8).

2   Remove the bolts that secure the fly-wheel/driveplate to the crankshaft **(see illus-tration)**. If the crankshaft turns, wedge a screwdriver in the ring gear teeth to jam the flywheel/driveplate **(see illustration)**. **Note:** *If there is a retaining ring between the bolts and the driveplate, note which side faces the driveplate when removing it.*

3   Remove the flywheel/driveplate from the crankshaft. Since the flywheel/driveplate is fairly heavy, be sure to support it while removing the last bolt. **Caution:** *When removing a flywheel, wear gloves to protect your fingers - the edges of the ring gear teeth may be sharp.*

4   Clean the flywheel/driveplate to remove

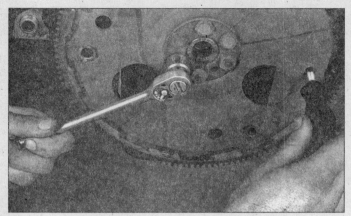

16.2b A large screwdriver wedged in one of the holes in the flywheel/driveplate can be used to keep the flywheel/driveplate from turning as the mounting bolts are removed

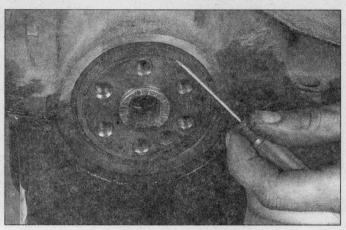

17.4 Carefully pry the old seal out

grease and oil. Inspect the surface for cracks, and check for cracked and broken ring gear teeth. Lay the driveplate on a flat surface to check for warpage.

5   Clean and inspect the mating surfaces of the flywheel/driveplate and the crankshaft. If the crankshaft rear seal is leaking, replace it before reinstalling the driveplate (see Section 17).

Installation

6   Position the flywheel/driveplate against the crankshaft. Be sure to align the marks made during removal. Note that some engines have an alignment dowel or staggered bolt holes to ensure correct installation. Before installing the bolts, apply thread locking compound to the threads and place the retaining ring in position on the flywheel/driveplate.

7   Wedge a screwdriver through the ring gear teeth to keep the flywheel/driveplate from turning as you tighten the bolts to the torque listed in this Chapter's Specifications. On vehicles equipped with automatic transaxles, if the front pump seal/O-ring is leaking, now would be a very good time to replace it.

8   The remainder of installation is the reverse of the removal procedure.

17   Rear main oil seal - replacement

Refer to illustration 17.4

1   Remove the transaxle (see Chapter 7).
2   Remove the flywheel/driveplate (see Section 16).
3   Inspect the oil seal, as well as the oil pan and engine block surface for signs of leakage. Sometimes an oil pan gasket leak can appear to be a rear oil seal leak.
4   Pry the oil seal from the block with a screwdriver **(see illustration)**. Be careful not to nick or scratch the crankshaft or the seal bore. Thoroughly clean the seal bore in the block with a shop towel. Remove all traces of oil and dirt.
5   Lubricate the lips of the new seal with engine oil or multi-purpose grease. Install the seal over the end of the crankshaft (make sure the lips of the seal point toward the engine) and carefully tap it into place. A special aftermarket tool may be available at your

local auto parts store. The tool just fits the diameter of the seal and, used with a hammer, drives the seal in. **Note:** *Do not drive it in any further than the original seal was installed.*

6   Install the flywheel/driveplate (see Section 16).
7   Install the transaxle (see Chapter 7).

18   Powertrain mounts - check and replacement

Refer to illustrations 18.1a and 18.1b

This procedure is essentially the same for all engines except that the engine mount strut and bracket on four cylinder models is referred to as the passenger side lower engine mount on V6 engines **(see illustrations)**. Note that all V6 engines are not equipped with the passenger side lower mount. Refer to Chapter 2A and follow the procedures outlined there for all other mounts except the passenger side lower mount. Be sure to use the bolt torque listed in this Chapter's Specifications.

18.1a The passenger side lower engine mount (A) is between the front of the oil pan and the chassis - with the engine supported from above remove the bolts (arrows) holding the bracket to the oil pan . . .

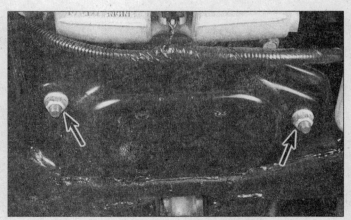

18.1b . . . and from below remove the nuts (arrows) on the crossmember, then unbolt the mount from the engine bracket and install the new mount

2B

Notes

Chapter 2 Part C
General engine overhaul procedures

Contents

2C

Specifications

2.4 liter OHC four-cylinder engines

General
VIN engine code	T
Displacement	146 cubic inches
Bore	3.54 inches
Stroke	3.70 inches
Cylinder numbers (drivebelt end-to-transaxle end)	1-2-3-4
Firing order	1-3-4-2
Cylinder compression pressure	
Minimum	100 psi
Maximum variation between cylinders	30-percent
Oil pressure	
At 900 rpm	10 psi minimum
At 3,000 rpm	30 psi minimum

Cylinder head
Warpage limit	0.010 inch (0.25 mm)

Valves and related components
Valve face angle	
Intake	46-degrees
Exhaust	45.5-degrees
Valve seat angle (intake and exhaust)	45-degrees
Valve margin width	1/32 inch (minimum)
Valve installed height*	0.9787 to 1.0024 inches (24.86 to 25.02 mm)
Valve stem-to-guide clearance	
Intake	0.0009 to 0.0025 inch (0.025 to 0.069 mm)
Exhaust	0.0015 to 0.0032 inch (0.038 to 0.081 mm)
Valve spring	
Free length	Not available
Installed height	1.4370 inches (36.49 mm)

*Measured from tip of stem to top of camshaft housing mounting surface.

2.4 liter OHC four-cylinder engines (continued)

Crankshaft and connecting rods

Crankshaft
Endplay	0.0034 to 0.0095 inch (0.087 to 0.243 mm)
Runout	
1997	
At center main journal	0.00098 inch (0.025 mm)
At flywheel flange	0.001 inch (0.030 mm)
1998 and later	
At center main journal	0.002 inch (0.050 mm)
At flywheel flange	0.002 inch (0.050 mm)
Main bearing journal	
Diameter	2.3622 to 2.3631 inches (60.000 to 60.024 mm)
Out-of-round limit	0.0002 inch (0.005 mm)
Taper limit	0.0003 inch (0.007 mm)
Main bearing oil clearance	0.0004 to 0.0023 inch (0.010 to 0.060 mm)
Connecting rod bearing journal	
Diameter	1.8887 to 1.8897 inches (47.975 to 48.00 mm)
Out-of-round	0.0002 inch (0.005 mm)
Taper limit	0.0002 inch (0.005 mm)
Connecting rod bearing oil clearance	
1997	0.0005 to 0.0020 inch (0.013 to 0.053 mm)
1998 and later	0.0004 to 0.0026 inch (0.010 to 0.068 mm)
Connecting rod side clearance (endplay)	0.0059 to 0.0177 inch (0.150 to 0.450 mm)
Rear oil seal journal	
Diameter	3.2210 to 3.2299 inches (81.96 to 82.04 mm)
Runout limit	0.002 inch (0.05 mm)

Engine block

Cylinder bore	
Diameter	3.5110 to 3.5435 inches (89.994 to 90.006 mm)
Out-of-round limit	0.0004 inch (0.010 mm)
Taper limit	0.0003 inch (0.008 mm)
Block deck warpage limit	If more than 0.010 inch (0.25 mm) must be removed, replace the block

Pistons and rings

Piston diameter	3.5420 to 3.5427 inches (89.968 to 89.984 mm)
Piston-to-bore clearance	
1997	0.0006 to 0.0018 inch (0.010 to 0.042 mm)
1998 and later	0.0006 to 0.0015 inch (0.010 to 0.038 mm)
Piston ring end gap	
Top compression ring	0.0060 to 0.0120 inch (0.15 to 0.30 mm)
Second compression ring	0.0119 to 0.0161 inch (0.30 to 0.41 mm)
Oil control ring	0.0098 to 0.0256 inch (0.25 to 0.65 mm)
Piston ring side clearance	
Top compression ring	0.0016 to 0.0031 inch (0.040 to 0.080 mm)
Second compression ring	0.0012 to 0.0028 inch (0.030 to 0.070 mm)

Camshaft

Lobe lift	
Intake	0.354 inch (9.00 mm)
Exhaust	0.346 inch (8.80 mm)
Journal diameter	
Number 1	1.5720 to 1.5728 inches (39.93 to 39.95 mm)
Numbers 2 through 5	1.3751 to 1.3760 inches (34.93 to 34.95 mm)
Endplay	0.0009 to 0.0088 inch (0.025 to 0.225 mm)

Torque specifications**

	Ft-lbs
Main bearing cap bolts	
Step 1	15
Step 2	Tighten an additional 90-degrees
Connecting rod cap nuts	
Step 1	18
Step 2	Tighten an additional 80-degrees

**Note: *Refer to Chapter 2A for additional torque specifications.*

3.1L and 3.4L V6 engines

General

VIN code	M or J
Displacement	
3.1L V6	191 cubic inches
3.4L V6	204 cubic inches
Bore and stroke	
3.1L V6	3.50 x 3.31 inches
3.4L V6	3.62 x 3.31 inches
Cylinder numbers (drivebelt end-to-transaxle end)	
Front bank (radiator side)	2-4-6
Rear bank	1-3-5
Firing order	1-2-3-4-5-6
Cylinder compression pressure	100 psi minimum
Maximum variation between cylinders	30-percent
Oil pressure	15 psi at 1100 rpm

Cylinder head

Warpage limit	0.005 inch (0.127 mm)*

If more than 0.010 inch (0.25 mm) must be removed, replace the head

Valves and related components

Valve face angle	45-degrees
Valve seat angle (intake and exhaust)	45-degrees
Valve margin width, minimum	
Intake	0.083 inch (2.10 mm)
Exhaust	0.106 inch (2.70 mm)
Valve stem-to-guide clearance	0.0010 to 0.0027 inch (0.026 to 0.068 mm)
Valve spring free length (intake and exhaust)	1.89 inches (48.5 mm)
Installed height	1.701 inches (43.2 mm)

Crankshaft and connecting rods

Connecting rod journal	
Diameter	1.9987 to 1.9994 inches (50.768 to 50.784 mm)
Bearing oil clearance	0.0007 to 0.0024 inch (0.018 to 0.062 mm)
Connecting rod side clearance (endplay)	0.007 to 0.017 inch (0.18 to 0.44 mm)
Main bearing journal	
Diameter	2.6473 to 2.6483 inches (67.239 to 67.257 mm)
Bearing oil clearance	0.0008 to 0.0025 inch (0.019 to 0.064
Taper/out-of-round limit	0.0002 inch (0.005 mm)
Crankshaft endplay (at thrust bearing)	0.002 to 0.008 inch (0.06 to 0.21 mm)

Engine block

Cylinder bore	
Out-of-round limit	0.0005 inch (0.014 mm)
Taper limit (thrust side)	0.0008 inch (0.020 mm)
Diameter	3.5046 to 3.5053 inches (89.016 to 89.034 mm)
Block deck warpage limit	If more than 0.010 inch (0.25 mm) must be removed, replace the block

Pistons and rings

Piston-to-bore clearance	0.0013 to 0.0027 inch (0.032 to 0.068 mm)
Piston ring end gap	
Top compression ring	0.006 to 0.014 inch (0.15 to 0.36 mm)
Second compression ring	0.0197 to 0.0280 inch (0.5 to 0.71 mm)
Oil control ring	0.0098 to 0.050 inch (0.25 to 1.27 mm)
Piston ring side clearance	
Compression ring	0.002 to 0.003 inch (0.05 to 0.085 mm)
Oil control ring	0.008 inch (0.20 mm)

Camshaft

Bearing journal diameter	1.868 to 1.869 inches (47.45 to 47.48 mm)
Bearing oil clearance	0.001 to 0.0039 inch (0.026 to 0.101 mm)
Lobe lift	0.2727 inch (6.9263 mm)

2C

3.1L and 3.4L V6 engines

Torque specifications***

	Ft-lbs (unless otherwise indicated)
Main bearing caps, bolts/studs	
Step 1	37
Step 2	Tighten an additional 77 degrees
Connecting rod caps	
Step 1	15
Step 2	Tighten an additional 75 degrees
Camshaft thrust plate screw	89 in-lbs

***Note: *Refer to Part B for additional torque specifications.*

1   General information - engine overhaul

Included in this portion of Chapter 2 are the general overhaul procedures for the cylinder head and internal engine components.

The information ranges from advice concerning preparation for an overhaul and the purchase of replacement parts to detailed, step-by-step procedures covering removal and installation of internal engine components and the inspection of parts.

The following Sections have been written based on the assumption that the engine has been removed from the vehicle. For information concerning in-vehicle engine repair, as well as removal and installation of the external components necessary for the overhaul, see Chapter 2A (2.4L four cylinder engine) or Chapter 2B (3.1L and 3.4L V6 engines), and Section 8 of this Chapter. The Specifications included in this Part are only those necessary for the inspection and overhaul procedures which follow. Refer to Chapter 2, Part A or Part B for additional Specifications.

It's not always easy to determine when, or if, an engine should be completely overhauled, as a number of factors must be considered.

High mileage is not necessarily an indication that an overhaul is needed, while low mileage doesn't preclude the need for an overhaul. Frequency of servicing is probably the most important consideration. An engine that's had regular and frequent oil and filter changes, as well as other required maintenance, will most likely give many thousands of miles of reliable service. Conversely, a neglected engine may require an overhaul very early in its life.

Excessive oil consumption is an indication that piston rings, valve seals and/or valve guides are in need of attention. Make sure that oil leaks aren't responsible before deciding that the rings and/or guides are bad. Perform a cylinder compression check to determine the extent of the work required (see Section 3). Also check the vacuum readings under various conditions (see Section 4).

Loss of power, rough running, knocking or metallic engine noises, excessive valve train noise and high fuel consumption rates may also point to the need for an overhaul,

especially if they're all present at the same time. If a complete tune-up doesn't remedy the situation, major mechanical work is the only solution.

An engine overhaul involves restoring the internal parts to the specifications of a new engine. During an overhaul, the piston rings are replaced and the cylinder walls are reconditioned (re-bored and/or honed). If a re-bore is done by an automotive machine shop, new oversize pistons will also be installed. The main bearings, connecting rod bearings and camshaft bearings are generally replaced with new ones and, if necessary, the crankshaft may be reground to restore the journals. Generally, the valves are serviced as well, since they're usually in less-than-perfect condition at this point. While the engine is being overhauled, other components, such as the starter and alternator, can be rebuilt as well. The end result should be a like new engine that will give many trouble free miles. **Note:** *Critical cooling system components such as the hoses, drivebelts, thermostat and water pump should be replaced with new parts when an engine is overhauled. The radiator should be checked carefully to ensure that it isn't clogged or leaking (see Chapter 3). If you purchase a rebuilt engine or short block, some rebuilders will not warranty their engines unless the radiator has been professionally flushed. Also, we don't recommend overhauling the oil pump - always install a new one when an engine is rebuilt.*

Before beginning the engine overhaul, read through the entire procedure to familiarize yourself with the scope and requirements of the job. Overhauling an engine isn't difficult, but it is time-consuming. Plan on the vehicle being tied up for a minimum of two weeks, especially if parts must be taken to an automotive machine shop for repair or reconditioning. Check on availability of parts and make sure that any necessary special tools and equipment are obtained in advance. Most work can be done with typical hand tools, although a number of precision measuring tools are required for inspecting parts to determine if they must be replaced. Often an automotive machine shop will handle the inspection of parts and offer advice concerning reconditioning and replacement. **Note:** *Always wait until the engine has been completely disassembled and all components, especially the engine block, have been*

inspected before deciding what service and repair operations must be performed by an automotive machine shop. Since the block's condition will be the major factor to consider when determining whether to overhaul the original engine or buy a rebuilt one, never purchase parts or have machine work done on other components until the block has been thoroughly inspected. As a general rule, time is the primary cost of an overhaul, so it doesn't pay to install worn or substandard parts.

As a final note, to ensure maximum life and minimum trouble from a rebuilt engine, everything must be assembled with care in a spotlessly-clean environment.

2   Oil Pressure check

Refer to illustrations 2.2a and 2.2b

1   Low engine oil pressure can be a sign of an engine in need of rebuilding. A "low oil pressure" indicator (often called an "idiot light") is not a test of the oiling system. Such indicators only come on when the oil pressure is dangerously low. Even a factory oil pressure gauge in the instrument panel is only a relative indication, although much better for driver information than a warning light. A better test is with a mechanical (not electrical) oil pressure gauge. When used in con-

2.2a On four cylinder engines, the oil pressure sending unit (arrow) is mounted to the front camshaft housing on the left (driver's side) of the engine compartment

2.2b On V6 engines, the oil pressure sending unit (arrow) is located next to the oil filter

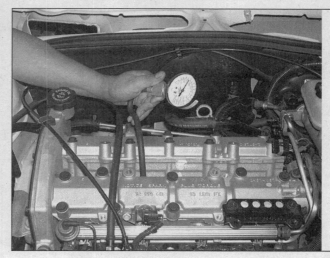

3.6 A compression gauge with a threaded fitting for the spark plug hole is preferred over the type that requires hand pressure to maintain the seal - be sure to open the throttle valve as far as possible during the compression check

junction with an accurate tachometer, an engine's oil pressure performance can be compared to the manufacturers Specifications.

2   Find the oil pressure indicator sending unit **(see illustrations)**.

3   Remove the oil pressure sending unit and install a fitting which will allow you to directly connect your hand-held, mechanical oil pressure gauge. Use Teflon tape or sealant on the threads of the adapter and the fitting on the end of your gauge's hose.

4   Connect an accurate tachometer to the engine, according to the tachometer manufacturer's instructions.

5   Check the oil pressure with the engine running (full operating temperature) at the specified engine speed, and compare it to this Chapter's Specifications. If it's extremely low, the bearings and/or oil pump are probably worn out.

3   Cylinder compression check

Refer to illustration 3.6

1   A compression check will tell you what mechanical condition the upper end (pistons, rings, valves, head gaskets) of the engine is in. Specifically, it can tell you if the compression is down due to leakage caused by worn piston rings, defective valves and seats or a blown head gasket. **Note:** *The engine must be at normal operating temperature and the battery must be fully charged for this check.*

2   Begin by cleaning the area around the spark plugs before you remove them. Compressed air should be used, if available, otherwise a small brush will work. The idea is to prevent dirt from getting into the cylinders as the compression check is being done.

3   Remove all of the spark plugs from the engine (see Chapter 1).

4   Block the throttle wide open.

5   Disable the ignition system by disconnecting the wires to the ignition control module (see Chapter 5). Also, disable the fuel

injection system by unplugging the electrical connector to the injector wiring harness or by removing the fuel pump fuse.

6   Install the compression gauge in the number one spark plug hole **(see illustration)**.

7   Crank the engine over at least seven compression strokes and watch the gauge. The compression should build up quickly in a healthy engine. Low compression on the first stroke, followed by gradually increasing pressure on successive strokes, indicates worn piston rings. A low compression reading on the first stroke, which doesn't build up during successive strokes, indicates leaking valves or a blown head gasket (a cracked head could also be the cause). Deposits on the undersides of the valve heads can also cause low compression. Record the highest gauge reading obtained.

8   Repeat the procedure for the remaining cylinders, turning the engine over for the same length of time for each cylinder, and compare the results to this Chapter's Specifications.

9   If the readings are below normal, add some engine oil (about three squirts from a plunger-type oil can) to each cylinder, through the spark plug hole, and repeat the test.

10   If the compression increases after the oil is added, the piston rings are definitely worn. If the compression doesn't increase significantly, the leakage is occurring at the valves or head gasket. Leakage past the valves may be caused by burned valve seats and/or faces or warped, cracked or bent valves.

11   If two adjacent cylinders have equally low compression, there's a strong possibility the head gasket between them is blown. The appearance of coolant in the combustion chambers or the crankcase would verify this condition.

12   If one cylinder is about 20-percent lower than the others, and the engine has a slightly rough idle, a worn exhaust lobe on the camshaft could be the cause.

13   If the compression is unusually high, the combustion chambers are probably coated

with carbon deposits. If that's the case, the cylinder heads should be removed and decarbonized.

14   If compression is way down or varies greatly between cylinders, it would be a good idea to have a leak-down test performed by an automotive repair shop. This test will pinpoint exactly where the leakage is occurring and how severe it is.

15   Installation of the fuses and the remaining components is the reverse of removal.

4   Vacuum gauge diagnostic checks

Refer to illustration 4.6

1   A vacuum gauge provides valuable information about what is going on in the engine at a low cost. You can check for worn rings or cylinder walls, leaking head or intake manifold gaskets, incorrect air/fuel adjustments, restricted exhaust, stuck or burned valves, weak valve springs, improper ignition or valve timing and ignition problems.

2   Unfortunately, vacuum gauge readings are easy to misinterpret, so they should be used in conjunction with other tests to confirm the diagnosis.

3   Both the gauge readings and the rate of needle movement are important for accurate interpretation. Most gauges measure vacuum in inches of mercury (in-Hg). As vacuum increases (or atmospheric pressure decreases), the reading will increase. Also, for every 1,000-foot increase in elevation above sea level, the gauge readings will decrease about one inch of mercury.

4   Connect the vacuum gauge directly to intake manifold vacuum, not to ported (above the plate) vacuum. Be sure no hoses are left disconnected during the test or false readings will result.

5   Before you begin the test, allow the engine to warm up completely. Block the wheels and set the parking brake. With the transmission in Park, start the engine and allow it to run at normal idle speed.

2C

6   Read the vacuum gauge; an average, healthy engine should normally produce about 17 to 22 inches of vacuum with a fairly steady needle. Refer to the following vacuum gauge readings and what they indicate about the engine's condition **(see illustration)**.

7   A low, steady reading usually indicates a leaking gasket between the intake manifold and carburetor or throttle body, a leaky vacuum hose, late ignition timing or incorrect camshaft timing. Eliminate all other possible causes, utilizing the tests provided in this Chapter before you remove the timing chain cover to check the timing marks.

8   If the reading is three to eight inches below normal and it fluctuates at that low reading, suspect an intake manifold gasket leak at an intake port.

9   If the needle has regular drops of about two to four inches at a steady rate, the valves are probably leaking. Perform a compression or leak-down test to confirm this.

10   An irregular drop or down-flick of the needle can be caused by a sticking valve or an ignition misfire. Perform a compression or leak-down test and read the spark plugs.

11   A rapid vibration of about four inches-Hg vibration at idle combined with exhaust smoke indicates worn valve guides. Perform a leak-down test to confirm this. If the rapid vibration occurs with an increase in engine speed, check for a leaking intake manifold gasket or head gasket, weak valve springs, burned valves or ignition misfire.

12   A slight fluctuation, say one inch up and down, may mean ignition problems. Check all the usual tune-up items and, if necessary, run the engine on an ignition analyzer.

13   If there is a large fluctuation, perform a compression or leak-down test to look for a weak or dead cylinder or a blown head gasket.

14   If the needle moves slowly through a wide range, check for a clogged PCV system, incorrect idle fuel mixture, throttle body or intake manifold gasket leaks.

15   Check for a slow return after revving the engine by quickly snapping the throttle open until the engine reaches about 2,500 rpm and let it shut. Normally the reading should drop to near zero, rise above normal idle reading (about 5 in-Hg over) and then return to the previous idle reading. If the vacuum returns slowly and doesn't peak when the throttle is snapped shut, the rings may be worn. If there is a long delay, look for a restricted exhaust system (often the muffler or catalytic converter). An easy way to check this is to temporarily disconnect the exhaust ahead of the suspected part and re-test.

5   Engine rebuilding alternatives

The home mechanic is faced with a number of options when performing an engine overhaul. The decision to replace the engine block, piston/connecting rod assemblies and crankshaft depends on a number of

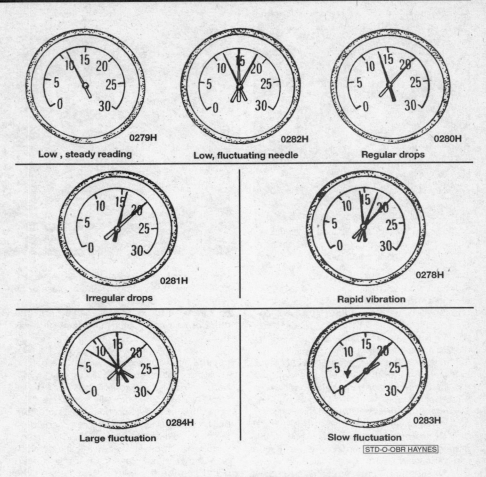

4.6 Typical vacuum gauge diagnostic readings

factors, with the number one consideration being the condition of the block. Other considerations are cost, access to machine shop facilities, parts availability, time required to complete the project and the extent of prior mechanical experience.

Some of the rebuilding alternatives include:

Individual parts - If the inspection procedures reveal the engine block and most engine components are in reusable condition, purchasing individual parts may be the most economical alternative. The block, crankshaft and piston/connecting rod assemblies should all be inspected carefully. Even if the block shows little wear, the cylinder bores should be surface-honed.

Short-block - A short-block consists of an engine block with a crankshaft and piston/connecting rod assemblies already installed. All new bearings are incorporated and all clearances will be correct. The existing camshaft, valve train components, cylinder heads and external parts can be bolted to the short block with little or no machine shop work necessary. Some rebuilding companies include a new timing chain, camshaft and lifters with their short-block assemblies.

Long-block - A long-block consists of a short block plus an oil pump, oil pan, cylinder heads, rocker arm covers, camshaft and valve train components, timing sprockets and chain and timing chain cover. All components are installed with new bearings, seals and gaskets incorporated throughout. The installation of manifolds and external parts is all that's necessary. Give careful thought to which alternative is best for you and discuss the situation with local automotive machine shops, auto parts dealers and experienced rebuilders before ordering or purchasing replacement parts.

6   Engine removal - methods and precautions

If you've decided the engine must be removed for overhaul or major repair work, several preliminary steps should be taken. Locating a suitable place to work is extremely important. Adequate work space, along with storage space for the vehicle, will be needed.

Cleaning the engine compartment and engine before beginning the removal procedure will help keep tools clean and organized.

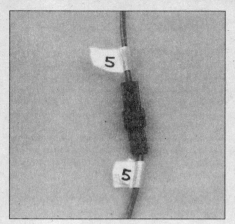

7.4 Label each wire before unplugging the connector

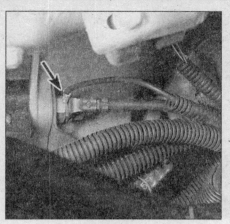

7.11 While the vehicle is raised, disconnect any wiring harnesses attached to the block

7.15 Paint or scribe alignment marks on the driveplate and the torque converter to ensure that the two components are still in balance when they're reassembled

2C

An engine hoist will also be necessary. Safety is of primary importance, considering the potential hazards involved in removing the engine from this vehicle.

If the engine is being removed by a novice, a helper should be available. Advice and aid from someone more experienced would also be helpful. There are many instances when one person cannot simultaneously perform all of the operations required when lifting the engine out of the vehicle.

Plan the operation ahead of time. Arrange for or obtain all of the tools and equipment you'll need prior to beginning the job. Some of the equipment necessary to perform engine removal and installation safely and with relative ease in addition to a hydraulic jack, jack stands and an engine hoist) are a complete sets of wrenches and sockets as described in the front of this manual, wooden blocks and plenty of rags and cleaning solvent for mopping up spilled oil, coolant and gasoline.

Plan for the vehicle to be out of use for quite a while. A machine shop will be required to perform some of the work which the do-it-yourselfer can't accomplish without special equipment. These shops often have a busy schedule, so it would be a good idea to consult them before removing the engine in order to accurately estimate the amount of time required to rebuild or repair components that may need work.

Always be extremely careful when removing and installing the engine. Serious injury can result from careless actions. Plan ahead, take your time and a job of this nature, although major, can be accomplished successfully. **Note:** *Because it may be some time before you reinstall the engine, it is very helpful to make sketches or take photos of various accessory mountings and wiring hookups before removing the engine.*

7   Engine - removal and installation

Warning 1: *The models covered by this manual are equipped with airbags. Always disable*

the airbag system before working in the vicinity of the impact sensors, steering column or instrument panel to avoid the possibility of accidental deployment of the airbag(s), which could cause personal injury (see Chapter 12).
Warning 2: *Gasoline is extremely flammable, so take extra precautions when you work on any part of the fuel system. Don't smoke or allow open flames or bare light bulbs near the work area, and don't work in a garage where a natural gas-type appliance (such as a water heater or a clothes dryer) is present. Since gasoline is carcinogenic, wear latex gloves when there's a possibility of being exposed to fuel, and, if you spill any fuel on your skin, rinse it off immediately with soap and water. Mop up any spills immediately and do not store fuel-soaked rags where they could ignite. The fuel system is under constant pressure, so, if any fuel lines are to be disconnected, the fuel pressure in the system must be relieved first. When you perform any kind of work on the fuel system, wear safety glasses and have a Class B type fire extinguisher on hand.*
Warning 3: *The air conditioning system is under high pressure - have a dealer service department or service station evacuate the system and recapture the refrigerant before disconnecting any of the hoses or fittings.*

Removal

Refer to illustrations 7.4, 7.11, 7.15 and 7.21
Note: *The procedure outlined below illustrates the necessary steps to remove the engine traditionally, i.e. from above with an engine hoist.*

1   Disconnect the cable from the negative terminal of the battery. **Caution:** *On models equipped with the Theftlock audio system, be sure the lockout feature is turned off before performing any procedure which requires disconnecting the battery (see the front of this manual).*
2   Cover the fenders and cowl. Special pads are available to protect the fenders, but an old bedspread or blanket will also work. Remove the hood (see Chapter 11).

3   Remove the air intake duct assembly (see Chapter 4).
4   To ensure correct reassembly, label each vacuum line, emission system hose, electrical connector, ground strap and fuel line, then disconnect them from the engine. Pieces of masking tape with numbers or letters written on them prevent confusion at assembly time **(see illustration)**. Or sketch the engine compartment routing of lines, hoses and wires.
5   Drain the cooling system (see Chapter 1). then label and remove the heater hoses and radiator hoses from the engine. Remove the coolant recovery tank (see Chapter 3).
6   Remove the cooling fans and radiator (see Chapter 3).
7   Disconnect the throttle and cruise control cables (see Chapter 4).
8   Remove the engine drivebelt (see Chapter 1). Unbolt the power steering pump and bracket and tie them out of the way (see Chapter 10).
9   Remove the alternator (see Chapter 5).
10   Disconnect and remove the cruise control module from the engine compartment (if equipped).
11   Raise the vehicle and suitably support it on jackstands. While raised, perform the disassembly procedures that can be done only from underneath, such as disconnecting the exhaust, disconnecting the transmission cooler lines (if equipped) where they are held by a clip to one of the oil pan bolts and disconnecting wiring harnesses attached to the block **(see illustration)**.
12   Detach the air conditioning compressor from its mounting bracket and position it aside (see Chapter 3).
13   Drain the engine oil and remove the filter (see Chapter 1).
14   Remove the starter (see Chapter 5). Remove the flywheel/driveplate access cover.
15   On automatic transaxle equipped vehicles, make an alignment mark between the driveplate and the torque converter **(see illustration)**, then rotate the crankshaft and

7.21 With the accessories tied out of the way and the engine mount through-bolts removed, raise the engine with the hoist

remove the converter bolts.

16   On four cylinder engines remove the crankshaft pulley (see Chapter 2A).

17   Support the engine from above with an engine hoist. Refer to the last section ("powertrain mounts") in Part A or B to remove the passenger side engine mounts (do not disconnect the transaxle mounts).

18   Remove the brace (if equipped) between the engine and transaxle and the lower bellhousing to-engine bolts. **Note:** *On four cylinder models it will be necessary to remove the lower bellhousing to oil pan bolts, nuts and stud(s). Refer to the transaxle removal procedure in Chapter 7A for fastener locations.*

19   Working at the back of the engine, remove the remaining bellhousing bolts.

20   Raise the engine and pull it forward to free it from the torque converter. **Note:** *On manual transaxle equipped vehicles it will be necessary to remove the transaxle from the vehicle (see Chapter 7A).*

21   Tie the wiring harnesses out of the way, raise the engine with the hoist, and with a combination of tilting, twisting and raising, move the engine around any obstructions and pull it out of the vehicle, raising it high enough to clear the front of the body **(see illustration)**.

22   Remove the driveplate or flywheel (see Part B) while the engine is out of the vehicle but still on the hoist, and mount the engine on an engine stand.

Installation

23   While the engine is out, check the powertrain (engine and transmission) mounts (see Chapter 2A or 2B). If they're worn or damaged, replace them.

24   Carefully lower the engine, twisting it to clear any harnesses or obstructions, until the converter snout lines up (automatic transaxle) or the input shaft lines up with clutch disc (manual transaxle) and the bellhousing-to-engine bolts can be inserted. **Caution:** *DO NOT use the transmission-to-engine bolts to force the transmission and engine together. Take great care when mating the engine to the transaxle, following the procedure outlined in Chapter 7a or 7B. Make sure the alignment marks you made on the driveplate and the torque converter during removal are*

lined up.

25   Install the driveplate-to-torque converter bolts (automatic transaxle) and tighten them to the torque listed in Chapter 7 Specifications.

26   Reinstall the remaining components in the reverse order of removal. Double-check to make sure everything is hooked up right, using the sketches or photos taken earlier to go by.

27   Add coolant, oil, power steering and transmission fluid as needed.

28   Run the engine and check for leaks and proper operation of all accessories, then install the hood and test drive the vehicle.

8   Engine overhaul - disassembly sequence

1   It's much easier to disassemble and work on the engine if it's mounted on a portable engine stand. A stand can often be rented quite cheaply from an equipment rental yard. Before it's mounted on a stand, the flywheel/driveplate should be removed from the engine.

2   If a stand isn't available, it's possible to disassemble the engine with it blocked up on the floor. Be extra careful not to tip or drop the engine when working without a stand.

3   If you're going to obtain a rebuilt engine, all external components must come off first, to be transferred to the replacement engine, just as they will if you're doing a complete engine overhaul yourself. These include:

Alternator and brackets
Emissions control components
Ignition coil/module assembly, spark plug wires and spark plugs
Thermostat and housing cover
Water pump
Engine front cover
Fuel injection components
Intake/exhaust manifolds
Valve covers (V6 engines)
Oil filter
Engine mounts
Flywheel/driveplate

Note: *When removing the external components from the engine, pay close attention to*

details that may be helpful or important during installation. Note the installed position of gaskets, seals, spacers, pins, brackets, washers, bolts and other small items.

4   If you're obtaining a short-block, then the cylinder heads, oil pan and oil pump will have to be removed as well. See *Engine rebuilding alternatives* for additional information regarding the different possibilities to be considered.

5   If you're planning a complete overhaul, the engine must be disassembled and the internal components removed in the following general order:

Four cylinder engine

Timing chain and sprockets
Timing chain housing
Camshaft and lifter housings
Camshaft and lifters
Cylinder head
Oil pan
Oil pump
Balance shaft housing
Piston/connecting rod assemblies
Rear main oil seal retainer
Crankshaft and main bearings

V6 engines

Rocker arms and pushrods
Valve lifters
Cylinder heads
Timing chain cover
Timing chain and sprockets
Camshaft
Oil pan
Oil pump
Piston/connecting rod assemblies
Crankshaft and main bearings

6   Before beginning the disassembly and overhaul procedures, make sure the following items are available. Also, refer to *Engine overhaul - reassembly sequence* for a list of tools and materials needed for engine reassembly.

Common hand tools
Small cardboard boxes or plastic bags for storing parts
Gasket scraper
Ridge reamer
Engine balancer puller
Micrometers
Telescoping gauges
Dial indicator set
Valve spring compressor
Cylinder surfacing hone
Piston ring groove-cleaning tool
Electric drill motor
Tap and die set
Wire brushes
Oil gallery brushes
Cleaning solvent

9   Cylinder head - disassembly

Refer to illustrations 9.2, 9.3 and 9.4
Note: *New and rebuilt cylinder heads are commonly available for most engines at deal-*

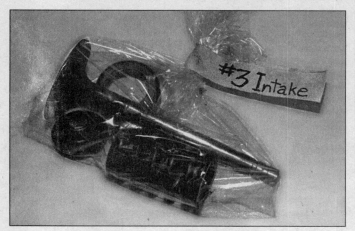

9.2 A small plastic bag, with an appropriate label, can be used to store the valve train components so they can be kept together and reinstalled in the original positions

9.3 Use a valve spring compressor to compress the spring, then remove the keepers from the valve stem

erships and auto parts stores. Due to the fact that some specialized tools are necessary for the disassembly and inspection procedures, and replacement parts aren't always readily available, it may be more practical and economical for the home mechanic to purchase replacement heads rather than taking the time to disassemble, inspect and recondition the originals.

1   Cylinder head disassembly involves removal of the intake and exhaust valves and related components. It is already assumed that the rocker arm components (V6 engines) and the camshaft housings (four cylinder engines) are removed from the cylinder head. If they're not already removed, label the parts and store them separately so they can be reinstalled in their original locations.

2   Before the valves are removed, arrange to label and store them, along with their related components, so they can be kept separate and reinstalled in their original locations **(see illustration)**.

3   Compress the springs on the first valve with a spring compressor and remove the keepers **(see illustration)**. Carefully release the valve spring compressor and remove the retainer, the spring and the spring seat (if used).

4   Pull the valve out of the head, then remove the oil seal from the guide. If the valve binds in the guide (won't pull through), push it back into the head and deburr the area around the keeper groove with a fine file or whetstone **(see illustration)**.

5   Repeat the procedure for the remaining valves. Remember to keep all the parts for each valve together so they can be reinstalled in the same locations.

6   Once the valves and related components have been removed and stored in an organized manner, the heads should be thoroughly cleaned and inspected. If a complete engine overhaul is being done, finish the engine disassembly procedures before beginning the cylinder head cleaning and inspection process.

9.4 If the valve won't pull through the guide, deburr the edge of the stem end and the area around the top of the keeper groove with a file or whetstone

10 Cylinder head - cleaning and inspection

1   Thorough cleaning of the cylinder heads and related valve train components, followed by a detailed inspection, will enable you to decide how much valve service work must be done during the engine overhaul. **Note:** *If the engine was severely overheated, the cylinder head is probably warped* (see Step 12).

Cleaning

2   Scrape all traces of old gasket material and sealant off the head gasket, intake manifold and exhaust manifold mating surfaces. Be very careful not to gouge the cylinder head. Special gasket-removal solvents that soften gaskets and make removal much easier are available at auto parts stores.

3   Remove all built-up scale from the coolant passages.

4   Run a stiff wire brush through the various holes to remove deposits that may have formed in them.

5   Run an appropriate-size tap into each of the threaded holes to remove corrosion and thread sealant that may be present. If com-

pressed air is available, use it to clear the holes of debris produced by this operation. **Warning:** *Wear eye protection when using compressed air!*

6   Clean the cylinder head with solvent and dry it thoroughly.

7   Compressed air will speed the drying process and ensure that all holes and recessed areas are clean. **Note:** *Decarbonizing chemicals are available and may prove very useful when cleaning cylinder heads and valve train components. They're very caustic and should be used with caution. Be sure to follow the instructions on the container.*

8   On V6 engines, clean the rocker arms, pivots, bolts and pushrods with solvent and dry them thoroughly (don't mix them up during the cleaning process). Compressed air will speed the drying process and can be used to clean out the oil passages.

9   Clean all the valve springs, keepers and retainers with solvent and dry them thoroughly. Do the components from one valve at a time to avoid mixing up the parts.

10   Scrape off any heavy deposits that may have formed on the valves, then use a motorized wire brush to remove deposits from the valve heads and stems. Again, make sure the valves don't get mixed up.

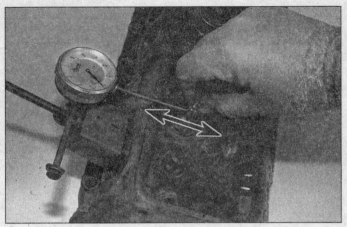

10.12 Check the cylinder head gasket surface for warpage by trying to slip a feeler gauge under the straightedge (see this Chapter's Specifications for the maximum warpage allowed and use a feeler gauge of that thickness)

10.14 A dial indicator can be used to determine the valve stem-to-guide clearance (move the valve stem as indicated by the arrows)

Inspection

Note: *Be sure to perform all of the following inspection procedures before concluding machine shop work is required. Make a list of the items that need attention.*

Cylinder head

Refer to illustrations 10.12 and 10.14

11   Inspect the head very carefully for cracks, evidence of coolant leakage and other damage. If cracks are found, check with an automotive machine shop concerning repair. If repair isn't possible, a new cylinder head must be obtained.

12   Using a straightedge and feeler gauge, check the head gasket mating surface for warpage **(see illustration)**. If the warpage exceeds the limit in this Chapter's Specifications, it can be resurfaced at an automotive machine shop. **Note:** *If the heads are resurfaced, the intake manifold flanges may also require machining.*

13   Examine the valve seats in each of the combustion chambers. If they're pitted, cracked or burned, the head will require valve

service that's beyond the scope of the home mechanic.

14   Check the valve stem-to-guide clearance by measuring the lateral movement of the valve stem with a dial indicator attached securely to the head **(see illustration)**. The valve must be in the guide and approximately 1/16-inch off the seat. The total valve stem movement indicated by the gauge needle must be divided by two to obtain the actual clearance. After this is done, if there's still some doubt regarding the condition of the valve guides, they should be checked by an automotive machine shop (the cost should be minimal).

Valves

Refer to illustrations 10.15 and 10.16

15   Carefully inspect each valve face for uneven wear, deformation, cracks, pits and burned areas. Check the valve stem for scuffing and galling and the neck for cracks. Rotate the valve and check for any obvious indication that it's bent. Look for pits and excessive wear on the end of the stem. The presence of any of these conditions **(see illustration)** indicates the need for valve service by an automotive machine shop.

16   Measure the margin width on each valve **(see illustration)**. Any valve with a margin narrower than specified in this Chapter will have to be replaced with a new one.

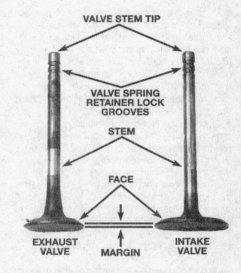

10.15 Check for valve wear at the points shown here

Valve components

Refer to illustrations 10.17 and 10.18

17   Check each valve spring for wear (on the ends) and pits. Measure the free length and compare it to this Chapter's Specifications **(see illustration)**. Any springs that are

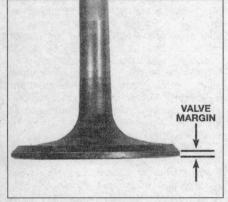

10.16 The margin width on the valve must be as specified (if no margin exists, the valve cannot be re-used)

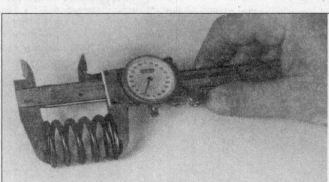

10.17 Measure the free length of each valve spring with a dial or vernier caliper

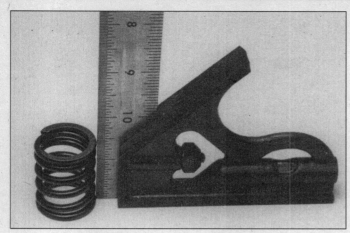

10.18 Check each valve spring for squareness

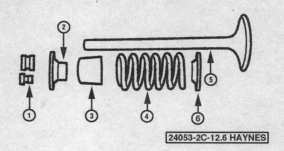

`24053-2C-12.6 HAYNES`

12.6 Typical valve components

1	*Keepers*	4	*Spring*
2	*Retainer*	5	*Valve*
3	*Oil seal*	6	*Valve spring seat or rotator*

2C

shorter than specified have sagged and shouldn't be re-used. The tension of all springs should be checked with a special fixture before deciding they're suitable for use in a rebuilt engine (take the springs to an automotive machine shop for this check).

18 Stand each spring on a flat surface and check it for squareness **(see illustration)**. If any of the springs are distorted or sagged, replace all of them with new parts. **Note:** *V6 engines are equipped with conical type valve springs which are smaller at the top. On these models it will be necessary to place the square along both sides of the spring (180 degrees apart) and note the measurements between the top of the spring and the square. if the spring is square the measurement will be the same on both sides.*

19 Check the spring retainers and keepers for obvious wear and cracks. Any questionable parts should be replaced with new ones, as extensive damage will occur if they fail during engine operation.

Valve actuating components

20 On V6 engines refer to Chapter 2B and inspect the rocker arms, pushrods and the valve lifters. On four cylinder engines refer to Chapter 2A and inspect the valve lifters.

All components

21 If the inspection process indicates the valve components are in generally poor condition and worn beyond the limits specified, which is usually the case in an engine that's being overhauled, reassemble the valves in the cylinder head (see Section 11 for valve servicing recommendations).

11 Valves - servicing

1 Because of the complex nature of the job and the special tools and equipment needed, servicing of the valves, the valve seats and the valve guides, commonly known as a valve job, should be done by a professional.

2 The home mechanic can remove and disassemble the head, do the initial cleaning and inspection, then reassemble and deliver it to an automotive machine shop for the actual service work. Doing the inspection will enable you to see what condition the head and valvetrain components are in and will ensure that you know what work and new parts are required when dealing with an automotive machine shop.

3 The automotive machine shop, will remove the valves and springs, recondition or replace the valves and valve seats, recondition the valve guides, check and replace the valve springs, spring retainers and keepers (as necessary), replace the valve seals with new ones, reassemble the valve components and make sure the installed spring height is correct. The cylinder head gasket surface will also be resurfaced if it's warped.

4 After the valve job has been performed by a professional, the head will be in like new condition. When the head is returned, be sure to clean it again before installation on the engine to remove any metal particles and abrasive grit that may still be present from the valve service or head resurfacing operations. Use compressed air, if available, to blow out all the oil holes and passages.

12 Cylinder head - reassembly

Refer to illustrations 12.6 and 12.7

1 Regardless of whether or not the head was sent to an automotive repair shop for valve servicing, make sure it's clean before beginning reassembly.

2 If the head was sent out for valve servicing, the valves and related components will already be in place. Begin the reassembly procedure with Step 8.

3 Beginning at one end of the head, lubricate and install the first valve. Apply moly-base grease or clean engine oil to the valve stem.

4 Install the spring seat and shims, if originally installed, before the valve seals.

5 Install new seals on each of the valve guides. Gently tap each seal into place until it's completely seated on the guide. Many

seal sets come with a plastic installer, but use hand pressure. Do not hammer on the seals or they could be driven down too far and subsequently leak. Don't twist or cock the seals during installation or they won't seal properly on the valve stems.

6 The valve components **(see illustration)** may be installed in the following order:

Four cylinder engine

Valves
Valve spring seat (exhaust valves only)
Valve rotator (intake valves only)
Valve stem seals
Valve spring shims (if any)
Valve springs
Retainers
Keepers

V6 engines

Valves
Valve spring seat
Valve stem seals
Valve spring shims (if any)
Valve springs
Retainers
Keepers

7 Compress the springs with a valve spring compressor and carefully install the keepers in

12.7 Apply a small dab of grease to each keeper as shown here before Installation - it'll hold them in place on the valve stem as the spring is released

13.2 Remove the bolt (arrow) and pull out the oil pump drive

13.4 Remove the retaining bolts and the camshaft thrust plate

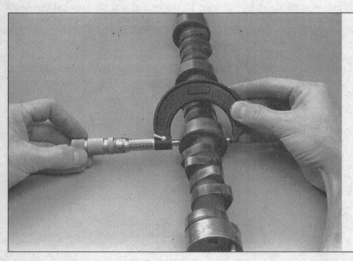

13.7a Measure the camshaft bearing journals with a micrometer

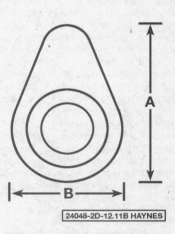

24048-2D-12.11B HAYNES

13.7b Measure the camshaft lobe maximum diameter (A) and the minimum (B) - subtract (B) from (A), the difference is the lobe lift

the groove, then slowly release the compressor and make sure the keepers seat properly. Apply a small dab of grease to each keeper to hold it in place if necessary **(see illustration)**. Tap the valve stem tips with a plastic hammer to seat the keepers, if necessary.

8   Repeat the procedure for the remaining valves. Be sure to return the components to their original locations - don't mix them up!

9   Check the installed valve spring height with a ruler graduated in 1/32-inch increments or a dial caliper. If the head was sent out for service work, the installed height should be correct (but don't automatically assume it is). The measurement is taken from the top of each spring seat or top shim to the bottom of the retainer. If the height is greater than specified in this Chapter, shims can be added under the springs to correct it. **Caution:** *Do not, under any circumstances, shim the springs to the point where the installed height is less than specified.*

10   On V6 engines, apply moly-base grease to the rocker arm faces and the pivots, then install the rocker arms and pivots on the cylinder heads. Tighten the bolts finger-tight.

13   Camshaft - removal and inspection

Removal

Four cylinder engine

1   Refer to Chapter 2A for the camshaft removal procedure on these engines. Then inspect the camshaft as described in Steps 6 through 12.

V6 engines

Refer to illustrations 13.2 and 13.4

2   Remove the bolt and clamp holding the oil pump drive and pull the oil pump drive straight up and out of the block **(see illustration)**.

3   Refer to Chapter 2B and remove the timing chain and sprockets (lifters should already be removed and stored in a marked container).

4   Remove the bolts holding the camshaft thrust plate to the block **(see illustration)**.

5   Slide the camshaft straight out of the engine, using a long bolt (with the same thread as the camshaft sprocket bolt) screwed into

the front of the camshaft as a "handle". Support the shaft near the block and be careful not to scrape or nick the bearings.

Inspection

Refer to illustrations 13.7a and 13.7b

6   After the camshaft has been removed, clean it with solvent and dry it, then inspect the bearing journals for uneven wear, pitting and evidence of seizure. If the journals are damaged, the camshaft bearings probably damaged as well. Both the shaft and bearings will have to be replaced.

7   Measure the bearing journals with a micrometer **(see illustration)** to determine whether they are excessively worn or out-of-round. Measure the camshaft lobes also to check for wear. Measure the camshaft lobes at their highest point, then subtract the measurement of the lobe at it's smallest diameter - the difference is the lobe lift **(see illustration)**.

14.1 A ridge reamer is required to remove the ridge from the top of each cylinder - do this before removing the pistons!

14.3 Check the connecting rod side clearance with a feeler gauge as shown

14.4 Mark the rod bearing caps in order from the front of the engine to the rear (one mark for the front cap, two for the second one and so on)

14.6 To prevent damage to the crankshaft journals and cylinder walls, slip sections of rubber or plastic hose over the rod bolts before removing the pistons/rods

8   Inspect the camshaft lobes for heat discoloration, score marks, chipped areas, pitting and uneven wear. If the lobes are in good condition and if the lobe lift measurements are as specified, you can reuse the camshaft.

9   On V6 engines, check the camshaft bearings in the block for wear and damage. Look for galling, pitting and discolored areas. On four cylinder engines the camshaft housing(s) are the actual bearing surface for the camshafts, so inspect the housing journals for damage and replace them if necessary.

10   The inside diameter of each bearing can be determined with a small hole gauge and outside micrometer or an inside micrometer. Subtract the camshaft bearing journal diameter(s) from the corresponding bearing inside diameter(s) to obtain the bearing oil clearance. If it's excessive, new bearings or housings will be required regardless of the condition of the originals.

11   On V6 engines, camshaft bearing replacement requires special tools and expertise that place it outside the scope of the home mechanic. Take the block to an automotive machine shop to ensure the job is done correctly.

14 Pistons and connecting rods - removal

Refer to illustrations 14.1, 14.3, 14.4 and 14.6
Note: *Prior to removing the piston/connecting rod assemblies, remove the cylinder heads, the oil pan, the oil pump and the balance shaft assembly (four cylinder engine only) by referring to the appropriate Sections in chapter 2 Part A or B.*

1   Use your fingernail to feel if a ridge has formed at the upper limit of ring travel (about 1/4-inch down from the top of each cylinder). If carbon deposits or cylinder wear have produced ridges, they must be completely removed with a special tool **(see illustration)**. Follow the manufacturer's instructions provided with the tool. Failure to remove the ridges before attempting to remove the piston/connecting rod assemblies may result in piston breakage.

2   After the cylinder ridges have been removed, turn the engine upside-down so the crankshaft is facing up. On V6 engines, remove the oil baffle plate bolted to the main caps.

3   Before the connecting rods are removed, check the endplay with feeler gauges. Slide them between the first connecting rod and the crankshaft throw until the play is removed **(see illustration)**. The endplay is equal to the thickness of the feeler gauge(s). If the endplay exceeds the service limit, new connecting rods will be required. If new rods (or a new crankshaft) are installed, the endplay may fall under the minimum specified in this Chapter (if it does, the rods will have to be machined to restore it - consult an automotive machine shop for advice if necessary). Repeat the procedure for the remaining connecting rods.

4   Check the connecting rods and caps for identification marks. If they aren't plainly marked, use a small center-punch **(see illustration)** to make the appropriate number of indentations on each rod and cap (1, 2, 3, etc., depending on the cylinder they're associated with).

5   Loosen each of the connecting rod cap nuts 1/2-turn at a time until they can be removed by hand. Remove the number one connecting rod cap and bearing insert. Don't drop the bearing insert out of the cap.

6   Slip a short length of plastic or rubber

15.3 Checking crankshaft endplay with a feeler gauge

15.4 The arrow on the main bearing cap indicates the front of the engine

hose over each connecting rod cap bolt to protect the crankshaft journal and cylinder wall as the piston is removed **(see illustration)**.

7   Remove the bearing insert and push the connecting rod/piston assembly out through the top of the engine. Use a wooden or plastic hammer handle to push on the upper bearing surface in the connecting rod. If resistance is felt, double-check to make sure all of the ridge was removed from the cylinder.

8   Repeat the procedure for the remaining cylinders.

9   After removal, reassemble the connecting rod caps and bearing inserts in their respective connecting rods and install the cap nuts finger tight. Leaving the old bearing inserts in place until reassembly will help prevent the connecting rod bearing surfaces from being accidentally nicked or gouged.

10   Don't separate the pistons from the connecting rods.

15 Crankshaft - removal

Refer to illustrations 15.3 and 15.4
Note: *The crankshaft can be removed only after the engine has been removed from the vehicle. It's assumed the flywheel/driveplate, timing chain, oil pan, oil pump, balance shaft assembly (four cylinder engine only) and piston/connecting rod assemblies have already been removed. On four cylinder engines, the rear main oil seal retainer must also be removed first.*

1   Before the crankshaft is removed, check the endplay. Mount a dial indicator with the stem in line with the crankshaft and touching one of the crank throws.

2   Push the crankshaft all the way to the rear and zero the dial indicator. Next, pry the crankshaft to the front as far as possible and check the reading on the dial indicator. The distance it moves is the endplay. If it's greater than listed in this Chapter's Specifications, check the crankshaft thrust surfaces for wear. If no wear is evident, new main bearings should correct the endplay.

3   If a dial indicator isn't available, feeler gauges can be used. Gently pry or push the crankshaft all the way to the front of the engine. Slip feeler gauges between the crankshaft and the front face of the thrust main bearing to determine the clearance **(see illustration)**. **Note:** *The thrust bearing is located at the number three main bearing cap on all engines.*

4   Check the main bearing caps to see if they're marked to indicate their locations. They should be numbered consecutively from the front of the engine to the rear. If they aren't, mark them with number stamping dies or a center-punch. Main bearing caps generally have a cast-in arrow, which points to the front of the engine **(see illustration)**. Loosen the main bearing cap bolts 1/4-turn at a time each, until they can be removed by hand. Note if any stud bolts are used and make sure they're returned to their original locations when the crankshaft is reinstalled. On four cylinder engines, mark the relationship of the balance shaft chain to the balance gear on the crankshaft so the chain (if reused) can be installed in its original direction. Premature wear and noise will occur if a "used" chain is installed in the opposite direction from which it was removed.

5   Gently tap the caps with a soft-face hammer, then separate them from the engine block. If necessary, use the bolts as levers to remove the caps. Try not to drop the bearing inserts if they come out with the caps.

6   Carefully lift the crankshaft straight out of the engine. It may be a good idea to have an assistant available, since the crankshaft is quite heavy. With the bearing inserts in place in the engine block and main bearing caps, return the caps to their respective locations on the engine block and tighten the bolts finger tight.

16 Engine block - cleaning

Refer to illustrations 16.4a, 16.4b, 16.8 and 16.10

1   Remove the main bearing caps and sep-

16.4a A hammer and large punch can be used to knock the core plugs sideways in their bores

arate the bearing inserts from the caps and the engine block. Tag the bearings, indicating which cylinder they were removed from and whether they were in the cap or the block, then set them aside.

2   Using a gasket scraper, remove all traces of gasket material from the engine block. Be very careful not to nick or gouge the gasket sealing surfaces.

3   Remove all of the covers and threaded oil gallery plugs from the block. The plugs are usually very tight - they may have to be drilled out and the holes retapped. Use new plugs when the engine is reassembled.

4   Remove the core plugs from the engine block. To do this, knock one side of the plugs into the block with a hammer and punch, then grasp them with large pliers and pull them out **(see illustrations)**.

5   If the engine is extremely dirty, it should be taken to an automotive machine shop to be cleaned. **Note:** *If the block is cleaned in a caustic-solution hot tank, this will ruin any bearing inserts left in the block, such as the camshaft bearings on V6 engines. If the engine is being rebuilt, these bearings should be replaced anyway.*

16.4b Pull the core plugs from the block with pliers

16.8 All bolt holes in the block - particularly the main bearing cap and head bolt holes - should be cleaned and restored with a tap (be sure to remove debris from the holes after this is done)

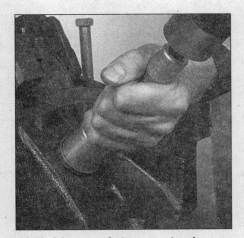

16.10 A large socket on an extension can be used to drive the new core plugs into the bores

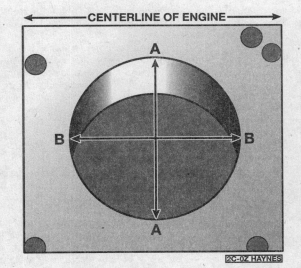

17.4a Measure the diameter of each cylinder at a right angle to the engine centerline (A), and parallel to the engine centerline (B) - out-of-round is the difference between A and B; taper is the difference between the diameter at the top of the cylinder and the diameter at the bottom of the cylinder

2C

6　After the block is returned, clean all oil holes and oil galleries one more time. Brushes specifically designed for this purpose are available at most auto parts stores. Flush the passages with warm water until the water runs clear, dry the block thoroughly and wipe all machined surfaces with a light, rust preventive oil. If you have access to compressed air, use it to speed the drying process and blow out all the oil holes and galleries. **Warning:** *Wear eye protection when using compressed air!*

7　If the block isn't extremely dirty or sludged up, you can do an adequate cleaning job with hot soapy water and a stiff brush. Take plenty of time and do a thorough job. Regardless of the cleaning method used, be sure to clean all oil holes and galleries very thoroughly, dry the block completely and coat all machined surfaces with light oil.

8　The threaded holes in the block must be clean to ensure accurate torque readings during reassembly. Run the proper size tap into each of the holes to remove rust, corrosion, thread sealant or sludge and restore damaged threads **(see illustration)**. If possible, use compressed air to clear the holes of debris produced by this operation. Now is a good time to clean the threads on the head bolts and the main bearing cap bolts as well.

9　Reinstall the main bearing caps and tighten the bolts finger tight.

10　After coating the sealing surfaces of the new core plugs with a non-hardening sealant (such as Permatex no. 2), install them in the engine block **(see illustration)**. Make sure they're driven in straight and seated properly or leakage could result. Special tools are available for this purpose, but a large socket, with an outside diameter that will just slip into the core plug, a 1/2-inch drive extension and a hammer will work just as well.

11　Apply non-hardening sealant (such as Permatex no. 2 or Teflon pipe sealant) to the new oil gallery plugs and thread them into the holes in the block. Make sure they're tightened securely.

12　If the engine isn't going to be reassembled right away, cover it with a large plastic trash bag to keep it clean.

17　Engine block - inspection

Refer to illustrations 17.4a, 17.4b and 17.4c
Note: *The manufacturer recommends checking the block deck for warpage and the main bearing bore concentricity and alignment. Since special measuring tools are needed, the checks should be done by an automotive machine shop.*

1　Before the block is inspected, it should be cleaned as described in Section 16.

2　Visually check the block for cracks, rust and corrosion. Look for stripped threads in the threaded holes. It's also a good idea to have the block checked for hidden cracks by an automotive machine shop that has the special equipment to do this type of work. If defects are found, have the block repaired, if possible, or replaced.

3　Check the cylinder bores for scuffing and scoring.

4　Check the cylinders for taper and out-of-round conditions as follows **(see illustrations):**

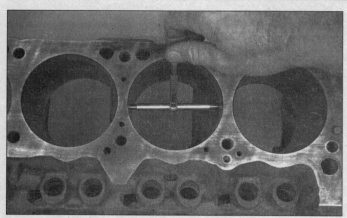

17.4b The ability to "feel" when the telescoping gauge is at the correct point will be developed over time, so work slowly and repeat the check until you're satisfied the bore measurement is accurate

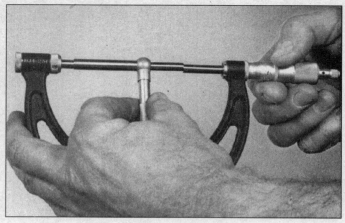

17.4c The gauge is then measured with a micrometer to determine the bore size

5   Measure the diameter of each cylinder at the top (just under the ridge area), center and bottom of the cylinder bore, parallel to the crankshaft axis.

6   Next, measure each cylinder's diameter at the same three locations perpendicular to the crankshaft axis.

7   The taper of each cylinder is the difference between the bore diameter at the top of the cylinder and the diameter at the bottom. The out-of-round specification of the cylinder bore is the difference between the parallel and perpendicular readings. Compare your results to this Chapter's Specifications.

8   If the cylinder walls are badly scuffed or scored, or if they're out-of-round or tapered beyond the limits given in this Chapter's Specifications, have the engine block rebored and honed at an automotive machine shop.

9   If a rebore is done, oversize pistons and rings will be required.

10   Using a precision straightedge and feeler gauge, check the block deck (the surface the cylinder heads mate with) for distortion as you did with the cylinder heads (see Section 10). If it's distorted beyond the specified limit, the block decks can be resurfaced by an automotive machine shop.

11   If the cylinders are in reasonably good condition and not worn to the outside of the limits, and if the piston-to-cylinder clearances can be maintained properly, they don't have to be rebored. Honing is all that's necessary (see Section 18).

18   Cylinder honing

Refer to illustrations 18.3a and 18.3b

1   Prior to engine reassembly, the cylinder bores must be honed so the new piston rings will seat correctly and provide the best possible combustion chamber seal. **Note:** *If you don't have the tools or don't want to tackle the honing operation, most automotive machine shops will do it for a reasonable fee.*

2   Before honing the cylinders, install the main bearing caps and tighten the bolts to the torque listed in this Chapter's Specifications.

3   Two types of cylinder hones are commonly available - the flex hone or "bottle brush" type and the more traditional surfacing hone with spring-loaded stones. Both will do the job, but for the less experienced mechanic the "bottle brush" hone will probably be easier to use. You'll also need some honing oil (kerosene will work if honing oil isn't available), rags and an electric drill motor. Proceed as follows:

a)   *Mount the hone in the drill motor, compress the stones and slip it into the first cylinder* **(see illustration)**. *Be sure to wear safety goggles or a face shield!*

b)   *Lubricate the cylinder with plenty of honing oil, turn on the drill and move the hone up-and-down in the cylinder at a pace that will produce a fine crosshatch pattern on the cylinder walls, and with the drill square and centered with the bore. Ideally, the crosshatch lines should intersect at approximately a 45-60-degree angle* **(see illustration)**. *Be sure*

to use plenty of lubricant and don't take off any more material than is absolutely necessary to produce the desired finish. **Note:** Piston ring manufacturers may specify a different crosshatch angle - read and follow any instructions included with the new rings.

c)   *Don't withdraw the hone from the cylinder while it's running. Instead, shut off the drill and continue moving the hone up-and-down in the cylinder until it comes to a complete stop, then compress the stones and withdraw the hone. If you're using a "bottle brush" type hone, stop the drill motor, then turn the chuck in the normal direction of rotation while withdrawing the hone from the cylinder.*

d)   *Wipe the oil out of the cylinder and repeat the procedure for the remaining cylinders.*

4   After the honing job is complete, chamfer the top edges of the cylinder bores with a

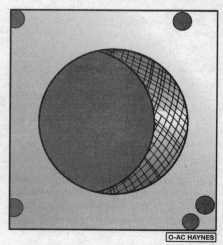

18.3a A "bottle brush" hone will produce a better cross hatch pattern when using a drill motor to hone the cylinders

18.3b The cylinder hone should leave a smooth, crosshatch pattern with the lines intersecting at approximately a 60-degree angle

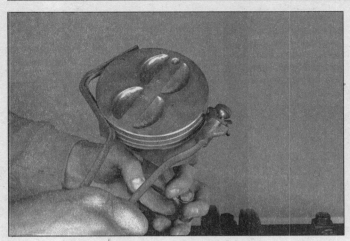

19.4a The piston ring grooves can be cleaned with a special tool, as shown here . . .

19.4b . . . or a section of broken ring

2C

small file so the rings won't catch when the pistons are installed. Be very careful not to nick the cylinder walls with the end of the file.

5 The entire engine block must be washed again very thoroughly with warm, soapy water to remove all traces of the abrasive grit produced during the honing operation. **Note:** *The bores can be considered clean when a lint-free white cloth - dampened with clean engine oil - used to wipe them out doesn't pick up any more honing residue, which will show up as gray areas on the cloth. Be sure to run a brush through all oil holes and galleries and flush them with running water.*

6 After rinsing, dry the block and apply a coat of light rust preventive oil to all machined surfaces. Wrap the block in a plastic trash bag to keep it clean and set it aside until reassembly.

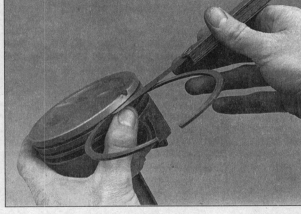

19.10 Check the ring side clearance with a feeler gauge at several points around the groove

19 Pistons and connecting rods - inspection

Refer to illustrations 19.4a, 19.4b, 19.10 and 19.11

1 Before the inspection process can be carried out, the piston/connecting rod assemblies must be cleaned and the original piston rings removed from the pistons. **Note:** *Always use new piston rings when the engine is reassembled.*

2 Using a piston ring installation tool, carefully remove the rings from the pistons. Be careful not to nick or gouge the pistons in the process.

3 Scrape all traces of carbon from the top of the piston. A hand-held wire brush or a piece of fine emery cloth can be used once the majority of the deposits have been scraped away. Do not, under any circumstances, use a wire brush mounted in a drill motor to remove deposits from the pistons. The piston material is soft and may be eroded away by the wire brush.

4 Use a piston ring groove-cleaning tool to remove carbon deposits from the ring grooves. If a tool isn't available, a piece broken off the old ring will do the job. Be very careful to remove only the carbon deposits - don't remove any metal and do not nick or scratch the sides of the ring grooves **(see illustrations)**.

5 Once the deposits have been removed, clean the piston/rod assemblies with solvent and dry them with compressed air (if available). **Warning:** *Wear eye protection. Make sure the oil return holes in the back sides of the ring grooves are clear.*

6 If the pistons and cylinder walls aren't damaged or worn excessively, and if the engine block isn't rebored, new pistons won't be necessary. Normal piston wear appears as even vertical wear on the piston thrust surfaces and slight looseness of the top ring in its groove. New piston rings, however, should always be used when an engine is rebuilt.

7 Carefully inspect each piston for cracks around the skirt, at the pin bosses and at the ring lands.

8 Look for scoring and scuffing on the thrust faces of the skirt, holes in the piston crown and burned areas at the edge of the crown. If the skirt is scored or scuffed, the engine may have been suffering from overheating and/or abnormal combustion, which

caused excessively high operating temperatures. The cooling and lubrication systems should be checked thoroughly. A hole in the piston crown is an indication that abnormal combustion (preignition) was occurring. Burned areas at the edge of the piston crown are usually evidence of spark knock (detonation). If any of the above problems exist, the causes must be corrected or the damage will occur again. The causes may include intake air leaks, incorrect fuel/air mixture, low octane fuel, ignition timing and EGR system malfunctions.

9 Corrosion of the piston, in the form of small pits, indicates coolant is leaking into the combustion chamber and/or the crankcase. Again, the cause must be corrected or the problem may persist in the rebuilt engine.

10 Measure the piston ring side clearance by laying a new piston ring in each ring groove and slipping a feeler gauge in beside it **(see illustration)**. Check the clearance at three or four locations around each groove. Be sure to use the correct ring for each groove - they are different. If the side clearance is greater than specified in this Chapter, new pistons will have to be used.

11 Check the piston-to-bore clearance by

19.11 Measure the piston diameter at a 90-degree angle to the piston pin and at the specified distance from the bottom of the piston skirt

20.1 The oil holes should be chamfered so sharp edges don't gouge or scratch the new bearings

measuring the bore (see Section 17) and the piston diameter. Make sure the pistons and bores are correctly matched. Measure the piston across the skirt, at a 90-degree angle to the piston pin **(see illustration)**. The measurement must be taken at a specific point to be accurate: The pistons are measured 0.45-inch from the bottom of the skirt, at right angles to the piston pin. Measure the cylinder bore 2.5-inches from the top for comparison with the piston measurement.

12   Subtract the piston diameter from the bore diameter to obtain the clearance. If it's greater than specified, the block will have to be rebored and new pistons and rings installed.

13   Check the piston-to-rod clearance by twisting the piston and rod in opposite directions. Any noticeable play indicates excessive wear, which must be corrected. The piston/connecting rod assemblies should be taken to an automotive machine shop to have the pistons and rods re-sized and new pins installed.

14   If the pistons must be removed from the connecting rods for any reason, they should be taken to an automotive machine shop.

While they are there, have the connecting rods checked for bend and twist, since automotive machine shops have special equipment for this purpose. **Note:** *Unless new pistons and/or connecting rods must be installed, do not disassemble the pistons and connecting rods.*

15   Check the connecting rods for cracks and other damage. Temporarily remove the rod caps, lift out the old bearing inserts, wipe the rod and cap bearing surfaces clean and inspect them for nicks, gouges and scratches. After checking the rods, replace the old bearings, slip the caps into place and tighten the nuts finger tight. **Note:** *If the engine is being rebuilt because of a connecting rod knock, be sure to install new or remanufactured connecting rods.*

20   Crankshaft - inspection

Refer to illustrations 20.1, 20.2, 20.5 and 20.7

1   Remove all burrs from the crankshaft oil holes with a stone, file or scraper **(see illustration)**.

2   Clean the crankshaft with solvent and

dry it with compressed air (if available). **Warning:** *Wear eye protection when using compressed air.* Be sure to clean the oil holes with a stiff brush **(see illustration)** and flush them with solvent.

3   Check the main and connecting rod bearing journals for uneven wear, scoring, pits and cracks.

4   Check the rest of the crankshaft for cracks and other damage. It should be Magnafluxed to reveal hidden cracks - an automotive machine shop will handle the procedure.

5   Using a micrometer, measure the diameter of the main and connecting rod journals and compare the results to this Chapter's Specifications **(see illustration)**. By measuring the diameter at a number of points around each journal's circumference, you'll be able to determine whether or not the journal is out-of-round. Take the measurement at each end of the journal, near the crank throws, to determine if the journal is tapered.

6   If the crankshaft journals are damaged, tapered, out-of-round or worn beyond the limits given in the Specifications, have the crankshaft reground by an automotive machine shop. Be sure to use the correct-

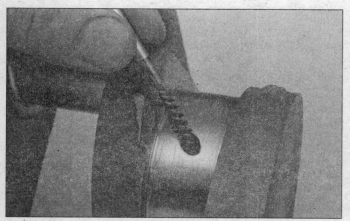

20.2 Use a wire or stiff plastic bristle brush to clean the oil passages in the crankshaft

20.5 Measure the diameter of each crankshaft journal at several points to detect taper and out-of-round conditions

20.7 If the seals have worn grooves in the crankshaft journals, or if the seal contact surfaces are nicked or scratched, the new seals will leak

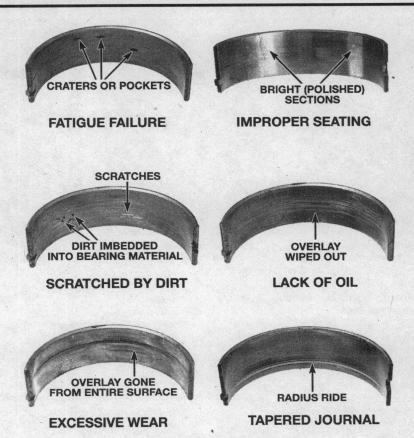

21.1 Typical bearing failures

size bearing inserts if the crankshaft is reconditioned.

7   Check the oil seal journals at each end of the crankshaft for wear and damage. If the seal has worn a groove in the journal, or if it's nicked or scratched **(see illustration)**, the new seal may leak when the engine is reassembled. In some cases, an automotive machine shop may be able to repair the journal by pressing on a thin sleeve. If repair isn't feasible, a new or different crankshaft should be installed.

8   Examine the main and rod bearing inserts (see Section 21).

21   Main and connecting rod bearings - inspection and selection

Refer to illustration 21.1

1   Even though the main and connecting rod bearings should be replaced with new ones during the engine overhaul, the old bearings should be retained for close examination, as they may reveal valuable information about the condition of the engine **(see illustration)**.

2   Bearing failure occurs because of lack of lubrication, the presence of dirt or other foreign particles, overloading the engine and corrosion. Regardless of the cause of bearing failure, it must be corrected before the engine is reassembled to prevent it from happening again.

3   When examining the bearings, remove them from the engine block, the main bearing caps, the connecting rods and the rod caps and lay them out on a clean surface in the same general position as their location in the engine. This will enable you to match any bearing problems with the corresponding crankshaft journal.

4   Dirt and other foreign particles get into the engine in a variety of ways. It may be left in the engine during assembly, or it may pass through filters or the PCV system. It may get into the oil, and from there into the bearings. Metal chips from machining operations and normal engine wear are often present. Abrasives are sometimes left in engine components after reconditioning, especially when parts aren't thoroughly cleaned using the proper cleaning methods. Whatever the source, these foreign objects often end up embedded in the soft bearing material and are easily recognized. Large particles won't embed in the bearing and will score or gouge the bearing and journal. The best prevention for this cause of bearing failure is to clean all parts thoroughly and keep everything spotlessly clean during engine assembly. Frequent and regular engine oil and filter changes are also recommended.

5   Lack of lubrication (or lubrication breakdown) has a number of interrelated causes. Excessive heat (which thins the oil), overloading (which squeezes the oil from the bearing face) and oil leakage or throw off (from excessive bearing clearances, worn oil pump or high engine speeds) all contribute to lubrication breakdown. Blocked oil passages, which usually are the result of misaligned oil holes in a bearing shell, will also oil starve a bearing and destroy it. When lack of lubrication is the cause of bearing failure, the bearing material is wiped or extruded from the steel backing of the bearing. Temperatures may increase to the point where the steel backing turns blue from overheating.

6   Driving habits can have a definite effect on bearing life. Low speed operation in too high a gear (lugging the engine) puts very high loads on bearings, which tends to squeeze out the oil film. These loads cause the bearings to flex, which produces fine cracks in the bearing face (fatigue failure). Eventually the bearing material will loosen in pieces and tear away from the steel backing. Short trip driving leads to corrosion of bearings because insufficient engine heat is produced to drive off the condensed water and corrosive gases. These products collect in the engine oil, forming acid and sludge. As the oil is carried to the engine bearings, the acid attacks and corrodes the bearing material.

7   Incorrect bearing installation during engine assembly will lead to bearing failure as well. Tight-fitting bearings leave insufficient oil clearance and will result in oil starvation. Dirt or foreign particles trapped behind a bearing insert result in high spots on the bearing which lead to failure.

22   Engine overhaul - reassembly sequence

1   Before beginning engine reassembly, make sure you have all the necessary new

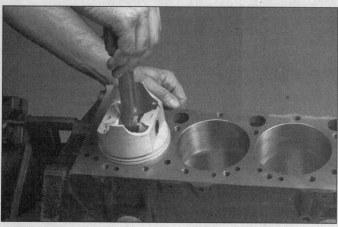

23.3 When checking piston ring end gap, the ring must be square in the cylinder bore (this is done by pushing the ring down with the top of a piston as shown)

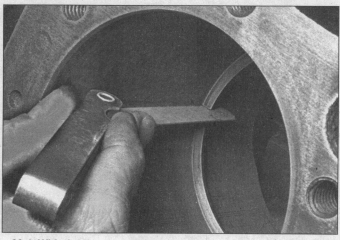

23.4 With the ring square in the cylinder, measure the end gap with a feeler gauge

parts, gaskets and seals as well as the following items on hand:

> Common hand tools
> Torque wrench (1/2-inch drive) with
>> angle-torque gauge
> Piston ring Installation tool
> Piston ring compressor
> Crankshaft balancer Installation tool
> Short lengths of rubber or plastic hose to
>> fit over connecting rod bolts
> Plastigage
> Feeler gauges
> Fine-tooth file
> New engine oil
> Engine assembly lube or moly-base
>> grease
> Gasket sealant
> Thread locking compound

2   In order to save time and avoid problems, engine reassembly must be done in the following general order:

Four cylinder engine

> Balance shaft drive chain on crankshaft
> Crankshaft and main bearings
> Piston/connecting rod assemblies
> Balance shaft housing assembly
> Oil pump
> Balance shaft driven gear
> Balance shaft drive chain tensioner
> Rear main oil seal
> Cylinder head
> Camshaft housings, valve lifters and
>> camshafts
> Timing chain housing
> Timing chain and sprockets
> Timing chain front cover
> Oil pan
> Flywheel/driveplate

V6 engines

> Crankshaft and main bearings
> Piston/connecting rod assemblies
> Rear main oil seal
> Camshaft
> Oil pump drive assembly
> Timing chain and sprockets

> Cylinder heads
> Valve lifters
> Rocker arms and pushrods
> Timing chain cover
> Oil pump
> Oil pan
> Driveplate

Assembled after engine installation

> Intake and exhaust manifolds
> Valve covers (V6 engines only)

23   Piston rings - installation

Refer to illustrations 23.3, 23.4, 23.5, 23.9a, 23.9b and 23.12

1   Before installing the new piston rings, the ring end gaps must be checked. It's assumed the piston ring side clearance has been checked and verified correct (see Section 19).

2   Lay out the piston/connecting rod assemblies and the new ring sets so the ring sets will be matched with the same piston and cylinder during the end gap measurement and engine assembly.

3   Insert the top (number one) ring into the first cylinder and square it up with the cylinder walls by pushing it in with the top of the piston **(see illustration)**. The ring should be near the bottom of the cylinder, at the lower limit of ring travel.

4   To measure the end gap, slip feeler gauges between the ends of the ring until a gauge equal to the gap width is found **(see illustration)**. The feeler gauge should slide between the ring ends with a slight amount of drag. Compare the measurement to this Chapter's Specifications. If the gap is larger or smaller than specified, double-check to make sure you have the correct rings before proceeding.

5   If the gap is too small, it must be enlarged or the ring ends may come in contact with each other during engine operation, which can cause serious engine damage. The end gap can be increased by filing the ring ends very carefully with a fine file. Mount the file in a vise equipped with soft jaws, slip the ring over the file with the ends contacting the file teeth and slowly move the ring to remove material from the ends. When performing this operation, file only from the outside in **(see illustration)**. **Note:** *When you have the end*

23.5 If the end gap is too small, clamp a file in a vise and file the ring ends (from the outside in only) to enlarge the gap slightly

gap correct, remove any burrs from the filed ends of the rings with a whetstone.

6   Excess end gap isn't critical unless it's greater than 0.040-inch. Again, double-check to make sure you have the correct rings for the engine. If the engine block has been bored oversize, necessitating oversize pistons, matching oversize rings are required.

7   Repeat the procedure for each ring that will be installed in the first cylinder and for each ring in the remaining cylinders. Remember to keep rings, pistons and cylinders matched up.

8   Once the ring end gaps have been checked/corrected, the rings can be installed on the pistons.

9   The oil control ring (lowest one on the piston) is usually installed first. It's composed of three separate components. Slip the spacer/expander into the groove **(see illustration)**. If an anti-rotation tang is used, make sure it's inserted into the drilled hole in the ring groove. Next, install the lower side rail. Don't use a piston ring installation tool on the oil ring side rails, as they may be damaged. Instead, place one end of the side rail into the

groove between the spacer/expander and the ring land, hold it firmly in place and slide a finger around the piston while pushing the rail into the groove **(see illustration)**. Next, install the upper side rail in the same manner.

10   After the three oil ring components have been installed, check to make sure both the upper and lower side rails can be turned smoothly in the ring groove.

11   The number two (middle) ring is installed next. It's usually stamped with a mark, which must face up, toward the top of the piston. **Note:** *Always follow the instructions printed on the ring package or box - different manufacturers may require different approaches. Don't mix up the top and middle rings, as they have different cross-sections.*

12   Use a piston ring installation tool and make sure the identification mark is facing the top of the piston, then slip the ring into the middle groove on the piston **(see illustration)**. Don't expand the ring any more than necessary to slide it over the piston.

13   Install the number one (top) ring in the same manner. Make sure the mark is facing up. Be careful not to confuse the number one and number two rings.

14   Repeat the procedure for the remaining pistons and rings.

---

24   Crankshaft - installation and main bearing oil clearance check

---

1   Crankshaft installation is the first step in engine reassembly. It's assumed at this point that the engine block and crankshaft have been cleaned, inspected and repaired or reconditioned.

2   Position the engine with the bottom facing up.

3   Remove the main bearing cap bolts and lift out the caps. Lay them out in the proper order to ensure correct installation.

4   If they're still in place, remove the original bearing inserts from the block and the

main bearing caps. Wipe the bearing surfaces of the block and caps with a clean, lint-free cloth. They must be kept spotlessly clean.

Main bearing oil clearance check

Refer to illustrations 24.11 and 24.15

Note: *Don't touch the faces of the new bearing inserts with your fingers. Oil and acids from your skin can etch the bearings.*

5   Clean the back sides of the new main bearing inserts and lay one in each main bearing saddle in the block. If one of the bearing inserts from each set has a large groove in it, make sure the grooved insert is installed in the block. Lay the other bearing from each set in the corresponding main bearing cap. Make sure the tab on the bearing insert fits into the recess in the block or cap, neither higher than the cap's edge nor lower. **Caution:** *The oil holes in the block must line up with the oil holes in the bearing inserts. Do not hammer the bearing into place and don't nick or gouge the bearing faces. No lubrication should be used at this time.*

6   The flanged thrust bearing must be installed in the number three main cap. **Caution:** *Some engines may have a oversize rear main bearing. Check your crankshaft for a marking on the last counterweight, and check the backside of the old bearing insert for a similar marking. If they are marked oversize, an oversize rear bearing will be required.*

7   Clean the faces of the bearings in the block and the crankshaft main bearing journals with a clean, lint-free cloth.

8   Check or clean the oil holes in the crankshaft, as any dirt here can go only one way - straight through the new bearings.

9   Once you're certain the crankshaft is clean, carefully lay it in position in the main bearings.

10   Before the crankshaft can be permanently installed, the main bearing oil clearance must be checked.

23.9a Installing the spacer/expander in the oil control ring groove

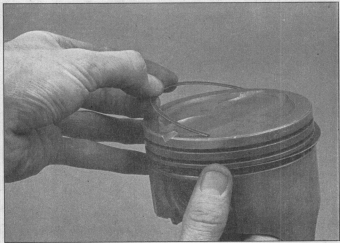

23.9b DO NOT use a piston ring installation tool when installing the oil ring side rails

23.12 Installing the compression rings with a ring expander - the mark (arrow) must face up

2C

24.11 Lay the Plastigage strips (arrow) on the main bearing journals, parallel to the crankshaft centerline

24.15 Measuring the width of the crushed Plastigage to determine the main bearing oil clearance (be sure to use the correct scale - standard and metric ones are included)

11   Cut several pieces of the appropriate size Plastigage (they should be slightly shorter than the width of the main bearings) and place one piece on each crankshaft main bearing journal, parallel with the journal axis **(see illustration)**.

12   Clean the faces of the bearings in the caps and install the caps in their original locations (don't mix them up) with the arrows pointing toward the front of the engine. Don't disturb the Plastigage.

13   Starting with the center main and working out toward the ends, tighten the main bearing cap bolts to the torque listed in this Chapter's Specifications in three steps. Don't rotate the crankshaft at any time during this operation, and do not tighten one cap completely - tighten all caps equally. Before tightening, the main caps should be seated using light taps with a brass or plastic mallet.

14   Remove the bolts/studs and carefully lift off the main bearing caps. Keep them in order. Don't disturb the Plastigage or rotate the crankshaft. If any of the main bearing caps are difficult to remove, tap them gently from side-to-side with a soft-face hammer to loosen them.

15   Compare the width of the crushed Plastigage on each journal to the scale printed on the Plastigage envelope to obtain the main bearing oil clearance **(see illustration)**. Check the Specifications to make sure it's correct.

16   If the clearance is not as specified, the bearing inserts may be the wrong size (which means different ones will be required). Before deciding different inserts are needed, make sure no dirt or oil was between the bearing inserts and the caps or block when the clearance was measured. If the Plastigage was wider at one end than the other, the journal may be tapered (see Section 20).

17   Carefully scrape all traces of the Plastigage material off the main bearing journals and/or the bearing faces. Use your fingernail or the edge of a credit card - don't nick or scratch the bearing faces.

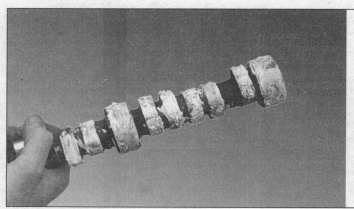

25.1 On V6 engines, coat the camshaft lobes and journals with assembly lube before installation

Final crankshaft installation

18   Carefully lift the crankshaft out of the engine.

19   Clean the bearing faces in the block, then apply a thin, uniform layer of moly-base grease or engine assembly lube to each of the bearing surfaces. Be sure to coat the thrust faces as well as the journal face of the thrust bearing.

20   Make sure the crankshaft journals are clean, then lay the crankshaft back in place in the block. **Note:** *On four cylinder engines it will be necessary to install the balance shaft drive chain onto the crankshaft drive gear before final installation of the crankshaft.*

21   Clean the faces of the bearings in the caps, then apply lubricant to them.

22   Install the caps in their original locations with the arrows pointing toward the front of the engine. **Note:** *On V6 engines, apply a small amount of RTV sealant between the rear main cap and engine block sealing surfaces.*

23   With all caps in place and bolts just started, tap the ends of the crankshaft forward and backward with a lead or brass hammer to line up the main bearing and crankshaft thrust surfaces.

24   Following the procedures outlined in Step 13, retighten all main bearing cap bolts

to the torque listed in this Chapter's Specifications, starting with the center main and working out toward the ends.

25   Rotate the crankshaft a number of times by hand to check for any obvious binding.

26   The final step is to check the crankshaft endplay with feeler gauges or a dial indicator as described in Section 15. The endplay should be correct if the crankshaft thrust faces aren't worn or damaged and new bearings have been installed.

25   Camshaft (V6 engines) - installation

Refer to illustration 25.1

1   Lubricate the camshaft bearing journals and cam lobes with a special camshaft installation lubricant **(see illustration)**.

2   Slide the camshaft into the engine, using a long bolt (the same thread as the camshaft sprocket bolt) screwed into the front of the camshaft as a "handle." Support the cam near the block and be careful not to scrape or nick the bearings. Install the camshaft retainer plate and tighten the bolts to the torque listed in this Chapter's Specifications.

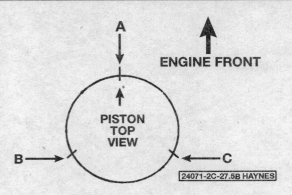

27.5a Ring end gap positions – four cylinder engine

A   Top compression ring gap
B   Second compression ring gap
C   Oil ring assembly gap

27.5b Ring end gap positions – V6 engines

A   Oil ring rail gaps
B   Second compression ring gap
C   Oil ring spacer gap (position between marks)
D   Top compression ring gap

3   Dip the gear portion of the oil pump drive in engine oil and insert it into the block. It should be flush with its mounting boss before inserting the retaining bolt. **Note:** *Position a new O-ring on the oil pump driveshaft before installation.*

4   Complete the installation of the timing chain and sprockets by referring to Chapter 2B.

26   Rear main oil seal - replacement

Refer to Chapter 2A or 2B for the rear main seal replacement procedure.

27   Pistons and connecting rods - installation and rod bearing oil clearance check

1   Before installing the piston/connecting rod assemblies, the cylinder walls must be perfectly clean, the top edge of each cylinder must be chamfered, and the crankshaft must be in place.

2   Remove the cap from the end of the number one connecting rod (check the marks made during removal). Remove the original bearing inserts and wipe the bearing surfaces of the connecting rod and cap with a clean, lint-free cloth. They must be kept spotlessly clean.

Piston installation and rod bearing oil clearance check

Refer to illustrations 27.5a, 27.5b, 27.11, 27.13 and 27.17

3   Clean the back side of the new upper bearing insert, then lay it in place in the connecting rod. Make sure the tab on the bearing fits into the recess in the rod. Don't hammer the bearing insert into place and be very careful not to nick or gouge the bearing face. Don't lubricate the bearing at this time.

4   Clean the back side of the other bearing insert and install it in the rod cap. Again, make sure the tab on the bearing fits into the recess in the cap, and don't apply any lubricant. It's critically important that the mating surfaces of the bearing and connecting rod are perfectly clean and oil free when they're assembled.

5   Stagger the piston ring gaps around the piston **(see illustrations)**.

6   Slip a section of plastic or rubber hose over each connecting rod cap bolt.

7   Lubricate the piston and rings with clean engine oil and attach a piston ring compressor to the piston. Leave the skirt protruding about 1/4-inch to guide the piston into the cylinder. The rings must be compressed until they're flush with the piston.

8   Rotate the crankshaft until the number one connecting rod journal is at BDC (bottom dead center) and apply a coat of engine oil to the cylinder walls.

9   With the mark or notch on top of the piston facing the front of the engine, gently insert the piston/connecting rod assembly into the number one cylinder bore and rest the bottom edge of the ring compressor on the engine block.

10   Tap the top edge of the ring compressor to make sure it's contacting the block around its entire circumference.

11   Gently tap on the top of the piston with the end of a wooden or plastic hammer handle **(see illustration)** while guiding the end of the connecting rod into place on the crankshaft journal. The piston rings may try to pop out of the ring compressor just before entering the cylinder bore, so keep some pressure down on the ring compressor. Work slowly, and if any resistance is felt as the piston enters the cylinder, stop immediately. Find out what's hanging up and fix it before proceeding. Do not, for any reason, force the piston into the cylinder - you might break a ring and/or the piston.

12   Once the piston/connecting rod assembly is installed, the connecting rod bearing oil clearance must be checked before the rod cap is permanently bolted in place.

13   Cut a piece of the appropriate size Plastigage slightly shorter than the width of the connecting rod bearing and lay it in place on the number one connecting rod journal, parallel with the journal axis **(see illustration)**.

27.11 Drive the piston into the cylinder bore with the end of a wooden or plastic hammer handle

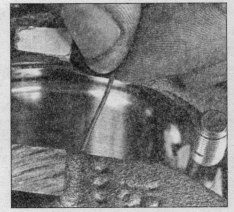

27.13 Lay the Plastigage strips on each rod bearing journal, parallel to the crankshaft centerline

14   Clean the connecting rod cap bearing face, remove the protective hoses from the connecting rod bolts and install the rod cap. Make sure the mating mark on the cap is on the same side as the mark on the connecting rod.

15   Install the nuts and tighten them to the torque listed in this Chapter's Specifications. Work up to it in three steps. **Note:** *Use a thin-wall socket to avoid erroneous torque readings that can result if the socket is wedged between the rod cap and nut. If the socket tends to wedge itself between the nut and the cap, lift up on it slightly until it no longer contacts the cap. Do not rotate the crankshaft at any time during this operation.*

16   Remove the nuts and detach the rod cap, being very careful not to disturb the Plastigage.

17   Compare the width of the crushed Plastigage to the scale printed on the Plastigage envelope to obtain the oil clearance **(see illustration)**. Compare it to this Chapter's Specifications to make sure the clearance is correct.

18   If the clearance is not as specified, the bearing inserts may be the wrong size (which means different ones will be required). Before deciding different inserts are needed, make sure no dirt or oil was between the bearing inserts and the connecting rod or cap when the clearance was measured. Also, recheck the journal diameter. If the Plastigage was wider at one end than the other, the journal may be tapered (see Section 20).

Final connecting rod installation

19   Carefully scrape all traces of the Plastigage material off the rod journal and/or bearing face. Be very careful not to scratch the bearing - use your fingernail or the edge of a credit card.

20   Make sure the bearing faces are perfectly clean, then apply a uniform layer of clean moly-base grease or engine assembly lube to both of them. You'll have to push the piston into the cylinder to expose the face of the bearing insert in the connecting rod - be sure to slip the protective hoses over the rod bolts first.

21   Slide the connecting rod back into place on the journal, remove the protective hoses from the rod cap bolts, install the rod cap and tighten the nuts to the torque listed in this Chapter's Specifications. Again, work up to

the torque in three steps.

22   Repeat the entire procedure for the remaining pistons/connecting rods.

23   The important points to remember are:

a) *Keep the back sides of the bearing inserts and the insides of the connecting rods and caps perfectly clean when assembling them.*

b) *Make sure you have the correct piston/rod assembly for each cylinder.*

c) *The arrow or mark on the piston must face the front of the engine.*

d) *Lubricate the cylinder walls with clean oil.*

e) *Lubricate the bearing faces when installing the rod caps after the oil clearance has been checked.*

24   After all the piston/connecting rod assemblies have been properly installed, rotate the crankshaft a number of times by hand to check for any obvious binding.

25   As a final step, the connecting rod endplay must be checked (see Section 14).

26   Compare the measured endplay to this Chapter's Specifications to make sure it's correct. If it was correct before disassembly and the original crankshaft and rods were reinstalled, it should still be right. If new rods or a new crankshaft were installed, the endplay may be inadequate. If so, the rods will have to be removed and taken to an automotive machine shop for re-sizing.

27   On four cylinder engines, complete the installation of the balance shaft housing assembly and driven sprocket by referring to Chapter 2A.

28   Initial start-up and break-in after overhaul

Warning: *Have a fire extinguisher handy when starting the engine for the first time.*

1   Once the engine has been installed in the vehicle, double-check the oil and coolant levels.

2   With the spark plugs out of the engine, disable the ignition system by disconnecting the wires to the ignition control module (see Chapter 5). Also, disable the fuel injection system by unplugging the electrical connector to the injector wiring harness or by removing the fuel pump fuse. Crank the engine until oil pressure registers on the gauge or the light goes out.

3   Install the spark plugs, hook up the plug

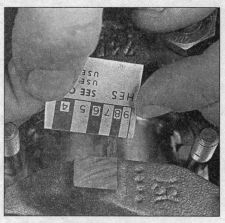

27.17 Measuring the width of the crushed Plastigage to determine the rod bearing oil clearance (be sure to use the correct scale - standard and metric ones are included)

wires and install the fuses.

4   Start the engine. It may take a few moments for the fuel system to build up pressure, but the engine should start without a great deal of effort. **Note:** *If the engine keeps backfiring, recheck the valve timing and spark plug wire routing.*

5   After the engine starts, it should be allowed to warm up to normal operating temperature. While the engine is warming up, make a thorough check for fuel, oil and coolant leaks.

6   Shut the engine off and recheck the engine oil and coolant levels.

7   Drive the vehicle to an area with no traffic, accelerate from 30 to 50 mph, then allow the vehicle to slow to 30 mph with the throttle closed. Repeat the procedure 10 or 12 times. This will load the piston rings and cause them to seat properly against the cylinder walls. Check again for oil and coolant leaks.

8   Drive the vehicle gently for the first 500 miles (no sustained high speeds) and keep a constant check on the oil level. It isn't unusual for an engine to use oil during the break-in period.

9   At approximately 500 to 600 miles, change the oil and filter.

10   For the next few hundred miles, drive the vehicle normally. Don't pamper it or abuse it.

11   After 2000 miles, change the oil and filter again and consider the engine broken in.

Chapter 3
Cooling, heating and air conditioning systems

Contents

Specifications

General

Coolant capacity	See Chapter 1
Coolant reservoir pressure cap rating	15 psi
Thermostat opening temperature	
Four cylinder engine	180-degrees F
V6 engines	195-degrees F
Refrigerant type	R-134a
Refrigerant capacity	1.75 pounds

Torque specifications

	Ft-lbs (unless otherwise indicated)
Transaxle oil cooler lines	22
Thermostat housing nuts/bolts	
Four cylinder engine	124 in-lbs
V6 engines	18
Thermostat inlet pipe-to-oil pan bolt (four cylinder engine)	18
Coolant outlet pipe-to-cylinder head (four cylinder engine	18
Water pump attaching bolts	
Four cylinder engine	
Water pump-to-timing chain housing nuts	19
Water pump cover-to-water pump housing bolts	124 in-lbs.
Water pump cover-to-engine block bolts	19
Radiator outlet pipe-to-water pump housing bolts	124 in-lbs.
V6 engines	89 in-lbs
Water pump pulley bolts (V6 engines)	18

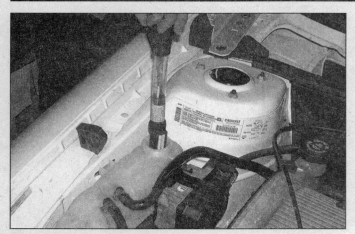

2.4 An inexpensive hydrometer can be used to test the condition of your coolant

3.7 Working from above, remove the coolant inlet pipe bolts-to-thermostat housing bolts (arrows) - exhaust manifold removed for clarity

1   General information

All vehicles covered by this manual employ a pressurized engine cooling system with thermostatically controlled coolant circulation. Coolant is drawn from the radiator by an impeller-type water pump mounted at the front of the block. The coolant is then circulated through the engine block, intake manifold and the cylinder head(s) before it's redirected back into the radiator.

A wax pellet type thermostat is located in the thermostat housing on the engine. During warm up, the closed thermostat prevents coolant from circulating through the radiator. When the engine reaches normal operating temperature, the thermostat opens and allows hot coolant to travel through the radiator, where it is cooled before returning to the engine.

The cooling system is pressurized by a spring-loaded coolant reservoir cap, which, by maintaining pressure, increases the boiling point of the coolant. If the coolant temperature goes above this increased boiling point, the extra pressure in the system forces the reservoir cap valve off its seat and allows the coolant to escape through the overflow tube into the coolant reservoir. When the system cools, the excess coolant is automatically drawn from the reservoir tank back into the radiator.

The coolant reservoir serves as both the point at which fresh coolant is added to the cooling system to maintain the proper fluid level and, as a holding tank for overheated coolant.

The heating system works by directing air through the heater core mounted in the dash and then to the interior of the vehicle by a system of ducts. Temperature is controlled by mixing heated air with fresh air, using a system of doors in the ducts, and a blower motor.

Air conditioning is an optional accessory, consisting of an evaporator core located under the dash, a condenser in front of the radiator, an accumulator in the engine compartment and a belt-driven compressor mounted at the front of the engine.

2   Antifreeze - general information

Refer to illustration 2.4

Warning: *Do not allow antifreeze to come in contact with your skin or painted surfaces of the vehicle. Rinse off spills immediately with plenty of water. Antifreeze is highly toxic if ingested. Never leave antifreeze lying around in an open container or in puddles on the floor; children and pets are attracted by it's sweet smell and may drink it. Check with local authorities about disposing of used antifreeze. Many communities have collection centers which will see that antifreeze is disposed of safely. Never dump used anti-freeze on the ground or pour it into drains.*

Caution: *The manufacturer recommends using only DEX-COOL coolant for these systems. DEX-COOL is a long-lasting coolant designed for 100,000 miles or 5 years. Never mix green-colored ethylene glycol anti-freeze and orange-colored "DEX-COOL" silicate-free coolant because doing so will destroy the efficiency of the "DEX-COOL".*

The cooling system should be filled with a water/ethylene glycol based antifreeze solution which will prevent freezing down to at least -20-degrees F (even lower in cold climates). It also provides protection against corrosion and increases the coolant boiling point.

The cooling system should be drained, flushed and refilled at least every other year (see Chapter 1). The use of antifreeze solutions for periods of longer than two years is likely to cause damage and encourage the formation of rust and scale in the system. However, these models are filled with a new, long-life "Dex-Cool coolant," which the manufacturer claims is good for five years.

Before adding antifreeze to the system, check all hose connections. Antifreeze can leak through very minute openings.

The exact mixture of antifreeze to water which you should use depends on the relative weather conditions. The mixture should contain at least 50-percent antifreeze, but should never contain more than 70-percent antifreeze. Consult the mixture ratio chart on the antifreeze container before adding coolant. Hydrometers are available at most auto parts stores to test the coolant **(see illustration)**. Always use antifreeze which meets the vehicle manufacturer's specifications.

3   Thermostat - check and replacement

Warning: *The engine must be completely cool when this procedure is performed.*
Note: *Don't drive the vehicle without a thermostat! The computer may stay in open loop mode and emissions and fuel economy will suffer.*

Check

1   Before assuming the thermostat is to blame for a cooling system problem, check the coolant level, drivebelt tension (see Chapter 1) and temperature gauge (or light) operation.
2   If the engine seems to be taking a long time to warm up (based on heater output or temperature gauge operation), the thermostat is probably stuck open. Replace the thermostat with a new one.
3   If the engine runs hot, use your hand to check the temperature of the upper radiator hose. If the hose isn't hot, but the engine is, the thermostat is probably stuck closed, preventing the coolant inside the engine from escaping to the radiator. Replace the thermostat. **Note:** *The thermostat and housing on four cylinder engines is located on the inlet side (before the water pump) and the thermostat and housing on V6 engines is located on*

3.8 Detach the bolts (arrows) and remove the engine-to-transaxle support brace

3.10 Detach the coolant inlet pipe from the lower radiator hose (A) and from the retaining stud on the oil pan (B)

3.11 Thermostat installation details - four cylinder engine

3.15 Upper radiator hose clamp at the thermostat housing cover - V6 engines

   A   *Radiator hose clamp*
   B   *Coolant reservoir (surge) tank line*

3.17 Thermostat housing cover bolts (arrows) - V6 engines

the outlet side (after the water pump). **Caution:** *Don't drive the vehicle without a thermostat. The computer may stay in open loop and emissions and fuel economy will suffer.*

4   If the upper radiator hose is hot, it means the coolant is flowing and the thermostat is open. Consult the *Troubleshooting* Section at the front of this manual for cooling system diagnosis.

Replacement

Four cylinder engine

Refer to illustrations 3.7, 3.8, 3.10 and 3.11

5   Drain the coolant from the radiator (see Chapter 1).

6   Remove the exhaust manifold heat shield (see Chapter 2A).

7   Remove coolant inlet pipe from the thermostat housing by accessing bolts through the exhaust manifold runners **(see illustration)**.

8   Raise the vehicle and secure it on jackstands. Remove the engine to transaxle support brace **(see illustration)**.

9   Disconnect the lower radiator hose from the coolant inlet pipe.

10   Remove the coolant inlet pipe to oil pan stud and carefully separate coolant inlet pipe from the thermostat housing **(see illustration)**. Slide the coolant pipe past the right of the transaxle to remove it.

11   Note how it's installed, then remove the thermostat from the inlet pipe **(see illustration)**. Be sure to use a replacement thermostat with the correct opening temperature (see this Chapter's Specifications).

V6 engines

Refer to illustrations 3.15, 3.17, 3.18a and 3.18b

12   Disconnect the cable from the negative terminal of the battery. **Caution:** *On models equipped with the Theftlock audio system, be sure the lockout feature is turned off before performing any procedure which requires disconnecting the battery (see the front of this manual).* Partially drain the cooling system. If the coolant is relatively new or in good condition, save it and reuse it. If it is to be replaced, see Section 2 for cautions about

proper handling of used antifreeze.

13   Remove the air cleaner and duct (see Chapter 4).

14   Follow the upper radiator hose to the engine to locate the thermostat housing cover. The thermostat is located at the end of the lower intake manifold on the driver's side of the engine compartment.

15   Loosen the hose clamp, then detach the hose and the coolant reservoir tank line from the thermostat housing cover **(see illustration)**. If the hose sticks, grasp it near the end with a pair of adjustable pliers and twist it to break the seal, then pull it off. If the hose is old or deteriorated, cut it off and install a new one.

16   If the outer surface of the cover fitting that mates with the hose is deteriorated (corroded, pitted, etc.) it may be damaged further by hose removal. If it is, the thermostat housing cover will have to be replaced.

17   Remove the bolts/nuts and detach the thermostat cover **(see illustration)**. If the cover is stuck, tap it with a soft-face hammer to jar it loose. Be prepared for some coolant to spill as the gasket seal is broken.

3

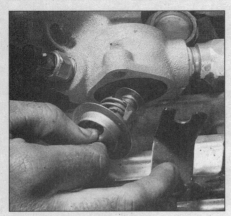

3.18a Thermostat installation details - V6 engines

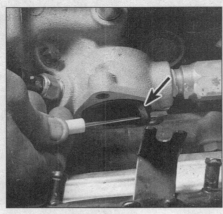

3.18b If the thermostat comes out without its rubber seal (arrow), pry it out of the housing with a small screwdriver

3.19 Install a new rubber seal around the thermostat

4.1 Disconnect the electrical connector from the fan and apply fused battery power and ground to test the fan

4.2 The cooling fan relays and fuses (arrows) are easily identifiable by viewing the decal on the inside of the underhood fuse/relay box cover - note that the No.1 fuse and relay control the left fan and the No.2 fuse and relay control the right fan

18   Note how it's installed (which end is facing up), then remove the thermostat **(see illustrations)**.

All engines

Refer to illustration 3.19

19   If a paper gasket was used, use a scraper or putty knife to remove all traces of old gasket material and sealant from the mating surfaces. **Note:** *Most of these models do not have a traditional gasket, but rather a rubber ring around the thermostat. If so, replace this ring and install the thermostat into the cover (V6 engines) or the coolant inlet pipe (four cylinder engines) without gasket sealant* **(see illustration)**.

20   Install the thermostat and make sure the correct end faces out - the spring is directed toward the engine.

21   If a traditional paper gasket was used, apply a thin coat of RTV sealant to both sides of the new gasket and position it on the engine side, over the thermostat, and make sure the gasket holes line up with the bolt holes in the housing.

22   Reattach the thermostat housing cover

(V6 engines) or the coolant inlet pipe (four cylinder engines) to the thermostat housing and tighten the bolts to the torque listed in this Chapter's Specifications. Now may be a good time to check and replace the hoses and clamps (see Chapter 1).

23   The remaining steps are the reverse of the removal procedure.

24   Refer to Chapter 1 and refill the system, then run the engine and check carefully for leaks.

25   Repeat steps 1 through 4 to be sure the repairs corrected the previous problem(s).

---

4   Engine cooling fan and circuit - check and component replacement

---

Warning: *Keep hands, tools and clothing away from the fan. To avoid injury or damage DO NOT operate the engine with a damaged fan. Do not attempt to repair fan blades - replace a damaged fan with a new one.*

Check

Refer to illustrations 4.1, 4.2 and 4.3

1   To test a fan motor, unplug the electrical connector at the motor and use fused jumper wires to connect battery power and ground directly to the fan **(see illustration)**. If the fan still doesn't work, replace the motor.

2   If the motor tests OK, check the cooling

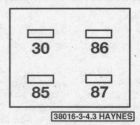

4.3 Test for continuity between relay terminals 30 and 87 - there should be no continuity until positive battery voltage is applied to terminal 85 and ground to terminal 86

4.10 Detach the air conditioning line support bracket (arrow)

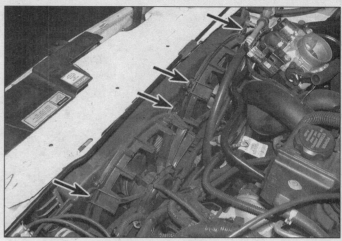

4.11 Detach the bolts securing the fans to the shroud (arrows indicate upper four bolts, lower four bolts not visible in this photo)

fan relay, located in the underhood fuse/relay panel **(see illustration)**.

3   Remove the cooling fan relay and test for continuity between terminals 30 and 87 **(see illustration)**. There should be no continuity.

4   Apply positive battery power to terminal 85, and ground terminal 86, and test for continuity again between 30 and 87. There should now be continuity; if not, replace the relay.

5   Test the number 30 terminal (on the fuse/relay panel). There should be power at all times, probing with a grounded test light. The number 85 terminal should have power only with the key ON.

6   If the relay and the fan motor are good, check the wiring from the relays to the PCM (computer) for open or short circuits. Refer to the wiring schematics at the end of Chapter 12.

7   If the circuit checks OK but the fan still don't come on, check the engine coolant temperature sensor (see Chapter 6).

Replacement

Refer to illustrations 4.10, 4.11, 4.13a and 4.13b

Warning: *Keep hands, tools and clothing away from the fan. To avoid injury or damage DO NOT operate the engine with a damaged fan. Do not attempt to repair fan blades - replace a damaged fan with a new one.*

8   Disconnect the cable from the negative terminal of the battery. **Caution:** *On models equipped with the Theftlock audio system, be sure the lockout feature is turned off before performing any procedure which requires disconnecting the battery (see the front of this manual).*

9   Disconnect the electrical connector from the fan motor.

10   If your removing the right fan, detach the A/C line support bracket from the fan **(see illustration)**.

11   Remove the bolts and detach the fan(s) from the fan shroud **(see illustration)**.

12   Pull the fan assembly outward to dislodge it from the fan shroud, then guide the

fan assembly out from the engine compartment, making sure that all wiring clips are disconnected. Be careful not to contact the radiator cooling fins.

13   If the fan blades or the fan motor are damaged, they can be replaced by removing the fan blade from the fan motor, then removing the fan motor from the fan housing **(see illustrations)**. **Note:** *It is extremely important to replace the original fan motor with the correct part because different models have different wattage ratings and insufficient cooling could occur if the original type fan motor is not installed back into the system.*

14   Installation is the reverse of removal.

---

5   Radiator and coolant reservoir - removal and installation

---

Warning 1: *The air conditioning system is under high pressure. DO NOT loosen any fittings or remove any components until after the system has been discharged. Air condi-*

3

4.13a Remove the nut (arrow) securing the fan blade to the fan motor . . .

4.13b . . . then remove the motor retaining screws (arrows) and separate the fan motor from the fan housing

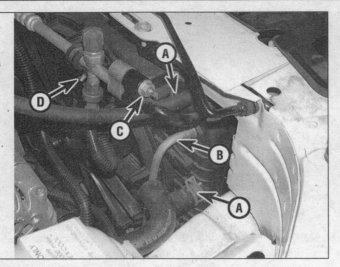

5.5 Remove the following components from the right side of the radiator

A   Coolant recovery hoses
B   Upper transaxle cooler line (if equipped)
C   Condenser inlet fitting
D   Discharge line support bracket

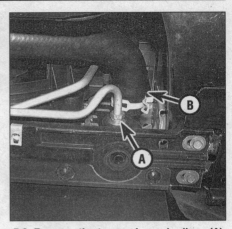

5.9 Remove the transaxle cooler lines (A) using a back-up wrench and a line wrench - to avoid damaging the radiator housing by twisting the cooler line if the fitting is tight or corroded - (B) is the location of the lower radiator hose

tioning refrigerant should be properly discharged into an EPA-approved container at a dealership service department or an automotive air conditioning repair facility. Always wear eye protection when disconnecting air conditioning system fittings.

Warning 2: The engine must be completely cool when this procedure is performed.

Radiator

Refer to illustrations 5.5, 5.9, 5.12a, 5.12b, 5.13 and 5.14

1   Have the air conditioning system discharged at dealer service department or other qualified repair facility (see **Warning** above).

2   Disconnect the cable from the negative terminal of the battery. **Caution:** On models equipped with the Theftlock audio system, be sure the lockout feature is turned off before performing any procedure which requires disconnecting the battery (see the front of this manual).

3   Drain the cooling system as described in Chapter 1. Refer to the coolant **Warning** in Section 2.

4   Remove the battery and the battery tray (see Chapter 5).

5   Disconnect the coolant recovery hoses, the upper automatic transaxle cooler line (if equipped), the condenser inlet fitting from the compressor discharge line and the discharge line support bracket from the right side of the radiator **(see illustration)**.

6   Disconnect the upper radiator hose from the radiator.

7   Raise the vehicle enough to allow removal of the radiator from below and support it securely on jackstands.

8   Remove the lower splash shield from the vehicle.

9   Remove the lower automatic transaxle cooler lines (if equipped) and the lower radiator hose **(see illustration)**. Be careful not to damage the lines or fittings. Plug the ends of the disconnected lines to prevent leakage and stop dirt from entering the system. Have a drip pan ready to catch any spills.

10   Unplug the cooling an electrical connectors.

11   Disconnect the evaporator line from the condenser outlet and the air conditioning lines from the radiator lower mounting panel.

5.12a Remove the following components from below the radiator

A   Evaporator line from condenser outlet line (if equipped)
B   Air conditioning lines (if equipped) from radiator support panel
C   Radiator support panel mounting bolts
D   Radiator support panel

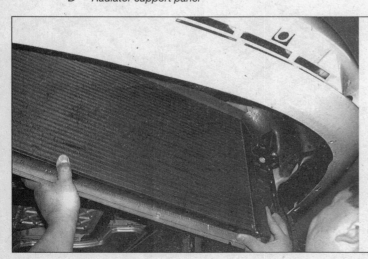

5.12b Carefully lower the radiator, condenser and cooling fans as an assembly from the vehicle

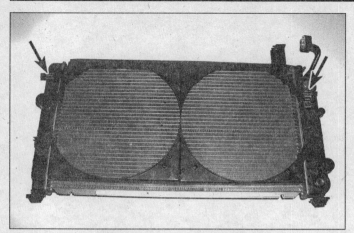

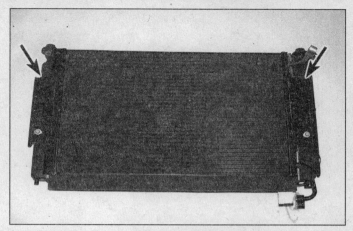

5.13 Use a flat bladed screwdriver to pry the fan shroud from the locking tabs (arrows) on the radiator

5.14 Remove the mounting screws (arrows) and separate the condenser from the radiator

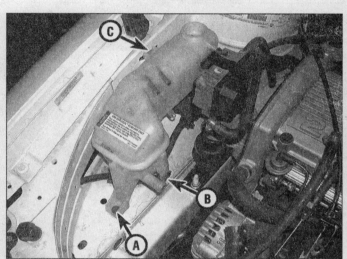

5.20 Remove the hose clamps and separate the hoses (arrows) from the coolant reservoir

5.21 Remove the reservoir mounting bolt (A) then lift the coolant reservoir from the notch on the inner fenderwell (B) and disconnect the coolant level sensor connector (C)

12   Detach the radiator lower mounting panel, then remove the radiator, condenser and cooling fans as an assembly from below the vehicle **(see illustrations)**.

13   Remove the cooling fans from the radiator, then pry the fan shroud from the plastic locking tabs on the radiator **(see illustration)**.

14   Remove the condenser mounting bolts and separate the condenser from the radiator **(see illustration)**.

15   Prior to installation of the radiator, replace any damaged hose clamps, radiator hoses and radiator mounts. If leaks have been noticed or there have been cooling problems, have the radiator cleaned and tested at a radiator shop.

16   Radiator installation is the reverse of removal. Be sure to seat the radiator securely in the upper mounts before installing the lower radiator support panel.

17   After installation, fill the system with the proper mixture of antifreeze, bleed the air from the cooling system as described in Chapter 1, and check the automatic transaxle fluid level. Also check the engine oil level, if equipped with an oil cooler (see Chapter 1).

Coolant reservoir

Refer to illustrations 5.20 and 5.21

18   Disconnect the cable from the negative terminal of the battery. **Caution:** *On models equipped with the Theftlock audio system, be sure the lockout feature is turned off before performing any procedure which requires disconnecting the battery (see the front of this manual).*

19   Drain the cooling system as described in Chapter 1 until the coolant reservoir is empty. Refer to the coolant **Warning** in Section 2.

20   Remove the coolant recovery hoses from the coolant reservoir **(see illustration)**.

21   Detach the reservoir mounting bolt, then lift the coolant reservoir from the notch on the inner fenderwell and disconnect the coolant level sensor connector **(see illustration)**.

22   Remove the coolant reservoir from the engine compartment.

23   Prior to installation make sure the reservoir is clean and free of debris which could be drawn into the radiator (wash it with soapy water and a brush if necessary, then rinse thoroughly).

24   Installation is the reverse of removal.

---

6   Water pump - check and replacement

Warning: *Wait until the engine is completely cool before starting this procedure.*

Check

Refer to illustration 6.2

1   Water pump failure can cause overheating and serious damage to the engine. There are three ways to check the operation of the water pump while it is installed on the engine. If any one of the following quick-checks indicates water pump problems, it should be replaced immediately.

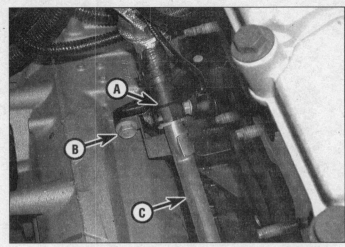

6.2 **The weep hole (arrow) is located on the top of the water pump (pump removed for clarity) (V6 engine shown)**

6.14 **Remove the quick connect fitting (A) and the heater outlet pipe bracket retaining bolt (B), then remove the heater outlet pipe (C) from the water pump cover**

2   A seal protects the water pump impeller shaft bearing from contamination by engine coolant. If this seal fails, a weep hole in the water pump snout will leak coolant **(see illustration)** (an inspection mirror can be used to look at the underside of the pump if the hole isn't on top). If the weep hole is leaking, shaft bearing failure will follow. Replace the water pump immediately.

3   The water pump impeller shaft bearing can also prematurely wear out. When the bearing wears out, it emits a high-pitched squealing sound. If such a noise is coming from the water pump during engine operation, the shaft bearing has failed - replace the water pump immediately. **Note:** *Do not confuse belt noise with bearing noise.*

4   To identify excessive bearing wear, grasp the water pump pulley (V6 engines only) and try to force it up-and-down or from side-to-side. If the pulley can be moved either horizontally or vertically, the bearing is nearing the end of its service life. Replace the water pump. Don't mistake drivebelt slippage, which causes a squealing sound, for water pump bearing failure.

5   It is possible for a water pump to be bad, even if it doesn't howl or leak water. Sometimes the fins on the back of the impeller can corrode away until the pump is no longer effective. The only way to check for this is to remove the pump for examination.

Replacement

6   Disconnect the cable from the negative terminal of the battery. **Caution:** *On models equipped with the Theftlock audio system, be sure the lockout feature is turned off before performing any procedure which requires disconnecting the battery (see the front of this manual).*

7   Drain the coolant (see Chapter 1).

8   Remove the serpentine drivebelt (see Chapter 1).

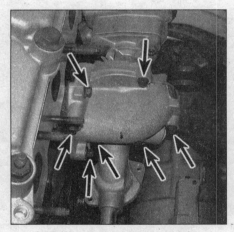

6.15 **Water pump cover retaining bolts (arrows)**

Four cylinder engine

Refer to illustrations 6.14, 6.15, 6.16a and 6.16b

Note: *Read the entire procedure before installing the water pump into the timing chain housing and the engine block to avoid an incorrect tightening sequence and leaky gasket mating surfaces.*

9   Raise the vehicle and support it securely on jackstands.

10   Remove the exhaust manifold from the engine (see Chapter 2A). **Caution:** *Do not rotate the flex coupling on the front exhaust pipe more than 4 degrees or damage to the flex coupling may occur.*

11   Remove the thermostat and the coolant inlet pipe from the water pump housing and separate the assembly from the engine (see Section 3).

12   Lower the vehicle. Remove the brake booster vacuum hose from the camshaft housing.

13   Remove the front timing chain cover and the timing chain tensioner from the engine

6.16a **Water pump retaining nuts (arrows indicate two of three retaining nuts)**

(see Chapter 2A, Section 6). **Caution:** *The timing chain tensioner must be removed to unload chain tension before removing the water pump from the housing otherwise the water pump will become jammed in the timing chain housing.*

14   Remove the heater outlet pipe from the water pump cover **(see illustration)**.

15   Remove the water pump cover bolts from the engine block and the water pump **(see illustration)**.

16   Remove the water pump retaining nuts and separate the water pump from the timing chain housing **(see illustrations)**.

17   Clean the sealing surfaces of all gasket material on the water pump, water pump cover and the timing chain housing. Wipe the mating surfaces with a rag saturated with lacquer thinner or acetone.

18   First, install the water pump cover and gasket to the water pump on the bench but leave the bolts finger tight. Apply a thin layer of RTV sealant to both sides of the new gasket and install the gasket on the water pump.

6.16b Pull the water pump from the timing chain housing

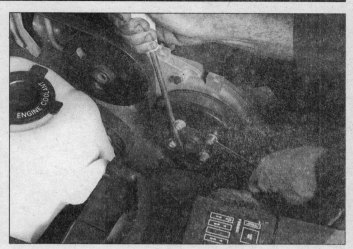

6.28 Use a pry bar or long screwdriver to hold the pulley in place while loosening the water pump pulley bolts

6.30 Water pump retaining bolts (arrows) - V6 engines

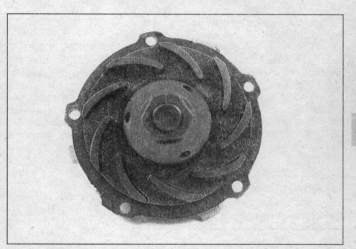

6.32 Inspect the impeller for damage and corrosion

This entire procedure must be performed in a timely manner before the RTV sealer sets up and dries.

19 Position the water pump and water pump cover assembly into the timing chain housing engaging the water pump sprocket in the timing chain, and install the nuts finger tight. Use caution to ensure that the gasket doesn't slip out of position. Leave the nuts loose to allow repositioning when the block bolts are installed in the next step.

20 Install the water pump cover to engine block bolts and tighten them by hand also.

21 Lubricate the O-ring on the heater outlet pipe with coolant and install it into the water pump cover. Hand tighten the bolts.

22 Now that the entire water pump housing assembly is installed and is fitted against the engine block, follow the tightening sequence to insure proper seating.

a) *Tighten the water pump-to-timing chain housing nuts to the torque listed in this Chapter's Specifications.*

b) *Tighten the water pump cover-to-water pump housing bolts to the torque listed in this Chapter's Specifications.*

c) *Tighten the water pump cover-to-engine block bolts to the torque listed in this Chapter's Specifications. Tighten the bottom bolts first then the top bolt.*

d) *Tighten the radiator outlet pipe-to-water pump housing to the torque listed in this Chapter's Specifications.*

23 Install the timing chain tensioner and cover (see Chapter 2A).

24 Install the exhaust manifold (see Chapter 2A).

25 Install the exhaust pipe to the exhaust manifold.

26 The remainder of the installation procedure is the reverse of removal. Add coolant to the specified level (see Chapter 1). Start the engine and check for the proper coolant level and the water pump and hoses for leaks. Bleed the cooling system of air as described in Chapter 1.

V6 engines
Refer to illustrations 6.28, 6.30 and 6.32

27 Refer to Section 5 and remove the coolant recovery tank.

28 Loosen the bolts on the water pump pulley, using a prybar to hold it **(see illustration)**.

29 Remove the small belt guard just above the water pump (if equipped).

30 Remove the bolts/nuts and detach the water pump from the engine **(see illustration)**. **Note:** *Most models have a locating mark at the top of the pump. If yours doesn't have one, make a mark yourself for help in orienting the pump during reassembly (if the same pump is to be reinstalled).*

31 Clean the fastener threads and any threaded holes in the engine to remove corrosion and sealant.

32 Compare the new pump to the old one to make sure they're identical. If using the existing pump, inspect the back of the pump for a broken or corroded impeller **(see illustration)**.

33 Remove all traces of old gasket material from the engine with a gasket scraper.

34 Clean the engine and water pump mating surfaces with lacquer thinner or acetone.

35 Carefully attach the pump and new gasket to the engine and start the bolts/nuts fin-

3

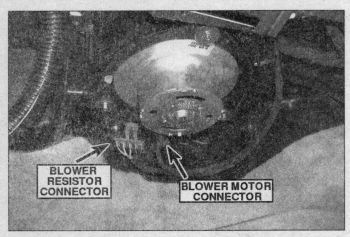

8.4 Connect a voltmeter to the heater blower motor connector by backprobing the connector using pins, and check the running voltage at each blower switch position

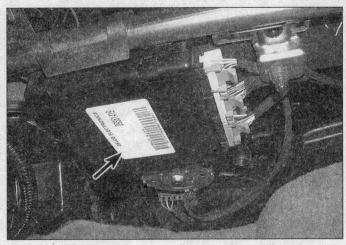

9.2 Slide the BCM (arrow) to the right, off the retaining bracket and set it aside

ger tight. Make sure the alignment mark is at the top.

36   Tighten the fasteners in 1/4-turn increments to the torque figure listed in this Chapter's Specifications. Don't overtighten them or the pump may be distorted.

37   Reinstall all parts removed for access to the pump.

38   Refill and bleed the cooling system (see Chapter 1). Run the engine and check for leaks.

7   Coolant temperature gauge sending unit - check and replacement

Check

1   The coolant temperature indicator system is composed of a temperature gauge or warning light mounted in the dash and a coolant temperature sensor mounted on the engine. This coolant temperature sensor doubles as an information sensor for the fuel and emissions systems (see Chapter 6) and as a sending unit for the temperature gauge.

2   If an overheating indication occurs, check the coolant level in the system and then make sure the wiring between the gauge and the sending unit is secure and all fuses are intact.

3   Check the operation of the coolant temperature sensor (see Chapter 6). If the sensor is defective, replace it with a new part of the same specification.

4   If the coolant temperature sensor is good, have the temperature gauge checked by a dealer service department. This test will require a scan tool to access the information as it is processed by the On Board computer.

Replacement

5   Refer to Chapter 6 for the engine coolant temperature sensor replacement procedure.

8   Blower motor and circuit - check

Refer to illustration 8.4

Warning: *The models covered by this manual are equipped with Supplemental Restraint Systems (SRS), more commonly known as airbags. Always disable the airbag system before working in the vicinity of any airbag system components to avoid the possibility of accidental deployment of the airbags, which could cause personal injury (see Chapter 12).*

1   Check the fuse (marked HVAC) and all connections in the circuit for looseness and corrosion. Make sure the battery is fully charged. **Note:** *The heater/blower relay is located in the fuse/relay center in the engine compartment and the HVAC fuse is located in the fuse panel under the dash.*

2   With the transaxle in Park or Neutral on manual transaxle equipped vehicles, set the parking brake securely, turn the ignition switch to the Run position. It isn't necessary to start the vehicle.

3   Remove the lower right dash insulator panel (below the glove box) for access to the blower motor.

4   Backprobe the blower motor electrical connector with two straight pins and connect a voltmeter to the blower motor connector and ground **(see illustration)**.

5   Move the blower switch through each of its positions and note the voltage readings. Changes in voltage indicate that the motor speeds will also vary as the switch is moved to the different positions.

6   If there is voltage present, but the blower motor does not operate, the blower motor is probably faulty. Disconnect the blower motor connector, then hook one side of the blower motor terminals to a chassis ground and the other to a fused source of battery voltage. If the blower doesn't operate, it is faulty.

7   If there was no voltage present at the blower motor at one or more speeds, and the motor itself tested OK, check the blower

motor resistor.

8   Disconnect the electrical connector from the blower motor resistor **(see illustration 8.4)**. With the ignition On, check for voltage at each of the terminals in the connector as the blower speed switch is moved to the different positions. If the voltmeter responds correctly to the switch and the blower is known to be good then the resistor is probably faulty. If there is no voltage present from the switch, then the switch, control panel or related wiring is probably faulty.

9   Test the blower motor relay for battery voltage and for correct relay operation. Also, check the blower fuse. The blower motor relay is identical to the cooling fan relay - follow the relay checks in Section 4, Steps 2 through 6. The blower relay is located in the fuse/relay center in the engine compartment. The fuse is located in the fuse/relay center on the right side of the instrument panel in the passenger compartment.

10   Follow the blower motor ground wire from the motor to the chassis and check the ground terminal for continuity to ground against the chassis metal.

9   Blower motor - removal and installation

Refer to illustrations 9.2, 9.3 and 9.4

Warning: *The models covered by this manual are equipped with Supplemental Restraint Systems (SRS), more commonly known as airbags. Always disable the airbag system before working in the vicinity of any airbag system components to avoid the possibility of accidental deployment of the airbags, which could cause personal injury (see Chapter 12).*

1   Remove the lower right dash insulator panel from below the glove box (see Chapter 11 if necessary).

2   Slide the Body Control Module (BCM) aside for access to the blower motor **(see illustration)**.

3   Remove the three screws from the

9.3 Remove the screws (arrows) retaining the blower motor to the housing

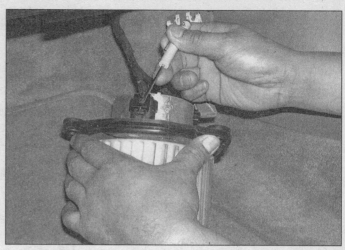

9.4 Lower the blower motor and fan assembly straight down and disconnect the electrical connector to remove it from the vehicle

10.3a Remove the screws (arrows) retaining the heater/air conditioning control assembly to the instrument panel

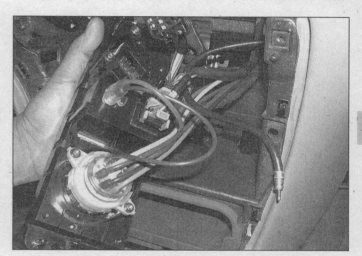

10.3b Pull the control assembly outward and disconnect the electrical connectors . . .

3

blower housing **(see illustration)**.

4   Pull the blower motor and fan straight down and disconnect the electrical connector from the blower motor **(see illustration)**.

5   If your replacing the blower motor with a new one, remove the fan from the blower motor.

6   Install the fan onto the new motor and install the blower motor into the heater housing.

10   Heater and air conditioning control assembly - removal and installation

Warning: *The models covered by this manual are equipped with Supplemental Restraint Systems (SRS), more commonly known as airbags. Always disable the airbag system before working in the vicinity of any airbag system components to avoid the possibility of accidental deployment of the airbags, which could cause personal injury (see Chapter 12).*

Removal

Refer to illustrations 10.3a, 10.3b and 10.3c

1   Disconnect the cable from the negative terminal of the battery. **Caution:** *On models equipped with the Theftlock audio system, be sure the lockout feature is turned off before performing any procedure which requires disconnecting the battery (see the front of this manual).*

2   Remove the main instrument panel bezel to allow access to the heater/air conditioning control mounting screws (see Chapter 11).

3   Remove the control assembly retaining screws and pull the unit from the dash. It can be pulled out just far enough to allow disconnecting the electrical connections and vacuum harness from the control head. Use a small screwdriver to release the clips and the vacuum harness **(see illustrations)**.

Installation

4   To install the control assembly, reverse

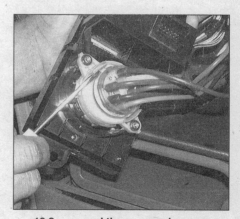

10.3c . . . and the vacuum harness

the removal procedure. **Caution:** *When reconnecting vacuum harness to the control assembly, do not use any lubricant to make them slip on easier; it can affect vacuum operation. If necessary, use a drop of plain water to make reconnection easier.*

11.3 Disconnect the heater core hoses (arrows) at the engine compartment firewall

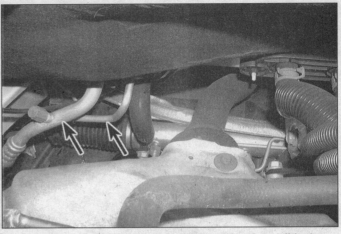

11.7 On Malibu and Cutlass models have the air conditioning system discharged by a dealership service department or an automotive air conditioning facility, then remove the air conditioning lines (arrows) from the evaporator core fittings at the firewall

11.8a Remove the heater core cover screws (arrows) - heater/air conditioning unit remove for clarity

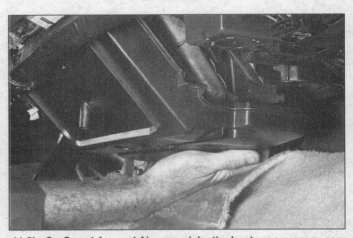

11.8b On Grand Am and Alero models, the heater core cover can be removed from inside the vehicle by gently separating it from the heater core housing

11   Heater core - removal and installation

Warning 1: *The models covered by this manual are equipped with Supplemental Restraint Systems (SRS), more commonly known as airbags. Always disable the airbag system before working in the vicinity of any airbag system components to avoid the possibility of accidental deployment of the airbags, which could cause personal injury (see Chapter 12).*
Warning 2: *The air conditioning system is under high pressure. DO NOT loosen any fittings or remove any components until after the system has been discharged. Air conditioning refrigerant should be properly discharged into an EPA-approved container at a dealership service department or an automotive air conditioning facility. Always wear eye protection when disconnecting air conditioning system fittings.*

Removal

Refer to illustrations 11.3, 11.7, 11.8a, 11.8b and 11.9

1   Disconnect the cable at the negative terminal of the battery. **Caution:** *On models equipped with the Theftlock audio system, be sure the lockout feature is turned off before performing any procedure which requires disconnecting the battery (see the front of this manual).*
2   Drain the cooling system (see Chapter 1).
3   Disconnect the heater hoses at the heater core inlet and outlet on the engine side of the firewall (passenger side) **(see illustration)** and plug the open fittings. If the hoses are stuck to the pipes, cut them off and replace them with new ones upon installation.
4   From the inside of the car, remove the lower right sound insulator panel from below the glove box, the lower left sound insulator

panel and the knee bolster from below the steering column. Then remove the center console (if equipped) and the glove box (see Chapter 11 if necessary).
5   Disconnect the drain tube from the heater core cover.
6   Remove the floor air duct from below the heater case.
7   On Malibu and Cutlass models have the air conditioning system discharged by a dealership service department or an automotive air conditioning facility (see **Warning** above) then remove the air conditioning lines from the evaporator core fittings at the firewall **(see illustration)**. It will also be necessary to remove the instrument panel from the passenger compartment. Refer to Chapter 11 and read through the entire instrument panel removal procedure before attempting to remove it. The instrument panel removal procedure is quite lengthy and can be particularly difficult for a beginner. Once the instrument panel cross beam is removed from the

11.9 Detach the heater core clamp (arrow) and remove the heater core from the housing

12.1 Check that the evaporator housing drain tube (arrow) at the firewall is clear of any blockage - the view here is from above the engine looking down

vehicle, the heating /air conditioning unit can be removed from the vehicle and set on a workbench. Then follow the remaining steps to remove the heater core from the heating /air conditioning unit. **Note:** *On Grand Am and Alero models, heater core removal does not require removal of the instrument panel or disconnecting the air conditioning lines. On these models there is sufficient space between the heater core cover and the floor of the vehicle to allow access to heater core cover screws.*

8 Remove the heater core cover **(see illustrations)**.

9 Remove the heater core clamp screw and clamp **(see illustration)**, then slide the heater core out carefully.

Installation

10 Installation is the reverse of removal. **Note:** *When reinstalling the heater core, make sure any original insulating/sealing materials are in place around the heater core pipes and around the core.*

11 Refill the cooling system (see Chapter 1).

12 Start the engine and check for proper operation.

---

12 Air conditioning and heating system - check and maintenance

Air conditioning system

Refer to illustration 12.1

Warning: *The air conditioning system is under high pressure. Do not loosen any hose fittings or remove any components until after the system has been discharged. Air conditioning refrigerant should be properly discharged into an EPA-approved recovery/recycling unit at a dealer service department or an automotive air conditioning repair facility. Always wear eye protection when disconnecting air conditioning system fittings.*

Caution 1: *All models covered by this manual use environmentally friendly R-134a. This refrigerant (and its appropriate refrigerant oils) are not compatible R-12 refrigerant system components and must never be mixed or the components will be damaged.*

Caution 2: *When replacing entire components, additional refrigerant oil should be added equal to the amount that is removed with the component being replaced. Be sure to read the can before adding any oil to the system, to make sure it is compatible with the R-134a system.*

1 The following maintenance checks should be performed on a regular basis to ensure that the air conditioning continues to operate at peak efficiency.

  a) *Inspect the condition of the compressor drivebelt. If it is worn or deteriorated, replace it (see Chapter 1).*

  b) *Check the drivebelt tension and, if necessary, adjust it (see Chapter 1).*

  c) *Inspect the system hoses. Look for cracks, bubbles, hardening and deterioration. Inspect the hoses and all fittings for oil bubbles or seepage. If there is any evidence of wear, damage or leakage, replace the hose(s).*

  d) *Inspect the condenser fins for leaves, bugs and any other foreign material that may have embedded itself in the fins. Use a "fin comb" or compressed air to remove debris from the condenser.*

  e) *Make sure the system has the correct refrigerant charge.*

  f) *If you hear water sloshing around in the dash area or have water dripping on the carpet, check the evaporator housing drain tube* **(see illustration)** *and insert a piece of wire into the opening to check for blockage.*

2 It's a good idea to operate the system for about ten minutes at least once a month. This is particularly important during the winter months because long term non-use can cause hardening, and subsequent failure, of

the seals. Note that using the Defrost function operates the compressor.

3 If the air conditioning system is not working properly, proceed to Step 6 and perform the general checks outlined below.

4 Because of the complexity of the air conditioning system and the special equipment necessary to service it, in-depth troubleshooting and repairs, beyond checking the refrigerant charge and the compressor clutch operation, are not included in this manual. However, simple checks and component replacement procedures are provided in this Chapter. For more complete information on the air conditioning system, refer to the *Haynes Automotive Heating and Air Conditioning Manual.*

5 The most common cause of poor cooling is simply a low system refrigerant charge. If a noticeable drop in system cooling ability occurs, one of the following quick checks will help you determine whether the refrigerant level is low. Should the system lose its cooling ability, the following procedure will help you pinpoint the cause.

Check

Refer to illustration 12.9

6 Warm the engine up to normal operating temperature.

7 Place the air conditioning temperature selector at the coldest setting and put the blower at the highest setting. Open the doors (to make sure the air conditioning system doesn't cycle off as soon as it cools the passenger compartment).

8 After the system reaches operating temperature, feel the two pipes connected to the evaporator at the firewall.

9 The pipe (thinner tubing) leading from the condenser outlet to the evaporator should be cold, and the evaporator outlet line (the thicker tubing that leads back to the compressor) should be slightly colder (3 to 10 degrees F colder). If the evaporator outlet is considerably warmer than the inlet, the system needs a charge. Insert a thermometer in

3

12.9 Insert a thermometer in the center duct while operating the air conditioning system - the output air should be 35-40 degrees F less than the ambient temperature, depending on humidity (but not lower than 40-degrees F)

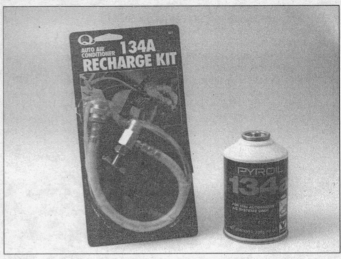

12.18 A basic charging kit for 134a systems is available at most auto parts stores - it must say 134a (not R-12) and so should the can of refrigerant

the center air distribution duct **(see illustration)** while operating the air conditioning system at its maximum setting - the temperature of the output air should be 35 to 40 degrees F below the ambient air temperature (down to approximately 40 degrees F). If the ambient (outside) air temperature is very high, say 110 degrees F, the duct air temperature may be as high as 60 degrees F, but generally the air conditioning is 35 to 40 degrees F cooler than the ambient air.

10   If the air isn't as cold as it used to be, the system probably needs a charge.

11   If the air is warm and the system doesn't seem to be operating properly, check the operation of the compressor clutch.

12   Have an assistant switch the air conditioning On while you observe the front of the compressor. The clutch will make an audible click and the center of the clutch should rotate.

13   If the clutch does not operate, check the appropriate fuses in the interior fuse panel.

14   Remove the compressor clutch (AC) relay from the engine compartment relay panel and test it (see Chapter 12). With the relay out and the ignition On, check for battery power at two of the relay terminal sockets (refer to the wiring diagrams for wire color designations to determine which terminals to check). There should be battery power with the key On, at the terminals for the relay control and power circuits.

15   Using a jumper wire, connect the terminals in the relay box from the relay power circuit to the terminal that leads to the compressor clutch (refer to the wiring diagrams for wire color designations to determine which terminals to connect). Listen for the clutch to click as you make the connection. If the clutch doesn't respond, disconnect the clutch connector at the compressor and check for battery voltage at the compressor clutch connector. Check for continuity to ground on the black wire terminal of the com-

pressor clutch connector. If power and ground are available and the clutch doesn't operate when connected, the compressor clutch is defective.

16   If the compressor clutch, relay and related circuits are good and the system is fully charged with refrigerant and the compressor does not operate under normal conditions, have the PCM and related circuits checked by a dealer service department or other properly equipped repair facility.

17   Further inspection or testing of the system is beyond the scope of the home mechanic and should be left to a professional.

Adding refrigerant

Refer to illustrations 12.18 and 12.21

Caution: *Make sure any refrigerant, refrigerant oil or replacement component your purchase is designated as compatible with environmentally friendly R-134a systems.*

18   Purchase an R-134a automotive charging kit at an auto parts store **(see illustration)**. A charging kit includes a 12-ounce can

of refrigerant, a tap valve and a short section of hose that can be attached between the tap valve and the system low side service valve. Because one can of refrigerant may not be sufficient to bring the system charge up to the proper level, it's a good idea to buy an additional can. **Warning:** *Never add more than two cans of refrigerant to the system.*

19   Hook up the charging kit by following the manufacturer's instructions. **Warning:** *DO NOT hook the charging kit hose to the system high side!* The fittings on the charging kit are designed to fit **only** on the low side of the system.

20   Back off the valve handle on the charging kit and screw the kit onto the refrigerant can, making sure first that the O-ring or rubber seal inside the threaded portion of the kit is in place. **Warning:** *Wear protective eyewear when dealing with pressurized refrigerant cans.*

21   Remove the dust cap from the low-side charging and attach the quick-connect fitting on the kit hose **(see illustration)**.

22   Warm up the engine and turn On the air

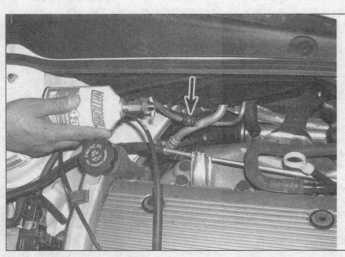

12.21 Attach the refrigerant kit to the low-side charging port (arrow) - it's near the right shock tower - the cap should be marked with an "L"

conditioning. Keep the charging kit hose away from the fan and other moving parts. **Note:** *The charging process requires the compressor to be running. If the clutch cycles off, you can put the air conditioning switch on High and leave the car doors open to keep the clutch on and compressor working.*

23. Turn the valve handle on the kit until the stem pierces the can, then back the handle out to release the refrigerant. You should be able to hear the rush of gas. Add refrigerant to the low side of the system, keeping the can upright at all times, but shaking it occasionally. Allow stabilization time between each addition. **Note:** *The charging process will go faster if you wrap the can with a hot-water-soaked shop rag to keep the can from freezing up.*

24. If you have an accurate thermometer, you can place it in the center air conditioning duct inside the vehicle and keep track of the output air temperature **(see illustration 12.9).** A charged system that is working properly should cool down to approximately 40-degrees F. If the ambient (outside) air temperature is very high, say 110 degrees F, the duct air temperature may be as high as 60 degrees F, but generally the air conditioning is 30-40 degrees F cooler than the ambient air.

25. When the can is empty, turn the valve handle to the closed position and release the connection from the low-side port. Replace the dust cap.

26. Remove the charging kit from the can and store the kit for future use with the piercing valve in the UP position, to prevent inadvertently piercing the can on the next use.

Heating systems

27. If the carpet under the heater core is damp, or if antifreeze vapor or steam is coming through the vents, the heater core is leaking. Remove it (see Section 12) and install a new unit (most radiator shops will not repair a leaking heater core).

28. If the air coming out of the heater vents isn't hot, the problem could stem from any of the following causes:

a) The thermostat is stuck open, preventing the engine coolant from warming up enough to carry heat to the heater core. Replace the thermostat (see Section 3).

b) There is a blockage in the system, preventing the flow of coolant through the heater core. Feel both heater hoses at the firewall. They should be hot. If one of them is cold, there is an obstruction in one of the hoses or in the heater core, or the heater control valve is shut. Detach the hoses and back flush the heater core with a water hose. If the heater core is clear but circulation is impeded, remove the two hoses and flush them out with a water hose.

c) If flushing fails to remove the blockage from the heater core, the core must be replaced (see Section 11).

Eliminating air conditioning odors

Refer to illustration 12.32

29. Unpleasant odors that often develop in air conditioning systems are caused by the growth of a fungus, usually on the surface of the evaporator core. The warm, humid environment there is a perfect breeding ground for mildew to develop.

30. The evaporator core on most vehicles is difficult to access, and factory dealerships have a lengthy, expensive process for eliminating the fungus by opening up the evaporator case and using a powerful disinfectant and rinse on the core until the fungus is gone. You can service your own system at home, but it takes something much stronger than basic household germ-killers or deodorizers.

31. Aerosol disinfectants for automotive air conditioning systems are available in most auto parts stores, but remember when shopping for them that the most effective treatments are also the most expensive. The basic procedure for using these sprays is to start by running the system in the RECIRC mode for ten minutes with the blower on its highest speed. Use the highest heat mode to dry out the system and keep the compressor from

engaging by disconnecting the wiring connector at the compressor (see Section 14).

32. The disinfectant can usually comes with a long spray hose. Remove the blower motor resistor (see Section 8), point the nozzle inside the hole and to the left towards the evaporator core, and spray according to the manufacturer's recommendations **(see illustration).** Try to cover the whole surface of the evaporator core, by aiming the spray up, down and sideways. Follow the manufacturer's recommendations for the length of spray and waiting time between applications.

33. Once the evaporator has been cleaned, the best way to prevent the mildew from coming back again is to make sure your evaporator housing drain tube is clear **(see illustration 12.1).**

Automatic heating and air conditioning systems

34. Some models are equipped with an optional automatic climate control system. This system has its own computer that receives inputs from various sensors in the heating and air conditioning system. This computer, like the PCM, has self diagnostic capabilities to help pinpoint problems or faults within the system. Vehicles equipped with automatic heating and air conditioning systems are very complex and considered beyond the scope of the home mechanic. Vehicles equipped with automatic heating and air conditioning systems should be taken to a dealer service department or other qualified facility for repair.

---

13 Air conditioning accumulator/drier - removal and installation

Removal

Refer to illustrations 13.4 and 13.5

Warning: *The air conditioning system is under high pressure. DO NOT loosen any fittings or remove any components until after the system has been discharged. Air conditioning refrigerant should be properly discharged into an EPA-approved container at a dealership service department or an automotive air conditioning repair facility. Always wear eye protection when disconnecting air conditioning system fittings.*

1. Have the air conditioning system discharged (see **Warning** above). Disconnect the cable from the negative terminal of the battery. **Caution:** *On models equipped with the Theftlock audio system, be sure the lock-out feature is turned off before performing any procedure which requires disconnecting the battery (see the front of this manual).*

2. Loosen right front wheel lug nuts, then raise the vehicle and support it securely on jackstands.

3. Remove the right front wheel and the inner fenderwell splash shield.

4. Disconnect the refrigerant inlet and out-

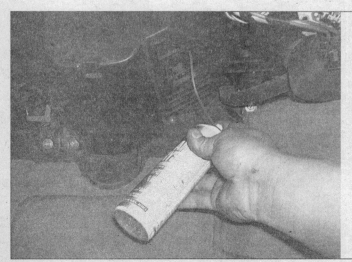

12.32 With the blower motor resistor removed, spray the disinfectant at the evaporator core

13.4 Disconnect the inlet and outlet lines - view here is from the inner fenderwell

13.5 Remove the mounting bracket bolt (arrow) - view here is through the headlight housing opening

14.7 Disconnect the electrical connector (A) and the retaining bolt (B) securing the refrigerant lines to the back of the compressor

14.8a Compressor mounting bolts (arrows) - four cylinder engine

let lines **(see illustration)**, using back-up wrenches. Cap or plug the open lines immediately to prevent the entry of dirt or moisture.

5   Loosen the clamp bolt **(see illustration)** on the mounting bracket and slide the accumulator/drier assembly up and out of the compartment. **Note:** *It may also be helpful to remove the right front headlight housing (see Chapter 12) to allow easier access to the mounting clamp bolt.*

Installation

6   If you are replacing the accumulator/drier with a new one, add one ounce of fresh refrigerant oil to the new unit (oil must be R-134a compatible).

7   Place the new accumulator/drier into position in the bracket.

8   Install the inlet and outlet lines, using clean refrigerant oil on the new O-rings. Tighten the mounting bolt securely.

9   Connect the cable to the negative terminal of the battery.

10   Have the system evacuated, recharged and leak tested by a dealership service department or an automotive air conditioning repair facility.

14 Air conditioning compressor - removal and installation

Note: *Whenever the compressor is replaced because of internal damage, the expansion (orifice) tube should also be replaced (see Section 16).*

Removal

Refer to illustrations 14.7, 14.8a and 14.8b

Warning: *The air conditioning system is under high pressure. DO NOT loosen any fittings or remove any components until after the system has been discharged. Air conditioning refrigerant should be properly discharged into an EPA-approved container at a dealership service department or an automotive air conditioning repair facility. Always wear eye protection when disconnecting air conditioning system fittings.*

Note: *The accumulator/drier (see Section 13) should be replaced whenever the compressor is replaced.*

1   Have the air conditioning system discharged (see **Warning** above). Disconnect

the cable from the negative terminal of the battery. **Caution:** *On models equipped with the Theftlock audio system, be sure the lockout feature is turned off before performing any procedure which requires disconnecting the battery (see the front of this manual).*

2   Clean the compressor thoroughly around the refrigerant line fittings.

3   Remove the serpentine drivebelt (see Chapter 1).

4   Raise the vehicle and support it securely on jackstands.

5   Remove the splash shield from below the engine (if equipped).

6   Disconnect the electrical connector from the air conditioning compressor.

7   Disconnect the suction and discharge lines from the compressor. Both lines are mounted to the back of the compressor with a plate secured by one bolt. Plug the open fittings to prevent the entry of dirt and moisture, and discard the seals between the plate and compressor **(see illustration)**.

8   Remove the compressor mounting bolts **(see illustrations)**. Detach the compressor from the mounting bracket and remove the compressor from the engine compartment.

14.8b Compressor mounting bolts (arrows) - V6 engines

16.3 Working from under the vehicle, disconnect the high pressure line from the bottom of the condenser (arrow)

Installation

9   If a new compressor is being installed, pour the oil from the old compressor into a graduated container and add that exact amount of new refrigerant oil to the new compressor. Also follow any directions included with the new compressor. **Note:** *Some replacement compressors come with refrigerant oil in them. Follow the directions with the compressor regarding the draining of excess oil prior to installation.* **Caution:** *The oil used must be labeled as compatible with R-134a refrigerant systems.*

10   Installation is the reverse of the disassembly. When installing the line fitting bolt to the compressor, use new seals lubricated with clean refrigerant oil, and tighten the bolt securely.

11   Reconnect the cable to the negative terminal of the battery.

12   Have the system evacuated, recharged and leak tested by a dealership service department or an automotive air conditioning repair facility.

15 Air conditioning condenser - removal and installation

This procedure is essentially the same procedure as removal of the radiator. Since the two components must be removed together as a unit, refer to Section 5 and follow the procedure outlined there.

16 Air conditioning expansion (orifice) tube - removal and installation

Refer to illustrations 16.3, 16.4 and 16.5

Warning : *The air conditioning system is under high pressure. DO NOT loosen any fittings or remove any components until after the system has been discharged. Air conditioning refrigerant should be properly discharged into an EPA-approved container at a dealership service department or an automotive air conditioning repair facility. Always wear eye protection when disconnecting air conditioning system fittings.*

1   Have the air conditioning system discharged and the refrigerant recovered (see **Warning** above). Disconnect the cable from the negative terminal of the battery. **Caution:** *On models equipped with the Theftlock audio system, be sure the lockout feature is turned off before performing any procedure which* requires disconnecting the battery (see the front of this manual).

2   Raise the front of the vehicle and support it securely on jackstands.

3   Disconnect the refrigerant high-pressure line at the fitting at the bottom of the condenser **(see illustration)**.

4   The expansion tube is a tube with a fixed-diameter orifice and a mesh filter at each end. When you separate the pipe at the fitting you will see one end of the orifice tube inside the pipe leading to the evaporator. Use needle-nose pliers to remove the orifice tube **(see illustration)**.

5   The orifice tube acts to meter the refrigerant, changing it from high-pressure liquid to low-pressure liquid. It is possible to reuse the orifice tube if **(see illustration)**:

a) *The screens aren't plugged with grit or foreign material*
b) *Neither screen is torn*
c) *The plastic housing over the screens is intact*
d) *The brass orifice inside the plastic housing is unrestricted*

6   Installation is the reverse of removal. Be sure to insert the expansion tube with the shorter end in first, toward the evaporator. **Caution:** *Always use a new O-ring when installing the expansion (orifice) tube.*

7   Retighten the fitting and refrigerant line, then have the system evacuated, recharged and leak-tested by the shop that discharged it.

16.4 Carefully remove the expansion tube using needle-nose pliers

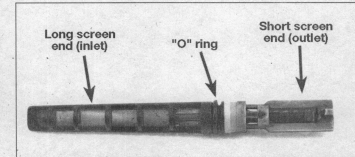

Long screen end (inlet)   "O" ring   Short screen end (outlet)

16.5 The expansion tube is equipped with a tapered mesh screen that must be cleaned and not have any holes or damage

3

Notes

Chapter 4
Fuel and exhaust systems

Contents

Specifications

General

Fuel pressure (key On, engine Off)
Four-cylinder engine
1997 and 1998 .... 41 to 47 psi
1999 and 2000
With EGR .... 41 to 47 psi
Without EGR .... 52 to 58 psi
V6 engines .... 41 to 47 psi
Fuel injector resistance
Four-cylinder engine
1997 and 1998 .... 1.95 to 2.3 ohms
1999 and 2000
With EGR .... 1.95 to 2.3 ohms
Without EGR .... 11.4 to 12.6 ohms
V6 engines .... 11.4 to 12.6 ohms

Torque specifications   **Ft-lbs** (unless otherwise indicated)

Throttle body mounting bolts
Four-cylinder engine .... 58 in-lbs
V6 engines 21
Fuel rail mounting bolts
Four-cylinder engine 18
V6 engines .... 89 in-lbs
Fuel tank mounting strap bolts 26

1   General information

Refer to illustrations 1.1a and 1.1b
Warning: *Gasoline is extremely flammable, so take extra precautions when you work on any part of the fuel system. Don't smoke or allow open flames or bare light bulbs near the* work area, and don't work in a garage where a natural gas-type appliance (such as a water heater or a clothes dryer) is present. Since gasoline is carcinogenic, wear latex gloves when there's a possibility of being exposed to fuel, and, if you spill any fuel on your skin, rinse it off immediately with soap and water. Mop up any spills immediately and do not *store fuel-soaked rags where they could ignite. The fuel system is under constant pressure, so, if any fuel lines are to be disconnected, the fuel pressure in the system must be relieved first. When you perform any kind of work on the fuel system, wear safety glasses and have a Class B type fire extinguisher on hand.*

All models covered by this manual are equipped with a sequential Multi Port Fuel Injection (MPFI) system **(see illustrations)**. This system uses timed impulses to sequentially inject the fuel directly into the intake ports of each cylinder. The injectors are controlled by the Powertrain Control Module (PCM). The PCM monitors various engine parameters and delivers the exact amount of fuel, in the correct sequence, into the intake ports.

All models are equipped with an electric fuel pump, mounted in the fuel tank. It is necessary to remove the fuel tank for access to the fuel pump. The fuel level sending unit is an integral component of the fuel pump module and it must be removed from the fuel tank in the same manner.

The exhaust system consists of exhaust

1.1a Typical fuel system components - four-cylinder engine

1	*Fuel pressure regulator*	*4*	*Fuel feed and return lines*	*6*	*Air filter housing*
2	*Accelerator cable*	*5*	*Fuel pump relay (inside Power*	*7*	*Throttle body*
3	*Fuel rail and injectors*		*Distribution Center)*		

1.1b Typical fuel system components - V6 engine

1	*Throttle body*	*4*	*Accelerator cable*
2	*Fuel pump relay (inside Power Distribution Center)*	*5*	*Fuel pressure regulator*
3	*Air filter housing*	*6*	*Fuel rail and injectors (under upper intake manifold plenum)*

manifolds, a catalytic converter, an exhaust pipe and a muffler. Each of these components is replaceable. For further information regarding the catalytic converter, refer to Chapter 6.

2   Fuel pressure relief procedure

Refer to illustration 2.4
Warning: *See the* **Warning** *in Section 1.*
Note: *After the fuel pressure has been relieved, it's a good idea to lay a shop towel over any fuel connection to be disassembled, to absorb the residual fuel that may leak out when servicing the fuel system.*

1   Before servicing any fuel system component, you must relieve the fuel pressure to minimize the risk of fire or personal injury.
2   Remove the fuel filler cap - this will relieve any pressure built up in the tank.

Four-cylinder models

3   Raise the vehicle and secure it on jackstands.
4   Disconnect the fuel pump electrical connector located near the fuel tank **(see illustration)**.
5   Attempt to start the engine. The engine should immediately stall. Continue to crank the engine for approximately three seconds.
6   Disconnect the cable from the negative battery terminal. **Caution:** *On models equipped with the Theftlock audio system, be sure the lockout feature is turned off before performing any procedure which requires disconnecting the battery (see the front of this manual).*
7   Place shop towels around the fuel fitting to be disconnected to absorb any residual fuel that may spill out.

V6 models

8   Disconnect the cable from the negative

2.4 Locate and disconnect the fuel pump electrical connector (arrow)

battery terminal. **Caution:** *On models equipped with the Theftlock audio system, be sure the lockout feature is turned off before performing any procedure which requires disconnecting the battery (see the front of this manual).*
9   Locate the test port on the fuel rail (see Section 3). Remove the cap and connect a fuel pressure gauge, equipped with a bleed-off valve and drain tube, to the test port. Relieve the fuel pressure by bleeding the fuel through the bleed-off valve and into an approved fuel container.
10   Place shop towels around the fuel fitting to be disconnected to absorb any residual fuel that may spill out.

3   Fuel pump/fuel pressure - check

Warning: *See the* **Warning** *in Section 1.*

Preliminary check

1   If you suspect insufficient fuel delivery, first inspect all fuel lines to ensure that the problem is not simply a leak in a line.

2   Set the parking brake and have an assistant turn the ignition switch to the ON position while you listen to the fuel pump (inside the fuel tank). You should hear a "whirring" sound, lasting for a couple of seconds indicating the fuel pump is operating. If the fuel pump is operating, proceed to the pressure check.
3   If there is no sound, check the fuel pump circuit, referring to Chapter 12 and the wiring diagrams. Check the related fuses, the fuel pump relay and the related wiring to ensure power is reaching the fuel pump connector. Check the ground circuit for continuity.
4   If the power and ground circuits are good and the fuel pump does not operate, replace the fuel pump (see Section 7).

Pressure check

Refer to illustrations 3.6a and 3.6b
Note: *In order to perform the fuel pressure test, you will need a fuel pressure gauge capable of measuring high fuel pressure. The fuel gauge must be equipped with the proper fittings or adapters required to attach it to the fuel line or fuel rail.*

5   Relieve the fuel pressure (see Section 2), but do not disconnect the negative battery cable. On four-cylinder models, reconnect the electrical connector to the fuel pump.
6   On V6 models, remove the cap from the fuel pressure test port on the fuel rail and attach a fuel pressure gauge **(see illustration)**. On four-cylinder models the fuel rail is not equipped with a test port; disconnect the fuel line quick-connect fitting at the point where the flexible fuel feed hose connects to the rigid metal line leading back to the fuel

4

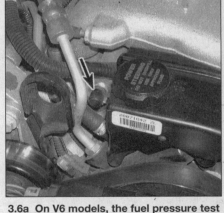

3.6a On V6 models, the fuel pressure test port is located on the fuel rail

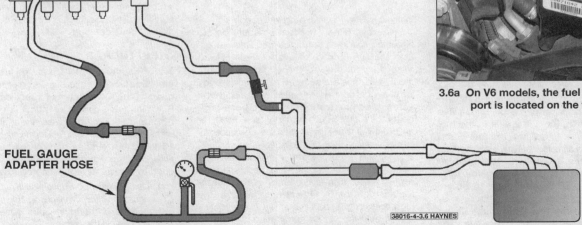

FUEL GAUGE
ADAPTER HOSE

38016-4-3.6 HAYNES

3.6b On four-cylinder models, a special adapter hose will be required to attach a fuel pressure gauge (the fuel rail isn't equipped with a test port)

tank. You will find this fitting behind the engine, near the firewall. Connect the special adapter hose to the fuel lines and connect a fuel pressure gauge to the adapter hose (**see illustration**).

7   Turn the ignition key On - the fuel pump should run for approximately two seconds then shut off. Note the pressure indicated on the gauge and compare your reading with the pressure listed in this Chapter's Specifications. Cycle the ignition key On and Off several times, if necessary, to obtain the highest reading.

8   If the fuel pressure is lower than specified, pinch-off the fuel return line. **Caution:** *Use special pliers designed specifically for pinching a rubber fuel line (available at most auto parts stores). Use of any other type pliers may damage the fuel line.* Turn the ignition key On and note the fuel pressure. **Caution:** *Do not allow the fuel pressure to rise above 65 psi or damage to the fuel pressure regulator may occur.* If the fuel pressure is now above the specified pressure, replace the fuel pressure regulator (see Section 14). If the fuel pressure is still lower than specified, check the fuel lines and the fuel filter for restrictions. If no restriction is found, remove the fuel pump module (see Section 7) and check the fuel strainer for restrictions, check the fuel flex pipe for leaks and check the fuel pump wiring for high resistance. If no problems are found, replace the fuel pump.

9   If the fuel pressure recorded in Step 7 is higher than specified, check the fuel return line for restrictions. If no restrictions are found, replace the fuel pressure regulator (see Section 14).

10   If the fuel pressure is within specifications, start the engine. **Warning:** *Make sure the fuel pressure gauge hose is positioned away from the engine drivebelt before starting the engine.* With the engine running, the fuel pressure should be 3 to 10 psi below the pressure recorded in Step 7. If it isn't, remove the vacuum hose from the fuel pressure regulator and verify there is 12 to 14 in-Hg of vacuum present at the hose. If vacuum is not present at the hose, check the hose for a restriction or a break. If vacuum is present, reconnect the hose to the fuel pressure regulator. If the fuel pressure regulator does not decrease the fuel pressure with vacuum applied, replace the fuel pressure regulator.

11   Now check the fuel system hold pressure. Turn the engine off and monitor the fuel pressure for five minutes; the fuel pressure should not drop more than 5 psi within five minutes. If it does, there is a leak in the fuel line, a fuel injector is leaking, the fuel pump module check valve is defective or the fuel pressure regulator is defective. To determine if the fuel injectors are leaking, cycle the ignition key On and Off several times to obtain the highest fuel pressure reading, then immediately pinch-off the fuel return hose. If the pressure drops below 5 psi within five minutes, the fuel pump is defective, a fuel injector (or injectors) is leaking, the main fuel line is leaking, or the fuel rail may be leaking (but

4.2 The metal fuel lines are secured to the underbody with a plastic retainer

4.11a To disconnect a plastic collar two-tab type fitting, squeeze the two tabs together and pull the lines apart

4.11b A special tool (available at most auto parts stores) is required to disconnect the metal collar type fitting . . .

an external leak such as this would be very apparent). If the fuel rail/injectors hold pressure with the return hose pinched-off, the fuel pressure regulator is defective.

4   Fuel lines and fittings - repair and replacement

Refer to illustrations 4.2, 4.11a, 4.11b and 4.11c

Warning: *See the* **Warning** *in Section 1.*

1   Always relieve the fuel pressure before servicing fuel lines or fittings (see Section 2).

2   Metal fuel feed and vapor lines extend from the fuel tank to the engine compartment. The lines are secured to the underbody with a plastic retainer and a sheet metal screw (**see illustration**). Flexible hose connects the metal lines to the fuel tank, fuel filter and fuel rail. Fuel lines must be occasionally inspected for leaks or damage.

3   In the event of any fuel line damage (metal or flexible lines) it is necessary to replace the damaged lines with factory replacement parts. Others may fail from the high pressures of this system.

4   If evidence of contamination is found in the system or fuel filter during disassembly, the line should be disconnected and blown out. Check the fuel strainer on the fuel pump

module for damage and deterioration.

5   Don't route fuel line or hose within four inches of any part of the exhaust system or within ten inches of the catalytic converter. Fuel line must never be allowed to chafe against the engine, body or frame. A minimum of 1/4-inch clearance must be maintained around a fuel line.

6   When replacing a fuel line, remove all fasteners attaching the fuel line to the vehicle body.

7   Because fuel lines used on fuel-injected vehicles are under high pressure, they require special consideration.

Steel tubing

8   If replacement of a steel fuel line or emission line is called for, use steel tubing meeting the manufacturers specification.

9   Don't use copper or aluminum tubing to replace steel tubing. These materials cannot withstand normal vehicle vibration.

10   Some fuel lines have threaded fittings with O-rings. Any time the fittings are loosened to service or replace components:

a) *Use a flare-nut wrench on the fitting nut and a backup wrench on the stationary portion of the fitting while loosening and tightening the fittings.*

b) *Check all O-rings for cuts, cracks and deterioration. Replace any that appear hardened, worn or damaged.*

4.11c . . . place the tool over the fuel line, insert it squarely into the fitting and pull the lines apart (the tool is not required to connect the lines)

5.8 Disconnect the EVAP hose (arrow) from the canister

5.9 Loosen the hose clamps and disconnect the fuel filler hose from the fuel tank

c) If the lines are replaced, always use original equipment parts, or parts that meet the original equipment standards.

Flexible hose

11   There are various methods of disconnecting the fittings, depending upon the type of quick-connect fitting installed on the fuel line (see illustrations). Clean any debris from around the fitting. Disconnect the fitting and carefully remove the fuel line from the vehicle. **Caution:** *The quick-connect fittings are not serviced separately. Do not attempt to repair these types of fuel lines in the event the fitting or line becomes damaged. Replace the entire fuel line as an assembly.*

12   Installation is the reverse of removal with the following additions:

a) Clean the quick-connect fittings with a lint-free cloth and apply clean engine oil the fittings.
b) After connecting a quick-connect fitting, check the integrity of the connection by attempting to pull the lines apart.
c) Use new O-rings at the threaded fittings (if equipped).
d) Cycle the ignition key On and Off several times and check for leaks at the fitting, before starting the engine.

5   Fuel tank - removal and installation

Refer to illustrations 5.8, 5.9 and 5.10
Warning: *See the* **Warning** *in Section 1.*
1   Remove the fuel tank filler cap to relieve fuel tank pressure.
2   Relieve the fuel system pressure (see Section 2).
3   Disconnect the cable from the negative battery terminal. **Caution:** *On models equipped with the Theftlock audio system, be sure the lockout feature is turned off before*

5.10 Remove the fuel tank strap bolts (arrows) and swing the straps forward

performing any procedure which requires disconnecting the battery (see the front of this manual).
4   Using a siphoning kit (available at most auto parts stores), siphon the fuel into an approved gasoline container. **Warning:** *Do not start the siphoning action by mouth!*
5   Raise the vehicle and support it securely on jackstands.
6   Remove the exhaust system support bolts and rubber insulators. Lower the exhaust system and rest it on the rear suspension. Remove the exhaust heat shield.
7   Disconnect the fuel pump module electrical connector from the body connector (see Section 2). Disconnect the fuel feed and return lines (see Section 4).
8   Disconnect the EVAP hose from the canister **(see illustration)**.
9   Loosen the hose clamps and disconnect the fuel filler hose from the fuel tank **(see illustration)**.
10   Position a transmission jack under the fuel tank and support the tank. Remove the fuel tank strap bolts and swing the straps forward **(see illustration)**.
11   Lower the jack and remove the tank from the vehicle.
12   Installation is the reverse of removal.

6   Fuel tank cleaning and repair - general information

1   The fuel tanks installed in the vehicles covered by this manual are not repairable. If the fuel tank becomes damaged, it must be replaced.
2   Cleaning the fuel tank (due to fuel contamination) should be performed by a professional with the proper training to carry out this critical and potentially dangerous work. Even after cleaning and flushing, explosive fumes may remain inside the fuel tank.
3   If the fuel tank is removed from the vehicle, it should not be placed in an area where sparks or open flames could ignite the fumes coming out of the tank. Be especially careful inside a garage where a natural gas-type appliance is located.

7   Fuel pump module - removal and installation

Refer to illustrations 7.4, 7.5 and 7.6
Warning: *See the* **Warning** *in Section 1.*
1   Relieve the fuel system pressure (see Section 2).

4

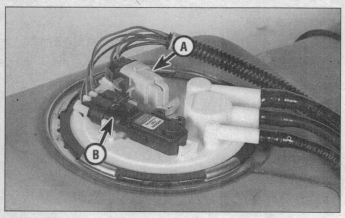

7.4 Disconnect the fuel pump/sending unit (A) and the fuel pressure sensor (B) electrical connectors from the fuel pump module

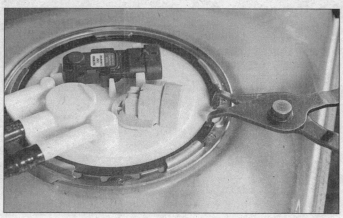

7.5 Using snap-ring pliers, remove the fuel pump module retaining ring from the slots in the fuel tank

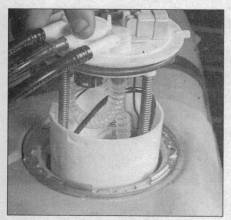

7.6 Carefully remove the fuel pump module from the tank and drain the fuel from the reservoir

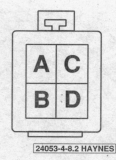

24053-4-8.2 HAYNES

8.2 Fuel pump module connector terminal identification

A   *Fuel level sending unit signal*
B   *Fuel pump 12-volt supply (from fuel pump relay)*
C   *Ground*
D   *Fuel level sending unit ground*

8.7 Disconnect the fuel pump/fuel level sending unit electrical connector from the fuel pump module

2   Disconnect the cable from the negative battery terminal. **Caution:** *On models equipped with the Theftlock audio system, be sure the lockout feature is turned off before performing any procedure which requires disconnecting the battery (see the front of this manual).*
3   Remove the fuel tank from the vehicle (see Section 5).
4   Disconnect the electrical connectors from the fuel pump module and remove the wiring harness **(see illustration)**.
5   While pressing down on the module, remove the fuel pump module retaining ring **(see illustration)**.
6   Remove the fuel pump module from the tank **(see illustration)**. Angle the assembly slightly to avoid damaging the fuel level sending unit float. **Warning:** *Some fuel may remain the module reservoir and spill as the module is removed. Have several shop towels ready and a drain pan near by to place the module in.*
7   The electric fuel pump is not serviced separately. In the event of failure, the complete assembly must be replaced. Transfer the fuel pressure sensor and fuel level sending unit to the new fuel pump module assem-

bly, if necessary (see Section 8).
8   Clean the fuel tank sealing surface and install a new O-ring on the fuel pump module.
9   Install the fuel pump module aligning the tab on the underside of the module with the notch in the fuel tank.
10   Press the fuel pump module down until seated and install the retaining ring. Make sure the retaining ring is seated in the slots.
11   The remainder of installation is the reverse of removal.

8   Fuel level sending unit - check and replacement

Warning: *See the* **Warning** *in Section 1.*

Check

Refer to illustration 8.2
1   Remove the fuel tank and the fuel pump module (see Sections 5 and 7).
2   Connect the probes of an ohmmeter to the two fuel level sensor terminals (A and D) of the fuel pump module electrical connector

(see illustration).
3   Position the float in the down (empty) position and note the reading on the ohmmeter.
4   Move the float up to the full position while watching the meter.
5   If the fuel level sending unit resistance does not change smoothly as the float travels

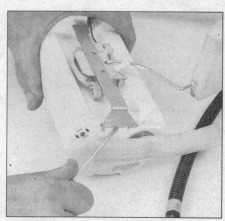

8.8 Carefully detach the sending unit bracket from the module (if equipped)

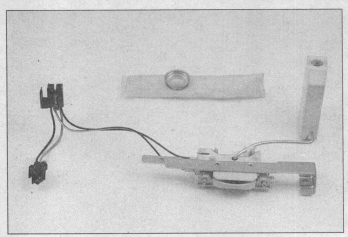

8.9 Fuel level sending unit removed from module

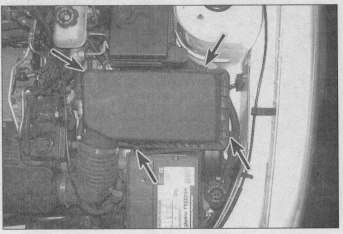

9.2 Loosen the screws (arrows) from the air filter cover and disconnect the air intake duct (four-cylinder shown, V6 similar)

9.3 Remove the air filter housing mounting nuts (arrows) (four-cylinder shown, V6 similar)

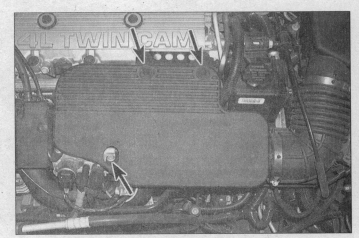

9.4 Remove the bolts retaining the resonator and loosen the hose clamp at the throttle body through the hole provided (arrows)

4

from empty to full, replace the fuel level sending unit assembly.

Replacement

Refer to illustrations 8.7, 8.8 and 8.9

6   Remove the fuel tank and the fuel pump module (see Sections 5 and 7).

7   Disconnect the fuel level sending unit electrical connector from the module cover **(see illustration)**.

8   Detach the sending unit bracket from the module, if equipped **(see illustration)**.

9   Slide the fuel level sending unit off the module **(see illustration)**. Note the routing of the wiring for installation.

10   Installation is the reverse of removal.

9   Air filter housing - removal and installation

Refer to illustrations 9.2, 9.3 and 9.4

1   On four-cylinder models, disconnect the electrical connector from the intake air temperature sensor. On V6 models, disconnect the electrical connectors from the intake air

temperature sensor and MAF sensor.

2   Loosen the hose clamp from the air intake duct and loosen the air filter cover screws **(see illustration)**. Remove the cover and the air filter element.

3   Remove the nuts from the mounting studs, pull the housing up and detach the duct from the inner fender **(see illustration)**. Remove the assembly from the engine compartment.

4   On a four-cylinder model, loosen the hose clamps, detach the fasteners and separate the air intake duct and resonator assembly from the throttle body **(see illustration)**. On V6 models, detach the intake duct from the throttle body.

5   Installation is the reverse of removal.

10   Accelerator cable - replacement

Refer to illustrations 10.2, 10.4, 10.5 and 10.6

1   Disconnect the cable from the negative battery terminal. **Caution:** *On models equipped with the Theftlock audio system, be sure the lockout feature is turned off before*

performing any procedure which requires disconnecting the battery (see the front of this manual).

2   Remove the throttle lever shield, if equipped **(see illustration)**.

3   On four-cylinder models equipped with

10.2 On four-cylinder models, remove the retainer (arrow) and the accelerator cable shield

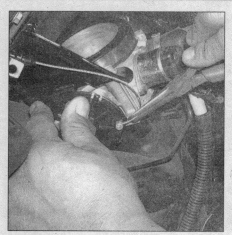

10.4 Rotate the throttle lever and pass the cable through the slot in the lever (four-cylinder shown, V6 similar)

10.5 Depress the locking tabs and remove the cable from the bracket (four-cylinder shown, V6 similar)

10.6 Pull the accelerator cable retainer out of the pedal and slide the cable through the slot (arrow)

cruise control, detach the cruise control cable from the throttle lever.

4   Rotate the throttle lever and separate the accelerator cable end from the throttle lever **(see illustration)**.

5   Depress the locking tabs on the cable housing and push the cable through the bracket **(see illustration)**. Detach the cable from the cable retainers.

6   Remove the trim panel from under the dash and detach the cable from the accelerator pedal **(see illustration)**.

7   Depress the locking tabs on the cable housing and push the cable through the firewall and into the engine compartment.

8   Remove the cable from the engine compartment.

9   Installation is the reverse of removal.

11   Fuel injection system - general information

The sequential Multi Port Fuel Injection (MPFI) system consists of three sub-systems: air intake, engine control and fuel delivery. The system uses a Powertrain Control Module (PCM) along with the sensors (coolant temperature sensor, throttle position sensor, manifold absolute pressure sensor, oxygen sensor, etc.) to determine the proper air/fuel ratio under all operating conditions.

The fuel injection system and the engine control system are closely linked in function and design. For additional information, refer to Chapter 6.

Air intake system

The air intake system consists of the air filter, the air intake ducts, the throttle body, the air intake plenum and the intake manifold.

When the engine is idling, the air/fuel ratio is controlled by the idle air control system, which consists of the Powertrain Control Module (PCM) and the idle air control valve. The idle air control valve is controlled by the

PCM and is opened and closed depending upon the running conditions of the engine (air conditioning system, power steering, cold and warm running etc.). This idle air control regulates the amount of airflow past the throttle plate and into the intake manifold, thus increasing or decreasing the engine idle speed. The PCM receives information from the sensors (vehicle speed, coolant temperature, air conditioning, power steering mode etc.) and adjusts the idle according to the demands of the engine and driver. Refer to Chapter 6 for information on the idle air control valve.

Emissions and engine control system

The emissions and engine control system is described in detail in Chapter 6.

Fuel delivery system

The fuel delivery system consists of these components: the fuel pump, the fuel pressure regulator, the fuel rail and the fuel injectors.

The fuel pump is an electric type. Fuel is drawn through an inlet screen into the pump, flows through the one-way valve, passes through the fuel filter and is delivered to the fuel rail and injectors. The pressure regulator maintains a constant fuel pressure to the injectors. Excess fuel is routed back to the fuel tank through the fuel pressure regulator.

The injectors are solenoid-actuated pintle types consisting of a solenoid, plunger, needle valve and housing. When current is applied to the solenoid coil, the needle valve raises and pressurized fuel sprays out the nozzle. The injection quantity is determined by the length of time the valve is open (the length of time during which current is supplied to the solenoid coils).

The fuel pump relay is located in the engine compartment Power Distribution Center. The fuel pump relay connects battery voltage to the fuel pump. If the PCM senses

there is NO signal from the camshaft or crankshaft sensors (as with the engine not running or cranking), the PCM will de-energize the relay.

12   Fuel injection system - check

Refer to illustrations 12.7, 12.8 and 12.10
Note: *The following procedure is based on the assumption that the fuel pressure is adequate (see Section 3).*

1   Check all electrical connectors that are related to the system. Check the ground wire connections on the intake manifold for tightness. Loose connectors and poor grounds can cause many problems that resemble more serious malfunctions.

2   Check to see that the battery is fully charged, as the control unit and sensors depend on an accurate supply voltage in order to properly meter the fuel.

3   Check the air filter element - a dirty or partially blocked filter will severely impede performance and economy (see Chapter 1).

4   Check the related fuses. If a blown fuse is found, replace it and see if it blows again. If it does, search for a wire shorted to ground in the harness.

5   Check the air intake duct to the intake manifold for leaks, which will result in an excessively lean mixture. Also check the condition of all vacuum hoses connected to the intake manifold and/or throttle body.

6   Remove the air intake duct from the throttle body and check for dirt, carbon or other residue build-up. If it's dirty, clean it with carburetor cleaner spray, a toothbrush and a shop towel. **Caution:** *Do not use a solvent containing Methyl Ethyl Ketone or damage to the throttle body may occur.*

7   With the engine running, place an automotive stethoscope against each injector, one at a time, and listen for a clicking sound, indicating operation **(see illustration)**. If you don't have a stethoscope, place the tip of a

12.7 Use a stethoscope to determine if the injectors are working properly - they should make a steady clicking sound that rises and falls with engine speed changes

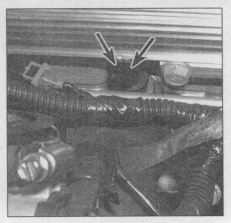

12.8 Measure the resistance of each injector across the two terminals of the injector

12.10 Install the "noid" light (available at most auto parts stores) into each injector electrical connector and confirm that it blinks when the engine is cranking

screwdriver against the injector and listen through the handle.

8   Disconnect the injector electrical connectors and measure the resistance of each injector **(see illustration)**. Compare the measurements with the resistance values listed in this Chapter's Specifications. **Note:** *On V6 models, follow the wiring harness from the fuel injectors to locate the fuel injector harness main connector and perform the tests there. Refer to the wiring diagrams at the end of Chapter 12 to determine the terminals for testing by wire color.*

9   Turn the ignition key On and check for battery voltage at the gray (four-cylinder) or pink (V6) wire terminal of one of the injector harness connectors. If battery voltage is not present, check the fuel pump/injector fuse, fuel pump relay (four-cylinder models only) and related wiring (see Chapter 12).

10   Install an injector test light ("noid" light) into each injector electrical connector, one at a time **(see illustration)**. Crank the engine over. Confirm that the light flashes evenly on each connector. This tests the PCM control of the injectors. If the light does not flash, have the PCM checked at a dealer service department or other properly equipped repair facility. **Note:** *On V6 models, follow the wiring harness from the fuel injectors to locate the fuel injector harness main connector and perform the tests there. Refer to the wiring diagrams at the end of Chapter 12 to determine the terminals for testing by wire color.*

11   The remainder of the engine control system checks can be found in Chapter 6.

13 Throttle body - removal and installation

Refer to illustrations 13.9a and 13.9b
Warning: *Wait until the engine is completely cool before beginning this procedure.*
1   Disconnect the cable from the negative battery terminal. **Caution:** *On models equipped with the Theftlock audio system, be*

sure the lockout feature is turned off before performing any procedure which requires disconnecting the battery (see the front of this manual).

2   On V6 models, clamp-off the coolant hoses attached to the throttle body.

3   Remove the air intake duct.

4   Disconnect the electrical connectors from the throttle body.

5   Label and detach the vacuum hoses from the throttle body.

6   Detach the accelerator cable (see Section 10) and if equipped, the cruise control cable. Remove the accelerator cable bracket.

7   On V6 models, detach the coolant hoses from the throttle body.

8   On four-cylinder models, remove the MAP sensor.

9   Remove the mounting bolts/nuts and remove the throttle body and gasket **(see illustrations)**.

10   Remove all traces of old gasket material from the throttle body and intake manifold and install a new gasket. **Caution:** *Do not use solvent or a sharp tool to clean the throttle body gasket surface or damage to the throttle*

body may occur.

11   Install the throttle body and tighten the bolts to the torque listed in this Chapter's Specifications.

12   The remainder of installation is the reverse of removal.

14 Fuel pressure regulator - replacement

Refer to illustrations 14.5a, 14.5b, 14.6a and 14.6b
Warning: *See the* **Warning** *in Section 1.*
1   Relieve the fuel system pressure (see Section 2).
2   Disconnect the cable from the negative battery terminal. **Caution:** *On models equipped with the Theftlock audio system, be sure the lockout feature is turned off before performing any procedure which requires disconnecting the battery (see the front of this manual).*
3   On four-cylinder models, remove the

13.9a Remove the bolts (arrows) and separate the throttle body from the intake manifold - four-cylinder model

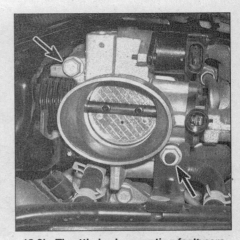

13.9b Throttle body mounting fasteners (arrows) - V6 models

4

14.5a Remove the fuel pressure regulator mounting screw (arrow) and withdraw the regulator with a twisting motion - four-cylinder model

14.5b Fuel pressure regulator details - V6 models

   A   Fuel return line
   B   Mounting screw

14.6a Replace the fuel pressure regulator-to-fuel rail O-ring . . .

fuel rail (see Section 15). On V6 models, remove the upper intake manifold (see Chapter 2B).

4   On V6 models, detach the fuel return line from the fuel pressure regulator. Remove

14.6b . . . and the return line inner O-ring

the vacuum hose from the port on the regulator.

5   Remove the pressure regulator mounting screw and detach the fuel pressure regulator **(see illustrations)**.

6   Be sure to replace all O-ring seals, lubricating them with a light film of engine oil **(see illustrations)**.

7   On V6 models, install the fuel return line before tightening the fuel pressure regulator mounting screw, otherwise a dangerous fuel leak may develop.

8   The remainder of installation is the reverse of removal.

15 Fuel rail and injectors - removal and installation

Warning: *See the* **Warning** *in Section 1.*

Removal

Refer to illustrations 15.6a, 15.6b, 15.6c, 15.7a, 15.7b, 15.7c, 15.8a and 15.8b

1   Relieve the fuel pressure (see Section 2).

2   Disconnect the cable from the negative battery terminal. **Caution:** *On models equipped with the Theftlock audio system, be sure the lockout feature is turned off before performing any procedure which requires disconnecting the battery (see the front of this manual).*

3   On V6 models, remove the upper intake manifold (see Chapter 2B). Cover the lower intake manifold ports to prevent any foreign objects from entering the engine. On four-cylinder models, remove the air intake duct and resonator (see Section 9).

4   Clearly label and remove any vacuum hoses or electrical wiring that will interfere with the fuel rail removal. Detach any wiring harness retainers from the fuel rail.

5   Disconnect the fuel injector electrical connectors and position the harness aside. **Note:** *Apply a numbered tag to each connector with the corresponding cylinder number.*

6   Disconnect the fuel inlet line from the fuel rail **(see illustration)**. Before separating the fuel line from the fuel rail, remove any fuel line bracket fasteners **(see illustration)**. On four-cylinder models, remove the bolts from the fuel return line bracket **(see illustration)**. On V6 models, disconnect the fuel return line

15.6a Disconnect the fuel inlet line from the fuel rail

15.6b Before separating the fuel line from the fuel rail, remove the nuts/bolts from any fuel line brackets

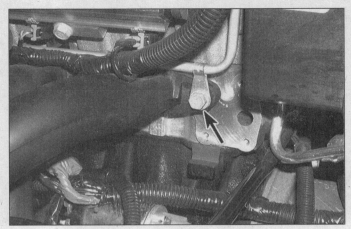

15.6c On four-cylinder models, remove the bolt from the fuel return line bracket

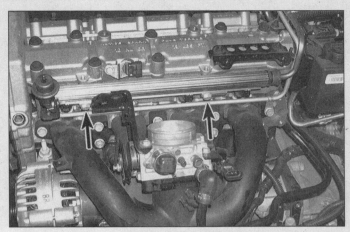

15.7a Remove the bolts (arrows) securing the fuel rail to the intake manifold (four-cylinder models)

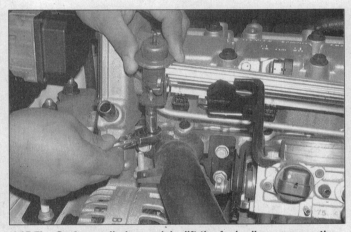

15.7b On four-cylinder models, lift the fuel rail up, remove the return line clamp screw and separate the return line from the fuel pressure regulator

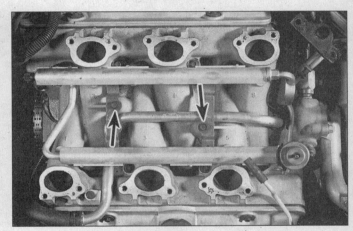

15.7c Fuel rail mounting bolts - V6 models

4

from the fuel pressure regulator (see Section 14).

7   Clean any debris from around the injectors. Remove the fuel rail mounting nuts/bolts **(see illustrations)**. Gently rock the fuel rail and injectors to loosen the injectors. On four-cylinder models, lift the fuel rail up and disconnect the fuel return line from the fuel pressure regulator. Remove the fuel rail and fuel injectors as an assembly.

8   Remove the retaining clip and remove the injector(s) from the fuel rail assembly **(see illustrations)**. Remove and discard the O-rings and seals. **Note:** *Whether you're replacing an injector or a leaking O-ring, it's a good idea to remove all the injectors from the fuel rail and replace all the O-rings.*

Installation

9   Coat the new seal rings (if equipped) with clean engine oil and slide them onto the injectors.

10   Coat the new O-rings with clean engine oil and install them on the injector(s), then insert each injector into its corresponding bore in the fuel rail. Install the injector retain-

15.8a Remove the injector retaining clip and pull the injector off the fuel rail

ing clip.

11   Install the injector and fuel rail assembly on the intake manifold. On four-cylinder models, install the fuel return line before seating the injectors. Fully seat the injectors, then tighten the fuel rail mounting nuts to the

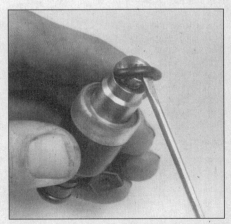

15.8b Carefully remove the O-rings from the injector

torque listed in this Chapter's Specifications.

12   Connect the fuel line (and fuel return line on V6 models) and make sure they're securely installed. Install the fuel line bracket fasteners, as required.

13   Connect the electrical connectors to

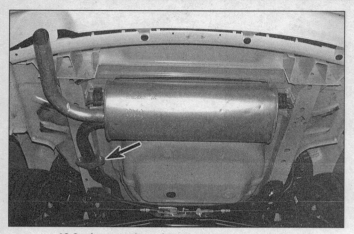

**16.2a Inspect the exhaust system connections
for leakage (arrow)**

16.2b Inspect the rubber insulators for damage

each injector, referring to the numbered tags.
14   The remainder of installation is the
reverse of removal.
15   After the injector/fuel rail assembly
installation is complete, turn the ignition
switch to On, but don't operate the starter
(this activates the fuel pump for about two
seconds, which builds up fuel pressure in the
fuel lines and the fuel rail). Repeat this about
two or three times, then check the fuel lines,
fuel rail and injectors for fuel leakage.

16   Exhaust system servicing - general information

Warning: *Inspection and repair of exhaust
system components should be done only
after enough time has elapsed after driving
the vehicle to allow the system components
to cool completely. Also, when working under
the vehicle, make sure it is securely supported on jackstands.*
1   The exhaust system consists of the
exhaust manifold(s), catalytic converter, muffler, resonators, the tailpipe and all connecting pipes, brackets, hangers and clamps. The
exhaust system is attached to the body with
mounting brackets and rubber hangers. If any
of the parts are improperly installed, excessive noise and vibration will be transmitted to
the body.

Muffler and pipes

Refer to illustrations 16.2a, and 16.2b
2   Conduct regular inspections of the
exhaust system to keep it safe and quiet.
Look for any damaged or bent parts, open
seams, holes, loose connections, excessive
corrosion or other defects which could allow
exhaust fumes to enter the vehicle **(see illus-**

trations). Also check the catalytic converter
when you inspect the exhaust system (see
below). Deteriorated exhaust system components should not be repaired; they should be
replaced with new parts.
3   If the exhaust system components are
extremely corroded or rusted together, welding equipment will probably be required to
remove them. The convenient way to accomplish this is to have a muffler repair shop
remove the corroded sections with a cutting
torch. If, however, you want to save money
by doing it yourself (and you don't have a
welding outfit with a cutting torch), simply cut
off the old components with a hacksaw. If
you have compressed air, special pneumatic
cutting chisels can also be used. If you do
decide to tackle the job at home, be sure to
wear safety goggles to protect your eyes
from metal chips and work gloves to protect
your hands.
4   Here are some simple guidelines to follow when repairing the exhaust system:

a) *Work from the back to the front when
removing exhaust system components.*
b) *Apply penetrating oil to the exhaust system component fasteners to make them
easier to remove.*
c) *Use new gaskets, hangers and clamps
when installing exhaust systems components.*
d) *Apply anti-seize compound to the
threads of all exhaust system fasteners
during reassembly.*
e) *Be sure to allow sufficient clearance
between newly installed parts and all
points on the underbody to avoid overheating the floor pan and possibly damaging the interior carpet and insulation.
Pay particularly close attention to the
catalytic converter and heat shield.*

**16.6 Inspect the catalytic converter and
heat shield for damage**

Catalytic converter

Refer to illustration 16.6
Warning: *The converter gets very hot during
operation. Make sure it has cooled down
before you touch it.*
Note: *See Chapter 6 for additional information on the catalytic converter.*
5   Periodically inspect the heat shield for
cracks, dents and loose or missing fasteners.
6   Inspect the converter for cracks or other
damage **(see illustration)**.
7   If the catalytic converter requires
replacement, refer to Chapter 6.

Chapter 5
Engine electrical systems

Contents

Specifications

General

Battery voltage
Engine off	12.0 to 12.6 volts
Engine running	13.5 to 14.5 volts

Torque specifications — Ft-lbs

Alternator mounting bolts	
Four-cylinder engine	37
V6 engines	
Bolts	37
Nuts	22
Ignition coil/module cover assembly mounting bolts (four-cylinder)	16
Starter mounting bolts	
Four-cylinder engine	66
V6 engines	32

1.1a Typical engine electrical system components - four-cylinder engine

1	Ignition coil/module assembly (under cover)	2	Power Distribution Center	4	Battery cable
		3	Battery	5	Alternator

1   General information and precautions

General information

Refer to illustrations 1.1a and 1.1b

The engine electrical systems include all ignition, charging and starting components **(see illustrations)**. Because of their engine-related functions, these components are discussed separately from body electrical devices such as the lights, the instruments, etc. (which are included in Chapter 12).

Precautions

Always observe the following precautions when working on the electrical system:

a) *Be extremely careful when servicing engine electrical components. They are easily damaged if checked, connected or handled improperly.*

b) *Never leave the ignition switched on for long periods of time when the engine is not running.*

c) *Never disconnect the battery cables while the engine is running.*

d) *Maintain correct polarity when connecting battery cables from another vehicle during jump starting - see the "Booster*

battery (jump) starting" section at the front of this manual.

e) *Always disconnect the negative battery cable before working on the electrical system.*

It's also a good idea to review the safety-related information regarding the engine electrical systems located in the *"Safety first!"* section at the front of this manual, before beginning any operation included in this Chapter.

Battery disconnection

Caution: *On models equipped with the Theft-lock audio system, be sure the lockout feature is turned off before performing any procedure which requires disconnecting the battery (see the front of this manual).*

Several systems on the vehicle require battery power to be available at all times, either to ensure their continued operation (such as the clock) or to maintain control unit memories (such as that in the engine management system's Powertrain Control Module [PCM]) which would be wiped out if the battery were to be disconnected. Therefore, whenever the battery is to be disconnected, first note the following to ensure that there are no unforeseen consequences of this action:

a) *First, on any vehicle with power door locks, it is a wise precaution to remove the key from the ignition and to keep it with you, so that it does not get locked inside if the power door locks should engage accidentally when the battery is reconnected!*

b) *The engine management system's PCM will lose the information stored in its memory when the battery is disconnected. This includes idling and operating values, and any fault codes detected (see Chapter 6). Whenever the battery is disconnected, the information relating to idle speed control and other operating values will have to be re-programmed into the unit's memory. The PCM does this by itself, but until then, there may be surging, hesitation, erratic idle and a generally inferior level of performance. To allow the PCM to relearn these values, start the engine and run it as close to idle speed as possible until it reaches its normal operating temperature, then run it for approximately two minutes at 1200 rpm. Next, drive the vehicle as far as necessary - approximately 5 miles of varied driving conditions is usually sufficient - to complete the relearning process.*

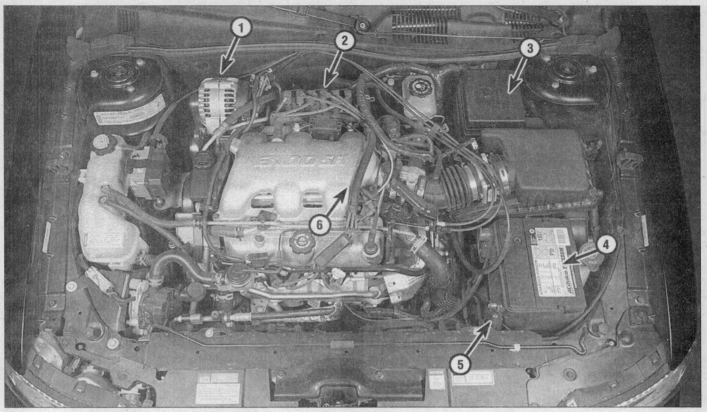

1.1b Typical engine electrical system components - V6 engines

1	*Alternator*	*3*	*Power Distribution Center*	*5*	*Battery cable*
2	*Ignition coils*	*4*	*Battery*	*6*	*Spark plug wires*

Devices known as "memory-savers" can be used to avoid some of the above problems. Precise details vary according to the device used. Typically, it is plugged into the cigarette lighter, and is connected by its own wires to a spare battery; the vehicle's own battery is then disconnected from the electrical system, leaving the "memory-saver" to pass sufficient current to maintain audio unit security codes and PCM memory values, and also to run permanently live circuits such as the clock, all the while isolating the battery in the event of a short-circuit occurring while work is carried out. **Warning:** *Some of these devices allow a considerable amount of current to pass, which can mean that many of the vehicle's systems are still operational when the main battery is disconnected. If a "memory-saver" is used, ensure that the circuit concerned is actually "dead" before carrying out any work on it!*

2   Battery - emergency jump starting

Refer to the *Booster battery (jump) starting* procedure at the front of this manual.

3   Battery - check and replacement

Warning: *Hydrogen gas is produced by the battery, so keep open flames and lighted cigarettes away from it at all times. Always wear eye protection when working around a battery. Rinse off spilled electrolyte immediately with large amounts of water.*

Check

Refer to illustrations 3.2 and 3.3

1   The battery's surface charge must be removed before accurate voltage measurements can be made. Turn On the high beams for ten seconds, then turn them Off, let the vehicle stand for two minutes. Remove the battery from the vehicle (see Steps 4 through 10).

2   Check the battery state of charge. Visually inspect the indicator eye on the top of the battery, if the indicator eye is clear, charge the battery as described in Chapter 1. Next perform an open voltage circuit test using a digital voltmeter **(see illustration)**. With the engine and all accessories Off, connect the negative probe of the voltmeter to the negative terminal of the battery and the positive probe to the positive terminal of the battery. The battery voltage should be 12.4 volts or more. If the battery is less than the specified voltage, charge the battery before proceeding to the next test. Do not proceed with the

3.2 To test the open circuit voltage of the battery, connect a voltmeter to the battery - a fully charged battery should measure at least 12.4 volts (depending on outside air temperature)

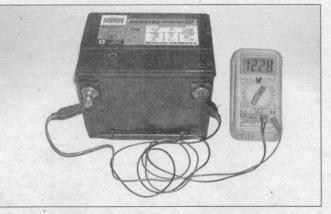

3.3 Connect a battery load tester to the battery and check the battery condition under load following the tool manufacturer's instructions

3.6 Remove the bolt (arrows) and detach the battery retainer

3.8 Inspect the tray, retainer brackets and related fasteners for corrosion or damage - if necessary, remove the bolts (arrows) and the battery tray

4.4a The battery cables are routed through the protective sheathing (arrow) to the power distribution center, starter solenoid and main engine ground point

battery load test unless the battery charge is correct.

3   Perform a battery load test. An accurate check of the battery condition can only be performed with a load tester (available at most auto parts stores). This test evaluates the ability of the battery to operate the starter and other accessories during periods of heavy amperage draw (load). Install a special battery load testing tool onto the terminals **(see illustration)**. Load test the battery according to the tool manufacturer's instructions. This tool utilizes a carbon pile to increase the load demand (amperage draw) on the battery. Maintain the load on the battery for 15 seconds or less and observe that the battery voltage does not drop below 9.6 volts. If the battery condition is weak or defective, the tool will indicate this condition immediately. **Note:** *Cold temperatures will cause the minimum voltage requirements to drop slightly. Follow the chart given in the tool manufacturer's instructions to compensate for cold climates. Minimum load voltage for freezing temperatures (32 degrees F) should be approximately 9.1 volts.*

Replacement

Refer to illustrations 3.6 and 3.8

4   Disconnect the cable from the negative battery terminal. **Caution:** *On models equipped with the Theftlock audio system, be sure the lockout feature is turned off before performing any procedure which requires disconnecting the battery (see the front of this manual).*

5   Disconnect the positive battery cable.

6   Remove the battery retainer bolt and retainer **(see illustration)**.

7   Lift the battery off the battery tray and remove the battery insulator. **Note:** *Battery handling tools are available at most auto parts stores for a reasonable price. They make it easier to remove and carry the battery.*

8   While the battery is removed, inspect the tray, retainer brackets and related fasten-

ers for corrosion or damage **(see illustration)**.

9   If corrosion is evident, remove the battery tray and use a baking soda/water solution to clean the corroded area to prevent further oxidation. Repaint the area as necessary using rust resistant paint.

10   Clean and service the battery and cables (see Chapter 1).

11   If you are replacing the battery, make sure you purchase one that is identical to yours, with the same dimensions, amperage rating, cold cranking amps rating, etc. Make sure it is fully charged prior to installation in the vehicle.

12   Installation is the reverse of removal. Connect the positive cable first and the negative cable last.

13   After connecting the cables to the battery, apply a light coating of petroleum jelly or grease to the connections to help prevent corrosion.

4   Battery cables - replacement

Refer to illustrations 4.4a, 4.4b and 4.4c
Caution: *On models equipped with the Theftlock audio system, be sure the lockout feature is turned off before performing any procedure which requires disconnecting the battery (see the front of this manual).*

1   Periodically inspect the entire length of each battery cable for damage, cracked or burned insulation and corrosion. Poor battery cable connections can cause starting problems and decreased engine performance.

2   Check the cable-to-terminal connections at the ends of the cables for cracks, loose wire strands and corrosion. The presence of white, fluffy deposits under the insulation at the cable terminal connection is a sign that the cable is corroded and should be replaced. Check the terminals for distortion, missing mounting bolts and corrosion.

3   When removing the cables, always disconnect the negative cable first and hook it up last or the battery may be shorted by the tool used to loosen the cable clamps. Even if

only the positive cable is being replaced, be sure to disconnect the negative cable first (see Chapter 1 for further information regarding battery cable maintenance).

4   Disconnect the old cables from the battery, then disconnect them at the opposite end. Detach the cables from the starter solenoid, power distribution center and ground terminals, as necessary **(see illustrations)**. Note the routing of each cable to ensure correct installation.

5   If you are replacing either or both of the battery cables, take them with you when buying new cables. It is vitally important that you replace the cables with identical parts. Cables have characteristics that make them easy to identify: positive cables are usually red and larger in cross-section; ground cables are usually black and smaller in cross-section.

6   Clean the threads of the starter solenoid or ground connection with a wire brush to remove rust and corrosion. Apply a light coat of battery terminal corrosion inhibitor or petroleum jelly to the threads to prevent future corrosion.

7   Attach the cable to the terminal and tighten the mounting nut/bolt securely.

4.4b One branch of the positive cable is connected to the power distribution center

4.4c Two branches of the negative cable are connected to the body

6.5 To use a calibrated ignition tester on a four-cylinder model: Bolt the cover to the engine with one of the mounting bolts and a spacer (arrow), connect the tester to a spark plug boot and use a jumper wire to connect the clip to a convenient ground, connect the remaining spark plug boots to either the spark plugs or ground them to the engine - crank the engine over, if there's enough power to fire the plug, bright blue sparks will be visible between the electrode tip and the tester body (weak sparks or intermittent sparks are the same as no sparks)

8   Before connecting a new cable to the battery, make sure that it reaches the battery without having to be stretched.

5   Ignition system - general information

All models are equipped with a distributorless ignition system (DIS). The ignition system consists of the battery, ignition coils, ignition control module, spark plug wires (V6 models only), spark plugs, camshaft position sensor, crankshaft position sensor and the Powertrain Control Module (PCM). The ignition control module and PCM control the ignition timing and spark advance characteristics for the engine. The ignition timing is not adjustable.

The DIS ignition systems use a "waste spark" method of spark distribution. Each cylinder is paired with its opposing cylinder in the firing order (1-4, 2-3 on four-cylinder models; 1-4, 2-5, 3-6 on V6 models) so one cylinder under compression fires simultaneously with its opposing cylinder, where the piston is on the exhaust stroke. Since the cylinder on the exhaust stroke requires very little of the available voltage to fire its plug, most of the voltage is used to fire the plug of the cylinder on the compression stroke. Conventional ignition coils have one end of the secondary winding connected to the engine ground. On DIS, neither end of the secondary winding is grounded - instead, one end of the coils secondary winding is directly attached to the spark plug and the other end is attached to the spark plug of the companion cylinder.

The crankshaft position sensor produces a signal voltage to indicate crankshaft position and crankshaft speed. This signal is used by the Ignition Control Module (ICM) during start up and passed on to the Powertrain Control Module (PCM) to control ignition timing.

The DIS system is also integrated with a knock sensor system. The system uses a knock sensor in conjunction with the Powertrain Control Module (PCM) to control spark timing. The knock sensor system allows the engine to use maximum spark advance without spark knock, which improves driveability and fuel economy.

6   Ignition system - check

Warning 1: *Because of the high voltage generated by the ignition system, extreme care should be taken whenever an operation is performed involving ignition components. This not only includes the ignition coil, but related components and test equipment.*
Warning 2: *The following procedure requires the engine to be cranked during testing, make sure the meter leads, loose clothing, long hair, etc. are away from the moving parts of the engine (drivebelt, cooling fan, etc.) before cranking the engine.*
Note: *Special test equipment is required to perform the following procedure. Read through the procedure and obtain the necessary equipment before proceeding.*
1   Before proceeding with the ignition system, check the following items:
  a) *Make sure the battery cable clamps, where they connect to the battery, are clean and tight.*
  b) *Test the condition of the battery (see Section 3). If it does not pass all the tests, replace it with a new battery.*
  c) *Check the external ignition coil and ignition control module wiring and connections.*
  d) *Check the related fuses inside the power distribution center (see Chapter 12). If they're burned, determine the cause and repair the circuit.*
2   If the engine turns over but won't start or has a severe misfire, make sure there is sufficient secondary ignition voltage to fire the spark plugs according to engine type as follows:

Four-cylinder models
Refer to illustrations 6.5, 6.7a, 6.7b, 6.7c, 6.8 and 6.9
Caution: *Because of the ignition system design on four-cylinder models, a special set of spark plug wires must be obtained or fabricated before an ignition system check can be*

performed. A test set can be fabricated from bulk spark plug wire and terminals available at most auto parts stores. As an alternative, individual jumper wires can be used to ground the spark plug terminals not being tested or four ignition system testers could be used. Whichever method you chose, make sure all the spark plug terminals are either connected to a spark plug or grounded to the engine block (and the cover is grounded) before cranking the engine or damage to the ignition control module may result.
3   Disable the fuel system by removing the fuel pump/injector fuse from the power distribution center (see Chapter 12).
4   Remove the bolts retaining the ignition coil/module cover (see Section 7). Carefully lift the cover up, turn the cover over and secure it to the engine using one of the bolts and a metal spacer. Use the bolt hole in the cover with the ground strap, the cover must be grounded to the engine block before proceeding. Do not disconnect the electrical connector from the ignition control module.
5   Attach a calibrated ignition system tester (available at most auto parts stores) to one of the spark plug boots. Connect the clip on the tester to a bolt or metal bracket on the engine **(see illustration)**. Ground the remaining spark plug terminals. Crank the engine and watch the end of the tester to see if a bright blue, well-defined spark occurs (weak spark or intermittent spark is the same as no spark).
6   If spark occurs, sufficient voltage is reaching the plug to fire it (repeat the check at the remaining spark plug terminals to verify that the ignition coils are good). If the ignition system is operating properly the problem lies

5

elsewhere; i.e. a mechanical or fuel system problem. However, the spark plugs may be fouled, so remove and check them as described in Chapter 1.

7   If no spark occurs, remove the ignition coil/module housing (see Section 7) and check the ignition coils. Using an ohmmeter, measure the primary and secondary resistance of the ignition coils **(see illustrations)**. Ignition coil primary resistance should be approximately 0.4 to 0.8 ohms and secondary resistance should be approximately 4,000 to 8,000 ohms. If it isn't, replace the defective ignition coil. If the ignition coils are good, remove the spark plug boot from the housing and check the terminals at both ends for damage and check the boot for an open or high resistance **(see illustration)**. Check the spring for damage and check the housing for continuity between the spark plug boot terminal and the inside coil terminal.

8   Check for battery voltage to the ignition coil from the ignition control module. Attach a 12 volt test light to the battery negative (-) terminal or other good ground. Disconnect the coil electrical connector from the ignition module and check for power at the positive (+) terminal at the ignition module **(see illustration)**. Battery voltage should be available with the ignition key On. If there is no battery voltage present at the coil connector, check for battery voltage at the pink wire terminal of the ignition module connector. If there is no battery voltage present at the module connector, check the wiring and/or circuit between the power distribution center and ignition control module (don't forget to check the fuses). Also check the black wire terminal for continuity to battery ground. **Note:** *Refer to the wiring diagrams at the end of Chapter 12 for wire color identification for testing and additional information on the circuits.*

9   Check for a trigger signal from the ignition control module. Attach the lead of a test light to the positive battery terminal and touch the probe of the test light to one of the coil trigger (-) terminals at the module **(see illustration)**. Crank the engine. The test light should blink with the engine cranking if a trig-

6.7a To check the primary resistance of the ignition coil, measure the resistance across the positive (+) terminal and each of the negative (-) terminals in turn

6.7b To check the secondary resistance, measure the resistance across the coil towers

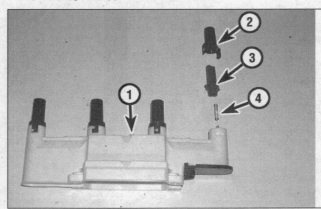

6.7c Ignition coil housing details

1   *Housing*
2   *Spark plug boot retainer*
3   *Spark plug boot*
4   *Spring*

ger signal is present. Check the other coil trigger terminal. If a trigger signal is present at the coil, the ignition control module and the crankshaft position sensor are functioning properly, check the ignition coils as described previously. If a trigger signal is not present at the coil terminals check the crankshaft position sensor (see Chapter 6). If the crankshaft position sensor is good, replace the ignition control module.

10   If all the components are good and there is no spark, have the PCM checked by a dealer service department or other qualified repair shop.

V6 models

Refer to illustrations 6.12, 6.14a, 6.14b, 6.15 and 6.17

11   Disable the fuel system by removing the fuel pump/injector fuse from the power distri-

6.8 Check for battery power at the positive (+) terminal of the ignition module coil connector

6.9 Using a test light connected to the positive battery terminal, check for a trigger signal on each of the negative (-) terminals

6.12 To use a calibrated ignition tester on a V6 model; disconnect a spark plug wire, connect the tester to the spark plug boot and clip the tester to a convenient ground - crank the engine over, if there's enough power to fire the plug, bright blue sparks will be visible between the electrode tip and the tester body (weak sparks or intermittent sparks are the same as no sparks)

6.14a Using an ohmmeter, check the resistance of the spark plug wire

6.14b Check the ignition coil secondary resistance across the two coil towers

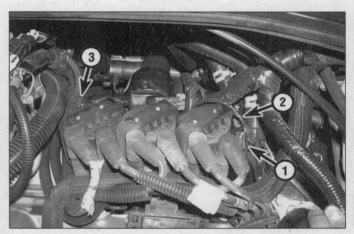

6.15 Disconnect the ignition supply and ground electrical connector from the ignition control module and check for battery voltage at the pink wire terminal

1   *Electronic spark control*   3   *Ignition supply and ground*
2   *Crankshaft position sensor*

bution center (see Chapter 12).

12   Disconnect a spark plug wire from one of the spark plugs and attach a calibrated ignition system tester (available at most auto parts stores) to the spark plug boot. Connect the clip on the tester to a bolt or metal bracket on the engine **(see illustration)**. Crank the engine and watch the end of the tester to see if a bright blue, well-defined spark occurs (weak spark or intermittent spark is the same as no spark).

13   If spark occurs, sufficient voltage is reaching the plug to fire it (repeat the check at the remaining spark plug wires to verify that the spark plug wires and ignition coils are good). If the ignition system is operating properly the problem lies elsewhere; i.e. a mechanical or fuel system problem. However, the spark plugs may be fouled, so remove and check them as described in Chapter 1.

14   If no spark occurs at one or more wires, remove the suspected spark plug wire from

the ignition coil and check the terminals at both ends for damage. Using an ohmmeter, check the wire for an open or high resistance **(see illustration)**. The resistance of a good spark plug wire should be approximately 600 ohms per foot. If the spark plug wires are good, remove both spark plug wires from the suspected coil and using an ohmmeter, measure the secondary resistance across the coil towers **(see illustration)**. Ignition coil secondary resistance should be approximately 5,000 to 7,000 ohms. If it isn't, replace the defective ignition coil.

15   If the engine won't start due to no spark, check for battery voltage to the ignition module from the ignition switch. Attach a 12 volt test light to the battery negative (-) terminal or other good ground. Disconnect the ignition power/ground electrical connector from the module and check for power at the pink wire terminal **(see illustration)**. Battery voltage should be available with the ignition key On. If there is no battery voltage present, check the

wiring and/or circuit between the power distribution center and ignition control module (don't forget to check the fuses). Also check the black wire terminal for continuity to battery ground. **Note:** *Refer to the wiring diagrams at the end of Chapter 12 for wire color identification for testing and additional information on the circuits.*

16   Check for a crankshaft position sensor signal. Disconnect the crankshaft position sensor connector (purple and yellow wires) from the ignition control module **(see illustration 6.15)**. Connect a digital voltmeter to the terminals in the connector (harness side) and set the meter to read AC volts. Crank the engine and note the voltage. The 7X crankshaft position sensor should produce a minimum of 200 millivolts with the engine cranking. If a crankshaft position sensor signal is not present, check the circuits from the ignition control module to the 7X crankshaft position sensor. If the circuits are good, replace the 7X crankshaft position sensor

5

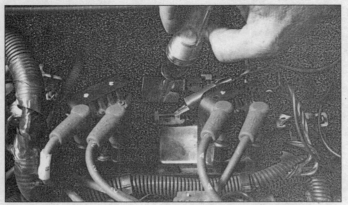

6.17 To check for a trigger signal from the ignition module, remove an ignition coil and connect a test light to the two terminals at the module

7.3a Disconnect the electrical connector from the ignition module

7.3b Remove the ignition coil/module cover mounting bolts (arrows) and pull the assembly straight up

7.4a Remove the ignition coil housing bolts (arrows) and remove the housing

(see Chapter 6).

17   Check for a trigger signal from the ignition control module. Remove one of the ignition coils from the ignition control module. Attach the lead of a test light to one of the coil terminals at the ignition control module and touch the probe of the test light to the other terminal **(see illustration)**. Crank the engine. The test light should blink with the engine cranking if a trigger signal is present. If a trigger signal is present at the coil, the ignition control module and 7X crankshaft position sensor are functioning properly; check the ignition coils as described previously. Check each pair of coil terminals, if necessary. If a trigger signal is not present at one or more of the coil terminal pairs and the 7X crankshaft position sensor is good, replace the ignition control module.

18   If all the components are good and there is no spark, have the PCM checked by a dealer service department or other qualified repair shop.

---

7   Ignition coil and ignition control module - removal and installation

---

1   Disconnect the cable from the negative battery terminal. **Caution:** *On models*

equipped with the Theftlock audio system, be sure the lockout feature is turned off before performing any procedure which requires disconnecting the battery (see the front of this manual).

Four-cylinder models

Refer to illustrations 7.3a, 7.3b, 7.4a and 7.4b

Removal

2   Detach the accelerator and cruise control cables from the retainer and position the cables aside. Remove the fuel line bracket bolt.

3   Disconnect the electrical connector from the ignition control module and remove the ignition coil/module cover mounting bolts **(see illustrations)**. Remove the coil/module cover assembly and place it upside down on a workbench. **Caution:** *If the spark plug boots adhere to the spark plugs, remove them with a spark plug boot removal tool and install them onto the housing secondary terminals. The boots must be installed on the housing before installation or damage may result.*

4   Remove the ignition coil housing screws and remove the housing **(see illustrations)**.

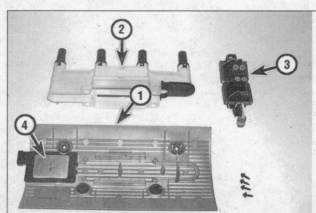

7.4b Ignition coil/module assembly details

1   *Ignition coil/module cover*
2   *Ignition coil housing*
3   *Ignition coils*
4   *Ignition control module*

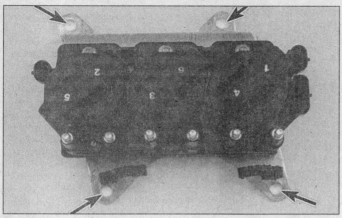

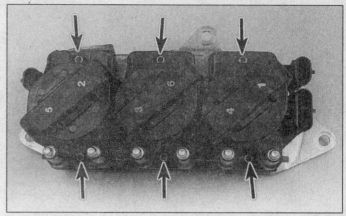

7.13 Remove the ignition coil/module bracket nuts/bolts (arrows) (removed for clarity)

7.14a Remove the ignition coil mounting screws (arrows) . . .

Disconnect the ignition coil electrical connector from the module and carefully remove the ignition coils.

5   Remove the ignition module mounting screws and remove the module.

6   Inspect the ignition coil contacts and seals for damage. Replace the damaged or defective parts as necessary. Do not wipe the silicone grease from the ignition control module if you plan on reusing the module.

Installation

7   Apply an even layer of silicone grease (provided with a new module) to the metal face of the ignition control module (this is essential for heat dissipation). Install the module to the cover.

8   Place the ignition coils in place on the cover and connect the electrical connector to the module.

9   Make sure the ground strap is in place and install the ignition coil housing. Make sure the spark plug boots and retainers are securely in place on the ignition coil housing secondary terminals.

10   Align the spark plug boots with the spark plugs and install the ignition coil/module assembly onto the engine. Apply thread-locking compound to the threads and install the mounting bolts (with the rubber on the insulator/washers down) and tighten the mounting bolts to the torque listed in this Chapter's Specifications.

11   The remainder of installation is the reverse of removal.

V6 models

Refer to illustrations 7.13, 7.14a and 7.14b

Removal

12   Label the spark plug wires corresponding to the cylinder numbers and remove them from the ignition coils. Disconnect the electrical connectors from the ignition module.

13   Remove the ignition coil/module mounting nuts/bolts and remove the assembly from the engine **(see illustration)**.

14   Remove the ignition coil mounting screws and remove the ignition coils from the module **(see illustrations)**.

Installation

15   Align the ignition coil with the blade terminals on the module and press the coil on until it's seated.

16   Install the mounting screws and tighten them securely.

17   Install the coil/module assembly onto the engine and tighten the fasteners securely. Connect the ignition module electrical connectors and install the spark plug wires in their proper locations.

8   Charging system - general information and precautions

The main components of the charging system are alternator (with an integral voltage regulator), the battery and the wiring connecting the components. The components work together to supply electrical power for the electrical system and the alternator maintains the battery in a charged condition. The alternator is driven by the drivebelt at the front of the engine.

All models are equipped with a model CS-130 alternator. CS type alternators should be considered non-serviceable and, if defective, exchanged as cores for new or rebuilt units.

The purpose of the voltage regulator is to limit the alternator voltage output to a preset value. This prevents power surges, circuit overloads, etc., during peak voltage output. On all models with which this manual is concerned, the voltage regulator is mounted inside the alternator housing.

The charging system doesn't ordinarily require periodic maintenance. However, the drivebelt, battery and wires and connections should be inspected at the intervals outlined in Chapter 1.

The dashboard warning light should come on when the ignition key is turned to START, then go off immediately after the engine has started. If it stays on or comes on when the engine is running, a charging system problem has occurred (see Section 9).

7.14b . . .and remove the coils from the module

Be very careful when making electrical circuit connections to a vehicle equipped with an alternator and note the following:

a) *When reconnecting wires to the alternator from the battery, be sure to note the polarity.*

b) *Before using arc welding equipment to repair any part of the vehicle, disconnect the wires from the alternator and the battery terminals.* **Caution:** *If the vehicle is equipped with a Theftlock audio system, make sure you have the correct activation code before disconnecting the battery.*

c) *Never start the engine with a battery charger connected.*

d) *Always disconnect both battery leads before using a battery charger.*

e) *The alternator is turned by an engine drivebelt which could cause serious injury if your hands, hair or clothes become entangled in it with the engine running.*

f) *Because the alternator is connected directly to the battery, it could arc or cause a fire if overloaded or shorted out.*

5

9.2 To measure battery voltage, attach the voltmeter leads to the battery terminals (engine OFF) - to measure charging voltage, start the engine

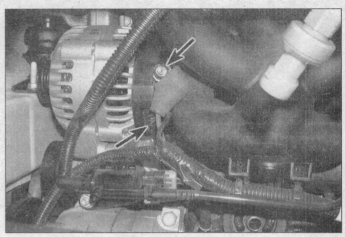

10.3 Disconnect the alternator electrical connections (arrows)

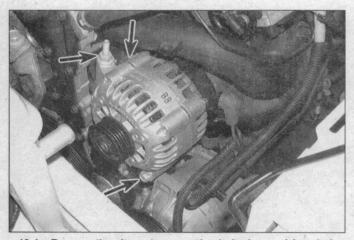

10.4a Remove the alternator mounting bolts (arrows) (one bolt not visible) - four-cylinder models

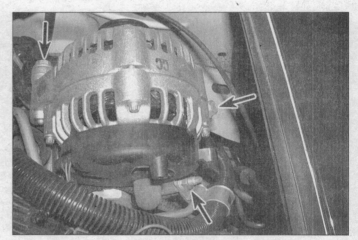

10.4b On V6 models, remove the nut and the power steering line clip from the stud before removing the alternator mounting bolts (arrows)

g) Wrap a plastic bag over the alternator and secure it with rubber bands before steam cleaning the engine.

9 Charging system - check

Refer to illustration 9.2
Note: *These vehicles are equipped with an On-Board Diagnostic (OBD) system that is useful for detecting charging system problems. Refer to Chapter 6 for the list of diagnostic codes and procedures for obtaining the codes.*

1   If a malfunction occurs in the charging circuit, do not immediately assume that the alternator is causing the problem. First check the following items:

a) *The battery cables where they connect to the battery. Make sure the connections are clean and tight.*
b) *The battery electrolyte specific gravity (by observing the charge indicator on*

the battery). If it is low, charge the battery.
c) *Check the external alternator wiring and connections.*
d) *Check the drivebelt condition and tension (see Chapter 1).*
e) *Check the alternator mounting bolts for tightness.*
f) *Run the engine and check the alternator for abnormal noise.*

2   Connect a voltmeter to the positive and negative battery terminals **(see illustration)**. Check the battery voltage with the engine off. It should be approximately 12.4 to 12.6 volts, if the battery is fully charged.

3   Start the engine and check the battery voltage again. It should now be greater than the voltage recorded in Step 2, but not more than 14.5 volts.

4   If the indicated voltage reading is less or more than the specified charging voltage, have the charging system checked at a dealer service department or other properly equipped repair facility.

10 Alternator - removal and installation

Refer to illustrations 10.3, 10.4a and 10.4b
1   Disconnect the cable from the negative battery terminal. **Caution:** *On models equipped with the Theftlock audio system, be sure the lockout feature is turned off before performing any procedure which requires disconnecting the battery (see the front of this manual).*
2   Remove the drivebelt (see Chapter 1).
3   Disconnect the output wire and the electrical connector from the alternator **(see illustration)**.
4   Remove the mounting bolts and remove the alternator from the engine **(see illustrations)**.
5   If you are replacing the alternator, take the old one with you when purchasing a replacement unit. Make sure the new/rebuilt unit looks identical to the old alternator. Look at the terminals - they should be the same in

number, size and location as the terminals on the old alternator. Finally, look at the identification numbers - they will be stamped into the housing or printed on a tag attached to the housing. Make sure the numbers are the same on both alternators.

6   Many new/rebuilt alternators do not have a pulley installed, so you may have to switch the pulley from the old unit to the new/rebuilt one. When buying an alternator, find out the shop's policy regarding pulleys; some shops will perform this service free of charge.

7   Installation is the reverse of removal. Tighten the mounting bolts to the torque listed in this Chapter's Specifications.

8   Install the drivebelt (see Chapter 1).

9   Check the charging voltage to verify proper operation of the alternator (see Section 9).

11   Starting system - general information and precautions

The starter motor assembly is a permanent magnet, planetary gear drive starter motor. The starter motor assembly is serviced as a complete unit. If any component of the starter motor fails, including the solenoid, the entire assembly must be replaced.

The sole function of the starting system is to turn over the engine quickly enough to allow it to start. The starting system consists of the battery, starter motor assembly and the wiring connecting the components.

When the ignition key is turned to the START position, the starter solenoid is actuated through the starter control circuit. The starter solenoid then connects the battery to the starter motor. The battery supplies the electrical energy to the starter motor, which does the actual work of cranking the engine.

Always observe the following precautions when working on the starting system:

a) *Excessive cranking of the starter motor can overheat it and cause serious damage. Never operate the starter motor for more than 15 seconds at a time without pausing to allow it to cool for at least two minutes.*

b) *The starter is connected directly to the battery and could arc or cause a fire if mishandled, overloaded or shorted.*

c) *Always detach the cable from the negative terminal of the battery before working on the starting system.*

12   Starter motor and circuit - check

Refer to illustration 12.4

1   If a malfunction occurs in the starting circuit, do not immediately assume that the starter is causing the problem. First, check the following items:

a) *Make sure the battery cable clamps, where they connect to the battery, are clean and tight.*

b) *Check the condition of the battery cables (see Section 4). Replace any defective battery cables with new parts.*

c) *Test the condition of the battery (see Section 3). If it does not pass all the tests, replace it with a new battery.*

d) *Check the starter motor wiring and connections.*

e) *Check the starter motor mounting bolts for tightness.*

f) *Check the related fuses in the engine compartment fuse box (see Chapter 12). If they're blown, determine the cause and repair the circuit.*

g) *Check the ignition switch circuit for correct operation (see Chapter 12).*

h) *Check the operation of the clutch start switch (manual transaxle) or the Park/Neutral position switch (automatic transaxle) (see Chapter 8 or 7B). These systems must operate correctly to provide battery voltage to the starter solenoid.*

2   If the starter does not activate when the ignition switch is turned to the start position, check for battery voltage to the starter solenoid. This will determine if the solenoid is receiving the correct voltage from the ignition switch. Install a 12-volt test light or a voltmeter to the starter solenoid terminal (purple wire). While an assistant turns the ignition switch to the start position, observe the test light or voltmeter. The test light should shine brightly or battery voltage should be indicated on the voltmeter. If voltage is not available to the starter solenoid, refer to the wiring diagrams in Chapter 12 and check the fuses and related wiring in series with the starting system. If voltage is available but there is no movement from the starter motor, remove the starter from the engine (see Section 14) and bench test the starter (see Step 4).

3   If the starter turns over slowly, check the starter cranking voltage and the current draw from the battery. This test must be performed with the starter assembly on the engine. Crank the engine over (for 10 seconds or less) and observe the battery voltage. It should not drop below 8.5 volts. Also, observe the current draw using an amp meter. Typically a starter amperage draw should not exceed 200 amps. If the starter motor amperage draw is excessive, have it tested by a dealer service department or other qualified repair shop. There are several conditions that may affect the starter cranking potential. The battery must be in good condition and the battery cold-cranking rating must not be under-rated for the particular application. Be sure to check the battery specifications carefully. The battery terminals and cables must be clean and not corroded. Also, in cases of extreme cold temperatures, make sure the battery and/or engine block is warmed before performing the tests.

4   If the starter is receiving voltage but does not activate, remove and check the starter motor assembly on the bench. Most likely the starter motor or solenoid is defective. In some rare cases, the engine may be seized so be sure to try and rotate the crankshaft pulley (see Chapter 2A or 2B) before proceeding. With the starter assembly mounted in a vise on the bench, install one jumper cable from the positive terminal of a test battery to the B+ terminal on the starter. Install another jumper cable from the negative terminal of the battery to the body of the starter **(see illustration)**. Install a starter switch and apply battery voltage to the solenoid S terminal (for 10 seconds or less) and observe the solenoid plunger, shift lever and overrunning clutch extend and rotate the pinion drive. If the pinion drive extends but does not rotate, the solenoid is operating but

5

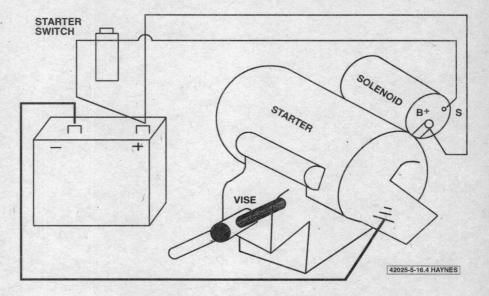

12.4 Starter motor bench testing details

the starter motor is defective. If there is no movement but the solenoid clicks, the solenoid and/or the starter motor is defective. If the solenoid plunger extends and rotates the pinion drive, the starter assembly is operating properly.

13   Starter motor - removal and installation

Refer to illustrations 13.5, 13.6a and 13.6b

1   Disconnect the cable from the negative battery terminal. **Caution:** *On models equipped with the Theftlock audio system, be sure the lockout feature is turned off before* performing any procedure which requires disconnecting the battery (see the front of this manual).

2   On four-cylinder models, remove the air intake duct and resonator (see Chapter 4). Remove the upper starter mounting bolt.

3   Raise the vehicle and support it securely on jackstands.

4   Remove the splash shield, if equipped.

5   Disconnect the wires from the terminals on the starter motor solenoid **(see illustration)**.

6   Remove the starter mounting bolts **(see illustrations)**. Carefully remove the starter from the vehicle.

7   Installation is the reverse of removal.

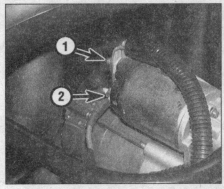

13.5 Remove the nuts and disconnect the battery cable (1) and the solenoid terminal (2) from the starter motor

13.6a Starter mounting bolts (arrows) - four-cylinder models

13.6b Starter mounting bolts (arrows) - V6 models

Chapter 6
Emissions and engine control systems

Contents

1   General information

Refer to illustrations 1.1a, 1.1b and 1.6

To prevent pollution of the atmosphere from incompletely burned and evaporating gases, and to maintain good driveability and fuel economy, a number of emission control systems are incorporated **(see illustrations)**. They include the:

Electronic engine control system
Crankcase ventilation system
Exhaust gas recirculation system
Evaporative emissions control system
Secondary air injection system
   (some 1999 and 2000 V6 models)
Catalytic converter

All of these systems are linked, directly or indirectly, to the emission control system.

The Sections in this Chapter include general descriptions, checking procedures

1.1a Typical emission and engine control system components - four-cylinder models

1	Camshaft position sensor	4	TPS and MAP sensor (on side of throttle body)
2	Oxygen sensor - upstream (on exhaust manifold)	5	Idle Air Control (IAC) valve
3	Intake Air Temperature (IAT) sensor	6	EVAP canister purge solenoid valve

1.1b Typical emission and engine control system components - V6 models

1 *Manifold absolute pressure sensor*	5 *Mass airflow sensor*	8 *AIR solenoid valve*
2 *EGR valve*	6 *Throttle position sensor (on side of*	9 *AIR check valve*
3 *Idle air control valve*	*throttle body)*	10 *AIR pump*
4 *Intake air temperature sensor*	7 *PCV valve*	

within the scope of the home mechanic and component replacement procedures (when possible) for each of the systems listed above.

Before assuming that an emissions control system is malfunctioning, check the fuel and ignition systems carefully. The diagnosis of some emission control devices requires specialized tools, equipment and training. If checking and servicing become too difficult or if a procedure is beyond your ability, con-

1.6 The Vehicle Emission Control Information (VECI) label is located in the engine compartment and contains information on the emission devices on your vehicle, vacuum line routing, etc.

sult a dealer service department or other properly equipped repair facility. Remember, the most frequent cause of emissions problems is simply a loose or broken vacuum hose or wire, so always check the hose and wiring connections first.

This doesn't mean, however, that emission control systems are particularly difficult to maintain and repair. You can quickly and easily perform many checks and do most of the regular maintenance at home with common tune-up and hand tools. **Note:** *Because of a Federally mandated warranty which covers the emission control system components, check with your dealer about warranty coverage before working on any emissions-related systems. Once the warranty has expired, you may wish to perform some of the component checks and/or replacement procedures in this Chapter to save money.*

Pay close attention to any special precautions outlined in this Chapter. It should be noted that the illustrations of the various systems may not exactly match the system installed on the vehicle you're working on because of changes made by the manufacturer during production or from year-to-year.

A Vehicle Emissions Control Information (VECI) label is located in the engine compartment **(see illustration)**. This label contains important emissions specifications and adjustment information, as well as a vacuum hose schematic with emissions components identified. When servicing the engine or emis-

sions systems, the VECI label in your particular vehicle should always be checked for up-to-date information.

---

2 On-Board Diagnostic system and trouble codes

Diagnostic tool information

Refer to illustrations 2.1 and 2.2

1   A digital multimeter is necessary for checking fuel injection and emission related components **(see illustration)**. A digital volt-ohmmeter is preferred over the older style analog multimeter for several reasons. The analog multimeter cannot display the volts-ohms or amps measurement in hundredths and thousandths increments. When working with electronic circuits which are often very low voltage, this accurate reading is most important. Another good reason for the digital multimeter is the high impedance circuit. The digital multimeter is equipped with a high resistance internal circuitry (10 million ohms). Because a voltmeter is hooked up in parallel with the circuit when testing, it is vital that none of the voltage being measured should be allowed to travel the parallel path set up by the meter itself. This dilemma does not show itself when measuring larger amounts of voltage (9 to 12 volt circuits) but if you are measuring a low voltage circuit such as the oxygen sensor signal voltage, a fraction of a

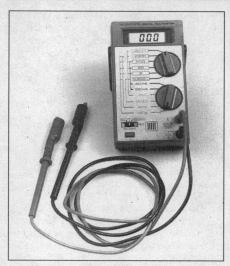

2.1 Digital multimeters can be used for testing all types of circuits; because of their high impedance, they are much more accurate than analog meters for measuring low-voltage computer circuits

2.2 Scanners like the Actron OBD II Diagnostic Tester, Actron Scantool and the AutoXray XP240 are powerful diagnostic aids - programmed with comprehensive diagnostic information, they can tell you just about anything you want to know about your engine management system

volt may be a significant amount when diagnosing a problem. However, there are several exceptions where using an analog voltmeter may be necessary to test certain sensors.

2   Hand-held scanners are the most powerful and versatile tools for analyzing engine management systems used on later model vehicles **(see illustration)**. Each brand scan tool must be examined carefully to match the year, make and model of the vehicle you are working on. Often interchangeable cartridges are available to access the particular manufacturer (Ford, GM, Chrysler, etc.). Some manufacturers will specify by continent (Asia, Europe, USA, etc.).

3   With the arrival of the Federally mandated emission control system (OBD-II), a specially designed scanner has been developed. Several tool manufacturers have released OBD-II scan tools for the home mechanic. Ask the parts salesman at a local auto parts store for additional information concerning availability and cost.

On-Board Diagnostic system general description

4   All models described in this manual are equipped with the second generation On-Board Diagnostic (OBD-II) system. The system consists of an onboard computer, known as the Powertrain Control Module (PCM), information sensors and output actuators.

5   The information sensors monitor various functions of the engine and send data to the PCM. Based on the data and the information programmed into the computer's memory, the PCM generates output signals to control various engine functions via control relays, solenoids and other output actuators. The PCM is specifically calibrated to optimize the emissions, fuel economy and driveability of the vehicle.

6   Because of a Federally mandated warranty which covers the emissions system components and because any owner-induced damage to the PCM, the sensors and/or the control devices may void the warranty, it isn't a good idea to attempt diagnosis or replacement of the PCM at home while the vehicle is under warranty. Take the vehicle to a dealer service department if the PCM or a system component malfunctions.

Information sensors

7   **Camshaft position sensor** - The camshaft position sensor provides information on camshaft position. The PCM uses this information, along with the crankshaft position sensor information, to control fuel injection synchronization.

8   **Crankshaft position sensor** - The crankshaft position sensor senses crankshaft position (TDC) during each engine revolution. The PCM uses this information to control ignition timing and fuel injection synchronization.

9   **Engine coolant temperature sensor** - The engine coolant temperature sensor senses engine coolant temperature. The PCM uses this information to control fuel injection duration and ignition timing.

10   **Intake air temperature sensor** - The intake air temperature senses the temperature of the air entering the intake manifold. The PCM uses this information to control fuel injection duration.

11   **Knock sensor** - The knock sensor is a piezoelectric element that detects the sound of engine detonation, or "pinging". The PCM uses the input signal from the knock sensor to recognize detonation and retard spark advance to avoid engine damage.

12   **Manifold absolute pressure sensor** - The manifold absolute pressure monitors intake manifold pressure and ambient barometric pressure. The PCM uses this input signal to determine engine load and adjusts fuel injection duration accordingly.

13   **Mass airflow sensor (V6 only)** - The mass airflow sensor measures the amount of air passing through the sensor body and ultimately entering the engine. The PCM uses this information to control fuel delivery.

14   **Oxygen sensor** - The oxygen sensors generate a voltage signal that varies with the difference between the oxygen content of the exhaust and the oxygen in the surrounding air. The PCM uses this information to determine if the fuel system is running rich or lean.

15   **Throttle position sensor** - The throttle position sensor senses throttle movement and position. This signal enables the PCM to determine when the throttle is closed, in a cruise position, or wide open. The PCM uses this information to control fuel delivery and ignition timing.

16   **Vehicle speed sensor** - The vehicle speed sensor provides information to the PCM to indicate vehicle speed.

17   **Miscellaneous PCM inputs** - In addition to the various sensors, the PCM monitors various switches and circuits to determine vehicle operating conditions. The switches and circuits include:
   a)   *Air conditioning system*
   b)   *Battery voltage*
   c)   *Brake On/Off switch*
   d)   *Cruise control system*
   e)   *EGR valve position*
   f)   *Engine oil level and pressure*
   g)   *EVAP system*
   h)   *Fuel level and fuel tank pressure*
   i)   *Ignition switch*
   j)   *Park/neutral position switch*
   k)   *Sensor signal and ground circuits*
   l)   *Transaxle controls*

Output actuators

18   **Air conditioning clutch relay** - The PCM controls the operation of the air conditioning compressor clutch with the air conditioning clutch relay.

19   **Check Engine light** - The PCM will illuminate the Check Engine light if a malfunction in the electronic engine control system occurs.

20   **Cruise control module** - The cruise control system operation is controlled by the PCM.

21   **Engine cooling fan relay** - The engine cooling fan is controlled by the PCM according to information received from the engine coolant temperature sensor.

6

22   **EGR valve** - The electronic EGR valve is controlled by the PCM. Ideal EGR flow is determined by the PCM and the EGR valve pintle position is adjusted accordingly.

23   **EVAP canister purge and vent valve solenoids** - The evaporative emission canister purge and vent valve solenoids are operated by the PCM to purge the fuel vapor canister and route fuel vapor to the intake manifold for combustion.

24   **Secondary air injection pump and vacuum valve/solenoid (some 1999 and 2000 V6 models** - The PCM operates the secondary air injection pump and opens the vacuum valve to inject fresh air into the exhaust stream, lower emission levels under certain operating conditions.

25   **Fuel injectors** - The PCM opens the fuel injectors individually in firing order sequence. The PCM also controls the time the injector is held open (pulse width). The pulse width of the injector (measured in milliseconds) determines the amount of fuel delivered. For more information on the fuel delivery system and the fuel injectors, including injector replacement, refer to Chapter 4.

26   **Fuel pump relay** - The fuel pump relay is activated by the PCM with the ignition switch in the Start or Run position. When the ignition switch is turned on, the relay is activated to supply initial line pressure to the system. For more information on fuel pump check and replacement, refer to Chapter 4.

27   **Idle air control valve** - The idle air control valve controls the amount of air allowed to bypass the throttle plate when the throttle valve is closed or at idle position. The more air allowed to bypass the throttle plate, the higher the idle speed. The idle air control valve opening and the resulting idle speed is controlled by the PCM.

28   **Ignition control module** - The PCM controls ignition timing through the ignition control module depending on engine operation conditions. Refer to Chapter 5 for more information on the ignition control module.

Obtaining diagnostic trouble codes

Refer to illustration 2.30

Note: *The diagnostic trouble codes on all models can only be extracted from the Powertrain Control Module (PCM) using a specialized scan tool. Have the vehicle diagnosed by a dealer service department or other qualified automotive repair facility if the proper scan tool is not available.*

29   The PCM will illuminate the CHECK ENGINE light (also known as the Malfunction Indicator Lamp) on the dash if it recognizes a fault in the system. The light will remain illuminated until the problem is repaired and the code is cleared or the PCM does not detect any malfunction for several consecutive drive cycles.

30   The diagnostic codes for the On-Board Diagnostic (OBD) system can only be extracted from the PCM using a scan tool. The scan tool is programmed to interface with the OBD system by plugging into the diagnostic connector **(see illustration)**. When used, the scan tool has the ability to diagnose in-depth driveability problems and it allows freeze frame data to be retrieved from the PCM stored memory. Freeze frame data is an OBD II PCM feature that records all related sensor and actuator activity on the PCM data stream whenever an engine con-

trol or emissions fault is detected and a trouble code is set. This ability to look at the circuit conditions and values when the malfunction occurs provides a valuable tool when trying to diagnose intermittent driveability problems. If the tool is not available and intermittent driveability problems exist, have the vehicle checked at a dealer service department or other qualified repair shop.

Clearing diagnostic trouble codes

31   After the system has been repaired, the codes must be cleared from the PCM memory. The preferred method is with a scan tool, but the codes can be cleared by disconnecting battery power from the PCM for a minimum of thirty seconds. Battery power can be disconnected from the PCM by removing the PCM fuse, disconnecting the PCM power connector near the positive battery terminal (if equipped) or by disconnecting the negative battery cable from the battery. **Caution:** *On models equipped with the Theftlock audio system, be sure the lockout feature is turned off before performing any procedure which requires disconnecting the battery (see the front of this manual).*

32   Always clear the codes from the PCM before starting the engine after a new electronic emission control component is installed onto the engine. The PCM stores the operating parameters of each sensor. The PCM may set a trouble code if a new sensor is allowed to operate before the parameters from the old sensor have been erased.

Diagnostic trouble code identification

33   The accompanying list of diagnostic trouble codes is a compilation of all the codes that may be encountered using a generic scan tool. Additional trouble codes may be available with the use of the manufacturer specific scan tool. Not all codes pertain to all models and not all codes will illuminate the Check Engine light when set. All models require a scan tool to access the diagnostic trouble codes.

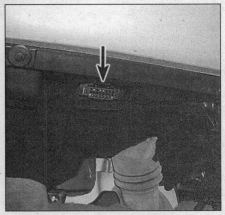

2.30 The diagnostic connector is typically located under the instrument panel

Code	Code Identification
P0101	Mass air flow sensor circuit, range or performance problem
P0102	Mass air flow sensor circuit, low input
P0103	Mass air flow sensor circuit, high input
P0105	Manifold absolute pressure sensor or throttle position sensor circuit malfunction
P0106	Manifold absolute pressure sensor circuit, range or performance problem
P0107	Manifold absolute pressure sensor circuit, low input
P0108	Manifold absolute pressure sensor circuit, high input
P0112	Intake air temperature circuit, low input
P0113	Intake air temperature circuit, high input
P0117	Engine coolant temperature circuit, low input
P0118	Engine coolant temperature circuit, high input
P0121	Throttle position sensor circuit, range or performance problem

Code	Code Identification
P0122	Throttle position sensor circuit, low input
P0123	Throttle position sensor circuit, high input
P0125	Insufficient coolant temperature for closed loop fuel control
P0131	Oxygen sensor circuit, low voltage (upstream sensor)
P0132	Oxygen sensor circuit, high voltage (upstream sensor)
P0133	Oxygen sensor circuit, slow response (upstream sensor)
P0134	Oxygen sensor circuit - no activity detected (upstream sensor)
P0135	Oxygen sensor heater circuit malfunction (upstream sensor no)
P0137	Oxygen sensor circuit, low voltage (downstream sensor)
P0138	Oxygen sensor circuit, high voltage (downstream sensor)
P0140	Oxygen sensor circuit - no activity detected (downstream sensor)
P0141	Oxygen sensor heater circuit malfunction (downstream sensor)
P0171	System too lean
P0172	System too rich
P0200	Injector circuit malfunction
P0201	Injector circuit malfunction - cylinder no. 1
P0202	Injector circuit malfunction - cylinder no. 2
P0203	Injector circuit malfunction - cylinder no. 3
P0204	Injector circuit malfunction - cylinder no. 4
P0205	Injector circuit malfunction - cylinder no. 5
P0206	Injector circuit malfunction - cylinder no. 6
P0218	Transaxle fluid over temperature
P0230	Fuel pump primary circuit malfunction
P0300	Random/multiple cylinder misfire detected
P0301	Cylinder no. 1 misfire detected
P0302	Cylinder no. 2 misfire detected
P0303	Cylinder no. 3 misfire detected
P0304	Cylinder no. 4 misfire detected
P0305	Cylinder no. 5 misfire detected
P0306	Cylinder no. 6 misfire detected
P0325	Knock sensor circuit malfunction
P0327	Knock sensor circuit, low input
P0335	Crankshaft position sensor circuit malfunction
P0336	Crankshaft position sensor circuit, range or performance problem
P0341	Camshaft position sensor circuit, range or performance problem
P0342	Camshaft position sensor circuit, low input
P0401	Exhaust gas recirculation, insufficient flow detected
P0403	Exhaust gas recirculation circuit malfunction
P0404	Exhaust gas recirculation circuit, range or performance problem
P0405	Exhaust gas recirculation sensor circuit low
P0410	Secondary air injection system
P0412	Secondary air injection solenoid control circuit
P0418	Secondary air injection pump relay control circuit
P0420	Catalyst system efficiency below threshold
P0440	Evaporative emission control system malfunction
P0442	Evaporative emission control system, small leak detected
P0443	Evaporative emission control system, purge control circuit malfunction
P0446	Evaporative emission control system, vent system performance
P0449	Evaporative emission control system, vent control circuit malfunction
P0452	Evaporative emission control system, pressure sensor low input
P0453	Evaporative emission control system, pressure sensor high input
P0460	Fuel level sensor circuit malfunction

6

Code	Code Identification
P0461	Fuel level sensor circuit, range or performance problem
P0462	Fuel level sensor circuit, low input
P0463	Fuel level sensor circuit, high input
P0480	Cooling fan control circuit malfunction
P0481	Cooling fan control circuit malfunction
P0502	Vehicle speed sensor circuit low output
P0503	Vehicle speed sensor signal intermittent
P0506	Idle control system, rpm lower than expected
P0507	Idle control system, rpm higher than expected
P0530	Air conditioning refrigerant pressure sensor, circuit malfunction
P0560	System voltage malfunction
P0562	System voltage low
P0563	System voltage high
P0601	Internal control module, memory error
P0602	Control module, programming error
P0620	Charging system performance
P0650	Malfunction Indicator Light (MIL) control circuit
P0654	Tachometer control circuit
P0705	Transmission range sensor, circuit malfunction (PRNDL input)
P0706	Transaxle range switch performance
P0711	Fluid temperature sensor circuit out-of-range
P0712	Fluid temperature sensor circuit low input
P0713	Fluid temperature sensor circuit high input
P0716	Input speed sensor circuit out-of-range
P0717	Input speed sensor circuit no signal
P0719	Torque converter clutch brake switch circuit low
P0724	Torque converter clutch brake switch circuit high
P0730	Incorrect gear ratio
P0741	Torque converter clutch system stuck off
P0742	Torque converter clutch system stuck on
P0748	Pressure control solenoid valve circuit
P0751	1-2 shift solenoid performance
P0753	1-2 shift solenoid circuit
P0756	2-3 shift solenoid performance
P0758	2-3 shift solenoid circuit
P1106	Manifold absolute pressure sensor circuit intermittent high voltage
P1107	Manifold absolute pressure sensor circuit intermittent low voltage
P1111	Intake air temperature sensor circuit intermittent high voltage
P1112	Intake air temperature sensor circuit intermittent low voltage
P1114	Engine coolant temperature sensor circuit intermittent low voltage
P1115	Engine coolant temperature sensor circuit intermittent high voltage
P1121	Throttle position sensor circuit intermittent high voltage
P1122	Throttle position sensor circuit intermittent low voltage
P1133	Oxygen sensor insufficient switching (upstream sensor)
P1134	Oxygen sensor transition time ratio (upstream sensor)
P1171	Fuel system lean during acceleration
P1200	Fuel injector control circuit
P1336	Crankshaft position sensor system variation not learned
P1350	By-pass line monitor
P1351	Ignition control circuit open
P1352	Ignition by-pass circuit open
P1361	Ignition control circuit not switching

Code	Code Identification
P1362	Ignition by-pass circuit shorted
P1374	Crankshaft position sensor 3X reference circuit
P1380	Electronic brake control module rough road sensing error
P1381	No serial data from electronic brake control module
P1404	EGR valve closed pintle position
P1406	EGR valve pintle position circuit
P1441	EVAP system flow during non-purge
P1554	Cruise control status circuit
P1573	Serial data communication failure with electronic brake and traction control module
P1601	Serial communication malfunction
P1602	Serial data communication failure with electronic brake control module
P1610	Serial data communication failure with body function controller
P1621	PCM memory performance
P1626	Serial data communication failure with vehicle theft deterent controller (no password)
P1630	Theft deterent PCM in learn mode
P1631	Theft deterent password incorrect
P1632	Theft deterent fuel disabled
P1635	5-volt reference circuit
P1639	5-volt reference circuit
P1641	Malfunction indicator light control circuit
P1651	Electric cooling fan relay control circuit
P1652	Electric cooling fan relay control circuit
P1654	Air conditioning relay control circuit
P1655	Evaporative emission control system, purge valve solenoid control circuit
P1662	Cruise control inhibit control circuit
P1665	Evaporative emission control system, vent valve solenoid control circuit
P1671	Malfunction indicator light control circuit
P1675	Evaporative emission control system, vent solenoid control circuit
P1676	Evaporative emission control system, purge valve solenoid control circuit
P1810	Pressure switch assembly malfunction
P1811	Long shift time
P1860	Torque converter clutch pulse width modulator solenoid circuit
P1887	Torque converter clutch release switch malfunction

---

3 Powertrain Control Module - (PCM) removal and installation

Refer to illustrations 3.3 and 3.4

Caution: *Avoid static electricity damage to the Powertrain Control Module (PCM) by grounding yourself to the body of the vehicle before touching the PCM and using a special anti-static pad to store the PCM on, once it is removed.*

Note 1: *Anytime the PCM is replaced with a new unit the PCM must be reprogrammed by a dealership service department with special equipment. A crankshaft position sensor variation relearn procedure and a vehicle anti-theft system password relearn procedure must be performed, as well. The following procedure pertains to removal and installation of the original PCM only. If the PCM must be replaced with a new unit, take the vehicle to a dealership service department.*

Note 2: *Anytime the battery is disconnected stored operating parameters may be lost from the PCM causing the engine to run rough for sometime while the PCM relearns the information.*

1   Disconnect the cable from the negative battery terminal. **Caution:** *On models equipped with the Theftlock audio system, be sure the lockout feature is turned off before performing any procedure which requires disconnecting the battery (see the front of this manual).*

2   The PCM is located under the instrument panel, near the steering column. Remove the insulation panel from beneath the instrument panel on the driver's side.

3   Loosen the PCM electrical connector retaining screws and carefully disconnect the electrical connectors from the PCM **(see Illustration)**.

3.3 Loosen the screws (arrows) and disconnect the electrical connectors from the PCM

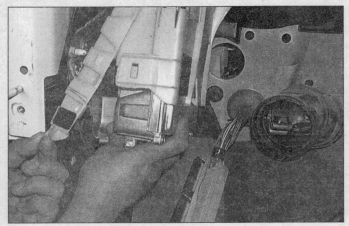

3.4 Slide the PCM out of the mounting bracket

4.2 The TPS (arrow) is located on the side of the throttle body

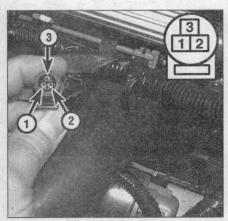

4.3 Disconnect the electrical connector from the TPS and check the voltage supply and ground circuits from the PCM at the harness connector

1   *5-volt supply*
2   *Sensor ground*
3   *Throttle position sensor signal*

4   Remove the PCM retainer and slide the PCM out of the bracket **(see illustration)**.
5   Installation is the reverse of removal.

4   Throttle Position Sensor (TPS) - check and replacement

1   The Throttle Position Sensor (TPS) is a variable potentiometer connected to the end of the throttle shaft on the throttle body. By monitoring the output voltage from the TPS, the PCM can determine fuel delivery based on throttle valve angle (driver demand). A broken or loose TPS can cause intermittent bursts of fuel from the injectors and an unstable idle because the PCM thinks the throttle is moving.

Check

Refer to illustrations 4.2, 4.3 and 4.4
Note: *Performing the following test will set a diagnostic trouble code and illuminate the*

4.4 To check the TPS, backprobe the dark blue wire terminal of the TPS connector with a voltmeter

Check Engine light. Clear the diagnostic trouble code after performing the tests and making the necessary repairs (see Section 2).
2   The Throttle Position Sensor (TPS) is located on the side of the throttle body **(see illustration)**. Check the terminals in the connector and the wires leading to the sensor for looseness and breaks. Repair as required.
3   Before checking the TPS, check the voltage supply and ground circuits from the PCM. Disconnect the electrical connector from the TPS and connect the positive lead of a voltmeter to the gray/black wire terminal and the negative lead to the black wire terminal at the harness connector **(see illustration)**. Turn the ignition key On - the voltage should read approximately 5.0 volts. If the voltage is incorrect, check the wiring from the TPS to the PCM. If the circuits are good, have the PCM checked at a dealer service department or other properly equipped repair facility.
4   To check the TPS operation, reconnect the connector to the TPS and using a suitable probe, backprobe the dark blue wire terminal of the TPS connector **(see illustration)** (see Chapter 12 for additional information on how to backprobe a connector). Connect the positive lead of a voltmeter to the probe and the

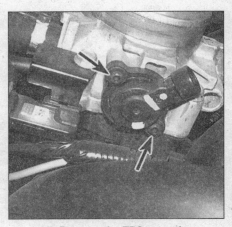

4.7 Remove the TPS mounting screws (arrows)

negative lead to a good engine ground point. Turn the ignition key On - with the throttle fully closed the voltage should read approximately 0.6 volts. Gradually open the throttle - the voltage should increase smoothly to approximately 4.5 volts at wide-open throttle. If the test results are incorrect, replace the TPS.

Replacement

Refer to illustration 4.7
5   On four-cylinder models, remove the air intake duct and resonator (see Chapter 4).
6   Disconnect the electrical connector from the TPS.
7   Remove the TPS mounting screws and remove the TPS from the throttle body **(see illustration)**.
8   Install a new O-ring on the TPS. With the throttle in the closed position, align the TPS with the throttle shaft and install the TPS. On four-cylinder models, install new mounting screws (included with a new TPS). On V6 models, apply a drop of thread sealing compound to the mounting screw threads. Tighten the screws securely.
9   The remainder of installation is the reverse of removal.

5.2 MAP sensor is location -
four-cylinder models

5.3 Disconnect the
electrical connector
from the MAP sensor
and check the
voltage supply and
ground circuits from
the PCM at the
harness connector

1   Sensor ground
2   MAP sensor
    signal
3   5-volt supply

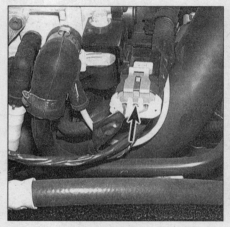

5.4 To check the MAP sensor, backprobe
the light green wire terminal of the MAP
sensor connector with a voltmeter

5.6a Remove the MAP sensor mounting
bolt (arrow) - four-cylinder models

5.6b Remove the MAP sensor mounting
screws (arrows) - V6 models

6

5   Manifold Absolute Pressure (MAP) sensor - check and replacement

1   The Manifold Absolute Pressure (MAP) sensor monitors the intake manifold pressure changes resulting from changes in engine load and speed and converts the information into a voltage output. The PCM receives information as a varying voltage signal from closed throttle (high vacuum) to wide open throttle (low vacuum). The PCM uses the MAP sensor to control fuel delivery and ignition timing.

Check

Refer to illustrations 5.2, 5.3 and 5.4
Note: *Performing the following test will set a diagnostic trouble code and illuminate the Check Engine light. Clear the diagnostic trouble code after performing the tests and making the necessary repairs (see Section 2).*
2   On four-cylinder models, the MAP sensor is located next to the throttle body **(see Illustration)**. On V6 models, the MAP sensor is located on the firewall side of the upper

intake manifold plenum **(see illustration 1.1b)**. Check the terminals in the connector and the wires leading to the sensor for looseness and breaks. Repair as required.
3   Before checking the MAP sensor, check the vacuum supply line. Make sure full manifold vacuum is present with the engine running. Check the voltage supply and ground circuits from the PCM. Disconnect the electrical connector from the MAP sensor and connect the positive lead of a voltmeter to the brown or gray wire terminal and the negative lead to the black wire terminal at the harness connector **(see illustration)**. Turn the ignition key On - the voltage should read approximately 5.0 volts. If the voltage is incorrect, check the wiring from the MAP sensor to the PCM. If the circuits are good, have the PCM checked at a dealer service department or other properly equipped repair facility.
4   To check the MAP sensor operation, reconnect the connector to the MAP sensor and using a suitable probe, backprobe the light green wire terminal of the MAP sensor connector **(see illustration)** (see Chapter 12 for additional information on how to backprobe a connector). Connect the positive lead of a voltmeter to the probe and the neg-

ative lead to a good engine ground point. Turn the ignition key On - with the engine not running the voltage should read 4.0 to 5.0 volts. Start the engine and allow it to idle - the voltage should decrease to approximately 0.5 to 2.0 volts. If the test results are incorrect, replace the MAP sensor.

Replacement

Refer to illustrations 5.6a and 5.6b
5   Disconnect the electrical connector and vacuum hose from the MAP sensor.
6   On four-cylinder models, remove the throttle body mounting bolt and remove the MAP sensor. On V6 models, remove the screws that retain the MAP sensor to the upper intake manifold plenum bracket **(see illustrations)**. Remove the MAP sensor.
7   Installation is the reverse of removal.

6   Mass Airflow (MAF) sensor - check and replacement

1   A Mass Airflow sensor (MAF) is used on V6 models only. The MAF sensor measures the amount of air passing through the sensor

6.2 The mass airflow sensor is located in the air intake duct

6.3 Disconnect the electrical connector from the MAF sensor and check the voltage supply and ground circuits at the harness connector

1   *MAF sensor signal*
2   *Sensor ground*
3   *12-volt supply*

6.6 Loosen the hose clamps (arrows) and remove the mass airflow sensor

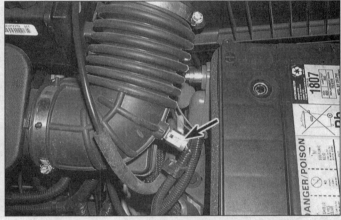

7.2 Intake air temperature sensor location

body and ultimately entering the engine through the throttle body. The PCM uses this information to control fuel delivery - the more air entering the engine (acceleration), the more fuel required.

Check

Refer to illustrations 6.2 and 6.3

2   The Mass Airflow Sensor (MAF) is located on the air intake duct **(see illustration)**. A scan tool is necessary to check the output of the MAF sensor (see Section 2). The scan tool displays the sensor output in grams per second. With the engine idling at normal operating temperature, the display should read approximately 4.0 to 6.0 grams per second. When the engine is accelerated the values should rise quickly and remain steady at a steady engine speed.

3   Before checking the MAF sensor operation, check the voltage supply and ground circuits. Disconnect the electrical connector from the MAF sensor and connect the positive lead of a voltmeter to the pink wire terminal and the negative lead to the black/white wire terminal at the harness connector **(see illustration)**. Turn the ignition key On - the voltage should read approximately 12.0 volts. If the voltage is incorrect, check the circuits

from the MAP sensor to the power distribution center and engine ground point (don't forget to check the fuses first). If the power and ground circuits are good, check the MAF sensor operation with a scan tool. If the MAF sensor does not respond as described in Step 2, replace the MAF sensor.

Replacement

Refer to illustration 6.6

4   Disconnect the electrical connector from the intake air temperature sensor and the MAF sensor.

5   Loosen the hose clamp securing the air intake duct to the MAF sensor and remove the duct.

6   Loosen the hose clamp retaining the MAF sensor to the air filter cover and remove the sensor **(see illustration)**.

7   Installation is the reverse of removal.

7   Intake Air Temperature (IAT) sensor - check and replacement

1   The intake air temperature sensor is a thermistor (a resistor which varies the value of its resistance in accordance with temperature changes). The change in the resistance

values will directly affect the voltage signal from the sensor to the PCM. As the sensor temperature INCREASES, the resistance values will DECREASE. As the sensor temperature DECREASES, the resistance values will INCREASE.

Check

Refer to illustrations 7.2, 7.3 and 7.4
Note: *Performing the following test will set a diagnostic trouble code and illuminate the Check Engine light. Clear the diagnostic trouble code after performing the tests and making the necessary repairs (see Section 2).*

2   The intake air temperature sensor is located in the air intake duct between the air filter housing and the throttle body **(see illustration)**. Check the terminals in the connector and the wires leading to the sensor for looseness and breaks. Repair as required.

3   Before checking the intake air temperature sensor, check the voltage supply and ground circuits from the PCM. Disconnect the electrical connector from the intake air temperature sensor and connect the positive lead a voltmeter to the tan wire terminal and the negative lead to the black or gray wire terminal at the harness connector **(see illustration)**. Turn the ignition key On - the volt-

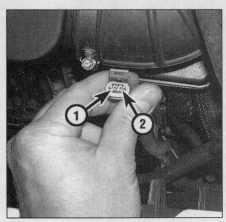

7.3 Disconnect the electrical connector from the intake air temperature sensor and check the voltage supply and ground circuits from the PCM at the harness connector

1   Intake air temperature sensor signal
2   Sensor ground

Temperature (degrees-F)	Resistance (ohms)
212	176
194	240
176	332
158	458
140	668
122	972
112	1182
104	1458
95	1800
86	2238
76	2795
68	3520
58	4450
50	5670
40	7280
32	9420

7.4 Intake air temperature sensor and engine coolant temperature sensor approximate temperature vs. resistance values

8.2a Engine coolant temperature sensor location - four-cylinder models

age should read approximately 5.0 volts. If the voltage is incorrect, check the wiring from the sensor to the PCM. If the circuits are good, have the PCM checked at a dealer service department or other properly equipped repair facility.

4   With the ignition switch OFF, disconnect the electrical connector from the intake air temperature sensor. Using an ohmmeter, measure the resistance between the two intake air temperature sensor terminals on the sensor while it is completely cold (50 to 80-degrees F). Reconnect the electrical connector to the sensor, start the engine and warm it up until it reaches operating temperature (180 to 200-degrees F), disconnect the connector and check the resistance again. Compare your measurements to the resistance chart (see illustration). If the sensor resistance test results are incorrect, replace the intake air temperature sensor. Note: A

more accurate check may be performed by removing the sensor and suspending the tip of the sensor in a container of water. Heat the water on the stove while you monitor the resistance of the sensor.

Replacement

5   Disconnect the electrical connector from the sensor.
6   Carefully remove the sensor from the air intake duct.
7   Installation is the reverse of removal.

8   Engine coolant temperature sensor - check and replacement

1   The engine coolant temperature sensor is a thermistor (a resistor which varies the value of its resistance in accordance with temperature changes). The change in the resistance values will directly affect the voltage signal from the sensor to the PCM. As the sensor temperature INCREASES, the resistance values will DECREASE. As the

sensor temperature DECREASES, the resistance values will INCREASE.

Check

Refer to illustrations 8.2a, 8.2b and 8.3
Note: Performing the following test will set a diagnostic trouble code and illuminate the Check Engine light. Clear the diagnostic trouble code after performing the tests and making the necessary repairs (see Section 2).
2   The engine coolant temperature sensor threads into a coolant passage near the coolant outlet (see illustrations). Check the terminals in the connector and the wires leading to the sensor for looseness and breaks. Repair as required.
3   Before checking the engine coolant temperature sensor, check the voltage supply and ground circuits from the PCM. Disconnect the electrical connector from the engine coolant temperature sensor and connect a voltmeter to the two terminals of the harness connector (see illustration). Turn the ignition key On - the voltage should read approximately 5.0 volts. If the voltage is incorrect, check the wiring from the engine coolant temperature sensor to the PCM. If

6

8.2b Engine coolant temperature sensor location - V6 models

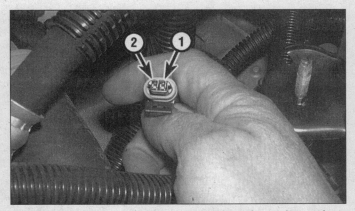

8.3 Disconnect the electrical connector from the engine coolant temperature sensor and check the voltage supply and ground circuits from the PCM at the harness connector

1   Engine coolant temperature sensor signal
2   Sensor ground

9.2a 7X crankshaft position sensor location - four-cylinder models

9.2b 7X crankshaft position sensor location - V6 models

the circuits are good, have the PCM checked at a dealer service department or other properly equipped repair facility.

4   With the ignition switch OFF, disconnect the electrical connector from the engine coolant temperature sensor. Using an ohmmeter, measure the resistance between the two terminals on the sensor while it is completely cold (50 to 80-degrees F). Reconnect the electrical connector to the sensor, start the engine and warm it up until it reaches operating temperature (180 to 200-degrees F), disconnect the connector and check the resistance again. Compare your measurements to the resistance chart **(see Illustration 7.4)**. If the sensor resistance test results are incorrect, replace the engine coolant temperature sensor. **Note:** *A more accurate check may be performed by removing the sensor and suspending the tip of the sensor in a container of water. Heat the water on the stove while you monitor the resistance of the sensor.*

Replacement

Warning: *Wait until the engine is completely cool before beginning this procedure.*

5   Partially drain the cooling system (see Chapter 1).

6   Disconnect the electrical connector from the sensor and carefully unscrew the sensor. **Caution:** *Handle the coolant sensor with care. Damage to this sensor will affect the operation of the entire fuel injection system.*

7   Before installing the new sensor, wrap the threads with Teflon sealing tape to prevent leakage and thread corrosion.

8   Installation is the reverse of removal.

9   Crankshaft position sensor - check and replacement

1   The crankshaft position sensor provides the ignition module and PCM with a crankshaft position signal. The ignition module uses the signal to determine the spark

sequence (firing order) for each cylinder. The PCM uses the signal to precisely control ignition timing and calculate engine speed (RPM). The signal is also used by the Onboard Diagnostic system for misfire detection. Four-cylinder models are equipped with a single 7X crankshaft position sensor. V6 models are equipped with two crankshaft position sensors; a 7X crankshaft position sensor and a 24X crankshaft position sensor. The 7X crankshaft position sensor is a magnetic inductive sensor triggered by seven slots cut into a reluctor ring on the crankshaft. The sensor tip is positioned approximately 0.050 inch from the reluctor ring. As the notches pass the sensor the magnetic field is altered, producing a pulsating voltage signal seven times per crankshaft revolution. The 24X crankshaft position sensor is a Hall effect device triggered by an interrupter ring behind the crankshaft pulley. The 24X crankshaft position sensor produces on-off pulses as the 24 blades and windows of the interrupter pass through the sensor's magnetic field. The ignition system will not operate if the PCM does not receive a 7X crankshaft position sensor input.

Check

Note: *Performing the following test will set a diagnostic trouble code and illuminate the Check Engine light. Clear the diagnostic trouble code after performing the tests and making the necessary repairs (see Section 2).*

7X crankshaft position sensor

Refer to illustrations 9.2a and 9.2b

2   The 7X crankshaft position sensor is located on the front (radiator) side of the engine block on four-cylinder models or on the rear (firewall) side of the engine block on V6 models **(see illustrations)**. Check the terminals in the connector and the wires leading to the sensor for looseness and breaks. Repair as required.

3   Disconnect the electrical connector from the 7X crankshaft position sensor. Connect a voltmeter to the two terminals of the

9.4 24X crankshaft position sensor location - V6 models

sensor and set the meter to read AC volts. **Note:** *The test connection can be made at the ignition control module electrical connector purple and yellow wire terminals, if desired.* Crank the engine and note the voltage. The 7X crankshaft position sensor should produce a minimum of 200 millivolts with the engine cranking. If a crankshaft position sensor signal is not present, replace the 7X crankshaft position sensor.

24X crankshaft position sensor (V6 models only)

Refer to illustrations 9.4 and 9.5

4   The 24X crankshaft position sensor is located on the front of the engine behind the crankshaft pulley **(see Illustration)**. Check the terminals in the connector and the wires leading to the sensor for looseness and breaks. Repair as required.

5   Before checking the crankshaft position sensor, check the voltage supply and ground circuits from the PCM. Disconnect the electrical connector and connect the positive lead of a voltmeter to the red/white wire terminal and the negative lead to the black wire termi-

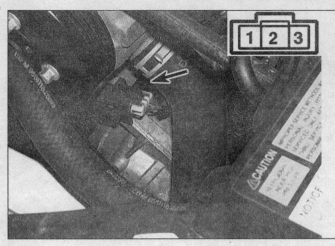

9.5 Disconnect the 24X crankshaft position sensor electrical connector (arrow) and check the voltage supply and ground circuits at the harness side of the connector

1 *12-volt supply*
2 *Sensor signal*
3 *Sensor ground*

10.2a Camshaft position sensor location - four-cylinder models

10.2b Camshaft position sensor location - V6 models

10.3 Disconnect the electrical connector from the camshaft position sensor and check the voltage supply and ground circuits at the harness connector

1 *Camshaft position sensor signal (four-cylinder); 12-volt supply (V6)*
2 *Sensor ground (four-cylinder); Camshaft position sensor signal (V6)*
3 *12-volt supply (four-cylinder); Sensor ground (V6)*

nal at the harness connector **(see illustration)**. Turn the ignition key On - the voltage should read approximately 12.0 volts. If the voltage is incorrect, check the wiring from the crankshaft position sensor to the PCM. If the circuits are good, have the PCM checked at a dealer service department or other properly equipped repair facility.

6   To check the 24X crankshaft position sensor operation, reconnect the connector to the crankshaft position sensor and using a suitable probe, backprobe the light blue/black wire terminal of the crankshaft position sensor connector (see Chapter 12 for additional information on how to back-probe a connector). Connect the positive lead of a voltmeter to the probe and the negative lead to a good engine ground point. Turn the ignition key On. The meter should indicate approximately 10.0 volts. Rotate the engine slowly with a breaker bar and socket attached to the crankshaft pulley center bolt while watching the meter. The voltage should fluctuate between 10.0 volts and zero as the blades and windows in the interupter pass the sensor. If the test results are incorrect, replace the 24X crankshaft position sensor. **Note:** *Rotate the engine slowly. Removing the spark plugs from the engine will make the crankshaft much easier to turn.*

Replacement

Note: *Anytime a crankshaft position sensor is disturbed, A Crankshaft Position Sensor Variation Learning Procedure should be performed or a false misfire diagnostic trouble code may be set. If after replacing the sensor, a false diagnostic trouble codes is set, take the vehicle to a dealership service department for the procedure.*

7X crankshaft position sensor

7   Disconnect the 7X crankshaft position sensor wiring harness connector.
8   Remove the 7X crankshaft position sensor mounting bolt and withdraw the sensor from the engine block.
9   Replace the O-ring and lightly lubricate it with clean engine oil.
10   Installation is the reverse of removal.

24X crankshaft position sensor (V6 models only)

11   Remove the drivebelt (see Chapter 1).
12   Raise the vehicle and support it securely on jackstands. Remove the crankshaft pulley (see Chapter 2B).
13   Remove the bolt and the sensor wiring harness retaining bracket.
14   Remove the 24X crankshaft position sensor mounting bolts. Remove the sensor and withdraw the wiring harness, noting the harness routing for installation.
15   Installation is the reverse of removal.

---

10 Camshaft position sensor - check and replacement

---

1   The camshaft position sensor, in conjunction with the crankshaft position sensor, determines the timing for the fuel injection on each cylinder. The camshaft position sensor is a Hall-Effect device triggered by a magnet on the camshaft.

Check

Refer to illustrations 10.2a, 10.3b, 10.3, 10.4a and 10.4b
Note: *Performing the following test will set a diagnostic trouble code and illuminate the Check Engine light. Clear the diagnostic trouble code after performing the tests and making the necessary repairs (see Section 2).*
2   On four-cylinder models, the camshaft position sensor is located at the left (drivers) side of the timing cover. On V6 models, the camshaft position sensor is located at the front of the engine block, below the power steering pump **(see Illustrations)**. Check the terminals in the connector and the wires leading to the sensor for looseness and breaks. Repair as required.
3   Before checking the camshaft position sensor, check the voltage supply and ground circuits from the PCM. Disconnect the electrical connector from the camshaft position sensor and connect the positive lead of a voltmeter to the red (four-cylinder) or red/white (V6) wire terminal and the negative

10.4a To check the camshaft position sensor, backprobe the brown/white wire terminal of the sensor connector with a voltmeter - four-cylinder model

10.4b On V6 models, the camshaft position sensor connector is located behind the alternator - backprobe the brown/white wire terminal of the sensor connector (harness-side) with a voltmeter

lead to the pink/black (four cylinder) or black (V6) wire terminal at the harness connector **(see illustration)**. Turn the ignition key On - the voltage should read approximately 12.0 volts. If the voltage is incorrect, check the wiring from the camshaft position sensor to the PCM. If the circuits are good, have the PCM checked at a dealer service department or other properly equipped repair facility.

4   To check the camshaft position sensor operation, reconnect the connector to the camshaft position sensor and using a suitable probe, backprobe the brown/white wire terminal of the camshaft position sensor connector **(see illustration)** (see Chapter 12 for additional information on how to backprobe a connector). Connect the positive lead of a voltmeter to the probe and the negative lead to a good engine ground point. Turn the ignition key On. The meter should indicate approximately 10.0 volts. Rotate the engine slowly with a breaker bar and socket attached to the crankshaft pulley center bolt while watching the meter. The voltage should remain a steady 10.0 volts then quickly drop to zero and back to 10.0 volts as the magnet passes the sensor **(see illustration)**. If the test results are incorrect, replace the camshaft position sensor. **Note:** *Rotate the engine slowly through at least two complete revolutions. Removing the spark plugs from the engine will make the crankshaft much easier to turn.*

Replacement

Four-cylinder models

5   Disconnect the electrical connector from the camshaft position sensor.
6   Remove the camshaft position sensor mounting bolt and withdraw the sensor from the camshaft cover.
7   Replace the O-ring and lightly lubricate it with clean engine oil.
8   Installation is the reverse of removal.

V6 models

9   Remove the drivebelt (see Chapter 1).

10   Without disconnecting the hoses, remove the power steering pump and position it aside (see Chapter 10).
11   Disconnect the electrical connector from the sensor harness, noting the harness routing for installation.
12   Remove the camshaft position sensor mounting bolt and withdraw the sensor from the engine block.
13   Installation is the reverse of removal.

---

11   Oxygen sensor - check and replacement

Note: *All models are equipped with two oxygen sensors; one upstream oxygen sensor and one downstream oxygen sensor.*

1   The oxygen in the exhaust reacts with the elements inside the oxygen sensor to produce a voltage output that varies from 0.1 volt (high oxygen, lean mixture) to 0.9 volt (low oxygen, rich mixture). The upstream oxygen sensor (mounted in the exhaust system before the catalytic converter) provides a feedback signal to the PCM that indicates the amount of leftover oxygen in the exhaust. The PCM monitors this variable voltage continuously to determine the required fuel injector pulse width and to control the engine air/fuel ratio. A mixture ratio of 14.7 parts air to 1 part fuel is the ideal ratio for minimum exhaust emissions, as well as the best combination of fuel economy and engine performance. Based on oxygen sensor signals, the PCM tries to maintain this air/fuel ratio of 14.7:1 at all times.
2   The downstream oxygen sensor (mounted in the exhaust system after the catalytic converter) has no effect on PCM control of the air/fuel ratio. However, the downstream sensor is identical to the upstream sensor and operates in the same way. The PCM uses the downstream signal to monitor the efficiency of the catalytic converter. A downstream oxygen sensor will produce a slower fluctuating voltage signal that reflects

the lower oxygen content in the post-catalyst exhaust.
3   An oxygen sensor produces no voltage when it is below its normal operating temperature of about 600-degrees F. During this warm-up period, the PCM operates in an open-loop fuel control mode. It does not use the oxygen sensor signal as a feedback indication of residual oxygen in the exhaust. Instead, the PCM controls fuel metering based on the inputs of other sensors and its own programs.
4   Proper operation of an oxygen sensor depends on four conditions:
   a)   ***Electrical*** - *The low voltages generated by the sensor require good, clean connections which should be checked whenever a sensor problem is suspected or indicated.*
   b)   ***Outside air supply*** - *The sensor needs air circulation to the internal portion of the sensor. Whenever the sensor is installed, make sure the air passages are not restricted.*
   c)   ***Proper operating temperature*** - *The PCM will not react to the sensor signal until the sensor reaches approximately 600-degrees F. This factor must be considered when evaluating the performance of the sensor.*
   d)   ***Unleaded fuel*** - *Unleaded fuel is essential for proper operation of the sensor.*
5   The PCM can detect several different oxygen sensor problems and set diagnostic trouble codes to indicate the specific fault (see Section 2). When an oxygen sensor fault occurs, the PCM will disregard the oxygen sensor signal voltage and revert to open-loop fuel control as described previously.

Check

Refer to illustrations 11.6a and 11.6b
Caution: *The oxygen sensor is very sensitive to excessive circuit loads and circuit damage of any kind. For safest testing, disconnect the oxygen sensor connector, install jumper wires between the two connectors and connect*

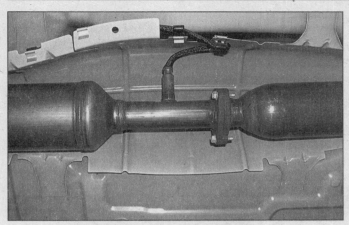

11.6a The downstream oxygen sensor is located in the exhaust pipe after the catalytic converter

11.6b Downstream oxygen sensor electrical connector location

your voltmeter to the jumper wires. If jumper wires aren't available, carefully backprobe the wires in the connector shell with suitable probes (such as T-pins). Do not puncture the oxygen sensor wires or try to backprobe the sensor itself. Use only a digital voltmeter to test an oxygen sensor.

Note: Performing the following test will set a diagnostic trouble code and illuminate the Check Engine light. Clear the diagnostic trouble code after performing the tests and making the necessary repairs (see Section 2).

6   Connect your voltmeter positive (+) lead to the purple or purple/white wire and connect the negative (-) lead to the tan or tan/white wire at the oxygen sensor connector **(see illustrations)**. If testing a single wire oxygen sensor, connect the negative lead to a good chassis ground. Turn the ignition ON but do not start the engine. The meter should read approximately 400 to 450 millivolts (0.40 to 0.45 volt). If it doesn't, trace and repair the circuit from the sensor to the PCM. **Note:** Refer to the wiring diagrams at the end of Chapter for additional information on the oxygen sensor circuits.

7   Start the engine and let it warm up to normal operating temperature; again check the oxygen sensor signal voltage.

a) Voltage from an upstream sensor should range from 100 to 900 millivolts (0.1 to 0.9 volt) and switch actively between high and low readings.

b) Voltage from a downstream sensor should also read between 100 to 900 millivolts (0.1 to 0.9 volt) but it should not switch actively. The downstream oxygen sensor voltage may stay toward the center of its range (about 400 millivolts) or stay for relatively longer periods of time at the upper or lower limits of the range.

8   Check the battery voltage supply and ground circuits to the oxygen sensor heater. Disconnect the electrical connector and connect the voltmeter negative (-) lead to the black wire terminal and the positive (+) lead to the brown wire terminal of the sensor connector. Turn the ignition ON, the meter should read approximately 12 volts. If battery

voltage is not present, check the power and ground circuits to the sensor (don't forget to check the fuses first).

9   Allow the oxygen sensor to cool and check the resistance of the oxygen sensor heater. With the connector disconnected. Connect an ohmmeter to the two oxygen sensor heater terminals of the connector (oxygen sensor side). The oxygen sensor pigtail is generally not color coded, but the heater wires are usually the white wires. The oxygen sensor heater resistance should be 3.0 to 10.0 ohms. If an open circuit or excessive resistance is indicated, replace the oxygen sensor. **Note:** If the tests indicate that a sensor is good, and not the cause of a driveability problem or diagnostic trouble code, check the wiring harness and connectors between the sensor and the PCM for an open or short circuit. If no problems are found, have the vehicle checked by a dealer service department or other qualified repair shop.

Replacement

Refer to illustration 11.13

10   The exhaust pipe contracts when cool, and the oxygen sensor may be hard to loosen when the engine is cold. To make sensor removal easier, start and run the engine for a minute or two; then shut it off. Be careful not to burn yourself during the following procedure. Also observe these guidelines when replacing an oxygen sensor.

a) The sensor has a permanently attached pigtail and electrical connector which should not be removed from the sensor. Damage or removal of the pigtail or electrical connector can harm operation of the sensor.

b) Keep grease, dirt and other contaminants away from the electrical connector and the louvered end of the sensor.

c) Do not use cleaning solvents of any kind on the oxygen sensor.

d) Do not drop or roughly handle the sensor.

11   If replacing the downstream oxygen sensor, raise the vehicle and place it securely on jackstands.

11.13 A special slotted socket, allowing clearance for the wiring harness, may be required for oxygen sensor removal (the tool is available at most auto parts stores)

6

12   Disconnect the electrical connector from the sensor.

13   Using a suitable wrench or specialized oxygen sensor socket, unscrew the sensor from the exhaust manifold **(see illustration)**.

14   Anti-seize compound must be used on the threads of the sensor to aid future removal. The threads of most new sensors will be coated with this compound. If not, be sure to apply anti-seize compound before installing the sensor.

15   Install the sensor and tighten it securely.

16   Reconnect the electrical connector to the sensor and lower the vehicle.

12 Knock sensor - check and replacement

1   The knock sensor detects abnormal vibration (spark knock or pinging) in the engine. The knock control system is designed to reduce spark knock during periods of heavy detonation. This allows the engine to use maximum spark advance to

12.2a Knock sensor location - four-cylinder models

12.2b Knock sensor location - V6 models

12.3 Knock sensor harness connector location - four-cylinder models with a two wire sensor

13.2 Vehicle speed sensor location

improve driveability. Knock sensors produce AC output voltage which increases with the severity of the knock. The signal is fed into the PCM and the timing is retarded to compensate for the severe detonation.

Check

Refer to illustrations 12.2a, 12.2b and 12.3

2   On four-cylinder models, the knock sensor is located on the rear (firewall) side of the engine block below the exhaust manifold **(see illustrations)**. On V6 models, the knock sensor is located on the front (radiator) side of the engine block above the oil filter **(see illustration)**. Four-cylinder models may be equipped with either a one wire or a two wire sensor.

3   Disconnect the electrical connector from the knock sensor. On four-cylinder models with a two wire sensor, follow the wiring harness from the knock sensor to locate the knock sensor harness connector **(see illustration)**. Single wire sensors contain an internal 100 K-ohm resistor; on two wire sensors, the 100 K-ohm resistor is in the PCM. Using an ohmmeter, measure the resistance between the terminal on the knock sensor and the engine block (single wire sen-

sor) or on a two wire sensor, between the two terminals of the sensor connector (harness side - not the sensor side). The resistance should be approximately 90 to 110 K-ohms. On a single wire sensor, if the resistance is not as specified, replace the knock sensor. On a two wire sensor (harness side), if the resistance is not as specified, check the wiring harness from the sensor connector to the PCM, including the PCM connector for looseness or damage. If the harness and connections are good, replace the PCM.

4   To check the sensor operation, reconnect the connector to the sensor. On models with a single wire sensor, backprobe the wire terminal at the electrical connector using a suitable probe (see Chapter 12 for additional information on how to backprobe a connector). On models with a two wire sensor, backprobe both terminals. On a single wire sensor, connect the positive lead of a voltmeter to the probe and the negative lead to a good engine ground point. On models with a two wire sensor, connect the positive lead to the dark blue wire and the negative lead to the gray wire. Set the voltmeter on the AC volts scale. Start the engine and check for an AC

voltage signal from the knock sensor. The AC voltage should increase as the engine speed increases. If a voltage signal is not present, replace the knock sensor.

Replacement

Warning: *The engine must be completely cool before beginning this procedure.*

5   Drain the cooling system (see Chapter 1).

6   Raise the vehicle and support it securely on jackstands.

7   Disconnect the electrical connector from the knock sensor.

8   Remove the sensor from the engine block.

9   Installation is the reverse of removal.

10   Refill the cooling system (see Chapter 1).

13   Vehicle Speed Sensor (VSS) - check and replacement

1   The Vehicle Speed Sensor (VSS) is a permanent magnet generator mounted on the transaxle. The sensor is triggered by a

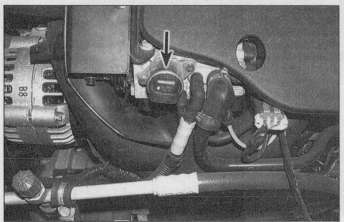

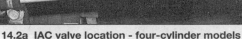

14.2a IAC valve location - four-cylinder models

14.2b IAC valve location - V6 models

toothed rotor on the transaxle output shaft. As the output shaft rotates, the sensor produces an AC voltage, the frequency of which is proportional to vehicle speed. The PCM uses the sensor input signal for several different engine and transmission control functions. The VSS signal also drives the speedometer on the instrument panel. A defective VSS can cause various driveability and transaxle problems.

Check

Refer to illustration 13.2

2   Raise the vehicle and support it securely on jackstands. Locate the vehicle speed sensor **(see illustration)**. Check the terminals in the connector and the wires leading to the sensor for looseness and breaks. Repair as required.

3   To check the VSS operation, backprobe the two wire terminals of the VSS connector using suitable probes (see Chapter 12 for additional information on how to backprobe a connector). Connect a voltmeter to the probes and set the meter on the AC volts scale. Turn the ignition key On. Rotate the right front tire by hand while watching the voltmeter. The sensor should produce a minimum of 0.5 volts and the voltage should increase as the transaxle output shaft rotates faster.

4   Turn the ignition key Off and disconnect the electrical connector from the sensor. Using an ohmmeter, measure the resistance across the two terminals of the sensor. The sensor resistance should be between 1500 and 1650 ohms at 68-degrees F. If the test results are incorrect, replace the vehicle speed sensor.

Replacement

5   Raise the vehicle and support it securely on jackstands.

6   Disconnect the electrical connector from the VSS.

7   Remove mounting bolt and withdraw the VSS from the transaxle case.

8   Replace the sensor O-ring.

9   Installation is the reverse of removal.

14 Idle Air Control (IAC) valve - check and replacement

1   The idle speed is controlled by the Idle Air Control (IAC) valve. The IAC valve regulates the air bypassing the throttle plate by moving the pintle in or out of the air passage. The IAC valve is controlled by the PCM, adjusting the idle speed depending upon the running conditions of the engine (air conditioning system, power steering, cold and warm running etc.). The engine idle speed is not adjustable on these models.

Check

Refer to illustrations 14.2a and 14.2b
Note: *Performing the following test will set a diagnostic trouble code and illuminate the Check Engine light. Clear the diagnostic trouble code after performing the tests and making the necessary repairs (see Section 2).*

2   The Idle Air Control (IAC) valve is located on the throttle body **(see illustrations)**. A scan tool is required for complete testing of the IAC valve and circuits. However, there are several tests the home mechanic can perform on the IAC system to verify operation but they are limited and are useful only in the case of definite IAC valve failure.

3   When the engine is started cold, the IAC valve should vary the idle as the engine begins to warm-up. Allow the engine to warm-up, then place a load on the engine by placing the transaxle in gear (automatic), turning the air conditioning on and/or operating the power steering. The idle should remain steady or increase slightly. If the engine stumbles or stalls, or if there are obvious signs that the IAC valve is not working, stop the engine and continue testing.

4   Disconnect the electrical connector from the IAC valve. Using an ohmmeter, measure the resistance across terminals 1 and 2 of the IAC valve, then measure the resistance across terminals 3 and 4 (the IAC valve terminals are numbered 1, 2, 3, 4, left-to-right) - the resistance should be about the same on both sets. If one or both checks indicate an

open circuit, replace the IAC valve.

5   If the IAC valve is good, have the PCM diagnosed by a dealer service department or other qualified repair shop.

Replacement

Refer to illustration 14.9

6   Disconnect the electrical connector from the IAC valve.

7   Remove the two mounting screws from the valve and withdraw it from the throttle body.

8   Inspect the IAC valve pintle and the air passage and valve seat in the throttle body for heavy carbon deposits. Clean the IAC valve with aerosol carburetor cleaner, a shop towel and a soft brush, if necessary. Do not submerge the IAC valve in any liquid cleaner. If the air passage requires further cleaning, remove the throttle body and clean it thoroughly.

9   If installing a new IAC valve, measure the distance from the tip of the IAC valve pintle to the mounting flange **(see illustration)**. If the distance is greater than 1-1/8 inch, press the pintle in by hand, as necessary. **Caution:** *Do not attempt to press the pintle in on a used IAC valve. The force required to move a*

6

14.9 Before installing a new Idle Air Control valve, measure the distance from the tip of the pintle to the mounting flange - press the pintle in until the distance is less than 1-1/8 inch

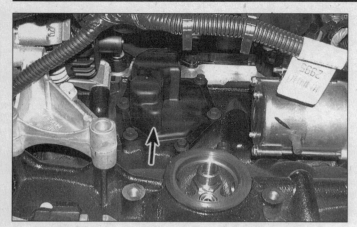

15.2 Oil/air separator location - four-cylinder models

15.11 PCV valve location - V6 models

pintle shaft with carbon build-up may damage the valve.

10   Install a new O-ring and lubricate it with clean engine oil.

11   Install the IAC valve and tighten the screws securely. Connect the electrical connector.

12   Cycle the ignition key On for ten seconds, then Off for ten seconds to reset the valve. Start the engine, allow it to reach operating temperature and check the idle operation.

15   Crankcase ventilation system

1   When the engine is running, a certain amount of the gasses produced during combustion escapes past the piston rings into the crankcase as blow-by gasses. The crankcase ventilation system is designed to reduce the resulting hydrocarbon emissions (HC) by routing the gasses and vapors from the crankcase into the intake manifold and combustion chambers, where they are consumed during engine operation. Two types of crankcase ventilation systems are used, depending on engine type.

Four-cylinder models

Refer to illustration 15.2

2   On four-cylinder models, crankcase vapors pass through a hose connected from the timing cover to the oil/air separator **(see Illustration)**. The oil/air separator separates the oil suspended in the blow-by gases and allows the oil to drain back into the crankcase. The crankcase vapors are drawn from the oil/air separator through a hose connected to the air intake resonator where they mix with the incoming air and are burned during the normal combustion process.

3   A plugged oil/air separator or hose will cause excessive crankcase pressures resulting in oil leaks and sludge build-up in the crankcase. Be sure to check the basic mechanical condition of the engine before condemning the oil/air separator (see Chapter 2C).

4   To remove the oil/air separator, remove the intake manifold (see Chapter 2A).

5   Remove the oil filter. Detach the hoses from the oil/air separator.

6   Remove the mounting bolts and remove the oil/air separator from the engine block.

7   Clean the engine block gasket surface and oil/air separator and install a new gasket.

8   Install the oil/air separator and tighten the mounting bolts securely.

9   Attach the hoses to the oil/air separator. Apply soapy water to the inside of the hoses to ease installation.

10   The remainder of installation is the reverse of removal.

V6 models

Refer to illustration 15.11

11   V6 models use a Positive Crankcase Ventilation (PCV) system. The main component of the Positive Crankcase Ventilation (PCV) system is the PCV valve **(see illustration)**. Fresh air flows from the air intake duct through a vent tube into the engine. Crankcase vapors are drawn from the crankcase by the PCV valve. To maintain idle quality and good driveability, the PCV valve restricts the flow when the intake manifold vacuum is high. When intake manifold vacuum is lower, maximum vapor flow is allowed through the valve.

Checking and replacement of the PCV valve is covered in Chapter 1.

16   Exhaust Gas Recirculation (EGR) system

1   The Exhaust Gas Recirculation (EGR) system is used to lower NOx (oxides of nitrogen) emission levels caused by high combustion temperatures. The EGR valve recirculates a small amount of exhaust gases into the intake manifold. The additional mixture lowers the temperature of combustion thereby reducing the formation of NOx compounds.

2   The EGR system consists of an electronic EGR valve and the PCM. The PCM controls the EGR flow rate by energizing the EGR valve solenoid coil, opening or closing the EGR passage in small increments. The PCM monitors the EGR valve pintle position with an EGR position sensor built into the EGR valve. This system allows for precise control of EGR flow, achieving optimum EGR flow depending on engine operating conditions.

Check

Refer to illustration 16.4

3   A scan tool is required for complete testing of the EGR valve, control system and circuits. However, there are several tests the home mechanic can perform on the system to verify operation but they are limited and are useful only in the case of definite system failure.

4   Disconnect the electrical connector

16.4 EGR valve harness connector terminal identification

1   EGR valve control (four-cylinder 12-volt supply - V6 ground)

2   Pintle position sensor ground

3   Pintle position sensor signal

4   Pintle position sensor 5-volt supply

5   Ground (four-cylinder); 12-volt supply (V6)

16.9 Remove the EGR valve mounting bolts (arrows)

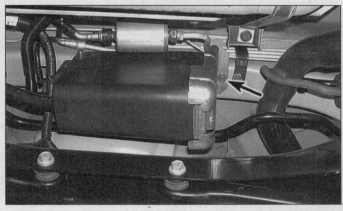

17.10 Remove the EVAP canister bracket mounting bolt (arrow)

from the EGR valve. Connect the negative lead of a voltmeter to a good engine ground point and probe terminal no. 1 of the EGR valve electrical connector (harness side) with the positive lead **(see illustration)**. Turn the ignition key On - on V6 models battery voltage should be indicated on the meter. If battery voltage is not indicated, check the circuits from the PCM to the EGR valve. If the circuits are good, have the PCM diagnosed by a dealer service department or other qualified repair shop. **Note:** *On four-cylinder models, battery voltage will only be indicated when the PCM is controlling the EGR valve.*

5   Check the voltage supply and ground circuits to the EGR valve position sensor. Disconnect the electrical connector from the EGR valve. Connect the positive lead of a voltmeter to terminal no. 4 of the EGR valve electrical connector (harness side). Connect the negative lead to terminal no. 2. Turn the ignition key On - approximately 5.0 volts should be indicated on the meter. If the 5.0 volt supply voltage is not present, check the circuits from the PCM to the EGR valve. If the circuits are good, have the PCM diagnosed by a dealer service department or other qualified repair shop.

6   To check the EGR valve, use an ohmmeter to check the continuity of the EGR valve position sensor and the EGR solenoid coil. Check for continuity across terminals 2 and 4 (EGR sensor) and across terminals 1 and 5 (EGR solenoid coil). If either check reveals an open circuit, replace the EGR valve. **Note:** *The EGR valve terminals are numbered 1, 2, 3, 4, 5, left-to-right).*

EGR valve replacement

Refer to illustration 16.9

7   On four-cylinder models, remove the air intake duct and resonator from the throttle body and air filter housing.
8   Disconnect the electrical connector from the EGR valve.
9   Remove the EGR valve mounting bolts **(see illustration)**. Remove the EGR valve and gaskets. Discard the gaskets.
10   Using a gasket scraper, clean the EGR valve gasket surfaces.
11   Installation is the reverse of removal.

17 Evaporative emissions control system

1   The fuel evaporative emissions control (EVAP) system absorbs fuel vapors from the fuel tank and, during engine operation, releases them into the engine intake system where they mix with the incoming air/fuel mixture. The main components of the evaporative emissions system are the canister (filled with activated charcoal to absorb fuel vapors), the purge valve, the vent valve, the fuel pressure sensor, the fuel tank and the vapor and purge lines.
2   After passing through a check valve, fuel tank vapor is carried through the vapor hose to the charcoal canister. The activated charcoal in the canister absorbs and stores the vapors. When a programmed set of conditions are met (engine running, warmed to a pre-set temperature, etc.), the PCM opens the purge valve and the vent valve. Fuel vapors from the canister are then drawn through the purge hose by intake manifold vacuum into the intake manifold and combustion chamber where they are consumed during normal engine operation.
3   The PCM regulates the rate of vapor flow from the canister to the intake manifold by controlling the duty cycle of the EVAP purge valve control solenoid. During cold running conditions and hot start time delay, the PCM does not energize the solenoid. After the engine has warmed up to the correct operating temperature, the PCM purges the vapors into the intake manifold according to the running conditions of the engine. The PCM will cycle (ON then OFF) the purge valve control solenoid about 5 to 10 times per second. The flow rate will be controlled by the pulse width, or length of time, the solenoid is allowed to be energized.
4   The system performs a self-diagnostic check when the engine is started cold. When the programmed conditions are met, the PCM opens the EVAP canister purge valve, leaving the vent valve closed. This action allows engine vacuum to draw a vacuum on the entire EVAP system. Once the proper vacuum level is reached, the PCM closes the

purge valve, sealing the system. The PCM then monitors the fuel tank pressure sensor and sets a diagnostic code if a leak is detected.

Check

Note: *The evaporative control system, like all emission control systems, is protected by a Federally-mandated warranty (5 years or 50,000 miles at the time this manual was written). The EVAP system probably won't fail during the service life of the vehicle; however, if it does, the hoses or charcoal canister are usually to blame.*
5   Always check the hoses first. A disconnected, damaged or missing hose is the most likely cause of a malfunctioning EVAP system. Refer to the Vacuum Hose Routing Diagram (attached to the radiator support) to determine whether the hoses are correctly routed and attached. Repair any damaged hoses or replace any missing hoses as necessary.
6   Check the related fuses and wiring to the purge and vent valves. Refer to the wiring diagrams at the end of Chapter 12, if necessary. The purge and vent valves are normally closed - no vapors will pass through the ports. When the PCM energizes the solenoid (by completing the circuit to ground), the valve opens and vapors flow through.
7   A scan tool is required to thoroughly check the system. If the above checks fail to identify the problem area, have the system diagnosed by a dealer service department or other qualified repair shop.

Component replacement

Refer to illustrations 17.10, 17.12a, 17.12b and 17.16

EVAP canister

8   The EVAP canister is attached to a bracket near the fuel tank.
9   Raise the vehicle and support it securely on jackstands.
10   Label and remove the hoses from the canister. Remove the bracket mounting bolt and remove the canister **(see illustration)**.
11   Installation is the reverse of removal.

6

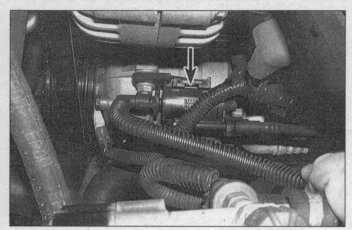

**17.12a EVAP purge valve/control solenoid location -
four-cylinder models**

**17.12b EVAP purge valve/control solenoid location -
V6 models**

Purge valve

12   On four-cylinder models, the purge valve is mounted on a bracket near the air conditioning compressor. On V6 models, the purge valve is attached to the rear (firewall side) cylinder head below the ignition coils **(see illustrations)**.

13   Disconnect the electrical connector. Label and remove the hoses from the purge valve.

14   Using a small screwdriver, depress the locking tab and remove the purge valve from the bracket.

15   Installation is the reverse of removal.

Vent valve

16   The vent valve is mounted on a bracket near the fuel tank **(see Illustration)**.

17   Raise the vehicle and support it securely on jackstands.

18   Disconnect the electrical connector. Remove the hose from the vent valve.

19   Remove the bracket mounting bolt and remove the vent valve. Depress the locking tab and remove the vent valve from the bracket.

20   Installation is the reverse of removal.

Fuel tank pressure sensor

21   The fuel tank pressure sensor is located on the fuel pump module.

22   Remove the fuel tank (see Chapter 4).

23   Disconnect the electrical connector from the fuel tank pressure sensor.

24   Release the retaining clip and remove the sensor from the top of the fuel pump module.

25   Installation is the reverse of removal.

---

18   Secondary air injection system

1   Some 1999 and 2000 V6 models are equipped with a secondary air injection (AIR) system. The secondary air injection system is used to reduce tailpipe emissions on initial engine start-up. The system uses an electric motor/pump assembly, vacuum valve/solenoid, check valve and tubing to inject fresh air directly into the exhaust manifolds. The fresh air (oxygen) reacts with the exhaust gas in the catalytic converter to reduce HC and CO levels. The air pump and solenoid are controlled by the PCM. During initial start-up, when the coolant temperature is between 50-

degrees F and 176-degrees F, the PCM will energize the vacuum valve/solenoid, opening the check valve and operate the air pump for approximate one minute. During normal operation, the check valve is closed to prevent exhaust backflow into the system.

Check

Refer to illustration 18.2

2   Check the air pump hoses and the vacuum hoses **(see Illustration)**. Repair any damaged hoses or replace any missing hoses as necessary. Check the vacuum source to the vacuum valve/solenoid. Intake manifold vacuum should be present with the engine running.

3   Check the related fuses and wiring to the air pump and vacuum valve/solenoid. Refer to the wiring diagrams at the end of Chapter 12, if necessary. The vacuum valve/solenoid is normally closed - no vacuum is applied to the valve. When the PCM energizes the solenoid (by completing the circuit to ground), the valve opens, vacuum is applied to the check valve, the check valve opens and air flows through the tube into the exhaust pipe.

**17.16 EVAP vent valve/control
solenoid location**

**18.2 Secondary air
injection component
locations**

1   *Air pump*
2   *Check valve*
3   *Vacuum
valve/solenoid*

18.11 Remove the mounting nut (arrow) from the vacuum valve/solenoid

19.7 Catalytic converter and related components

1   Catalytic converter
2   Downstream oxygen sensor
3   Flange bolts

4   A scan tool is required to thoroughly check the system. If the above checks fail to identify the problem area, have the system diagnosed by a dealer service department or other qualified repair shop.

Component replacement
Refer to illustration 18.11

Air pump
5   Disconnect the electrical connector and remove the hose from the air pump.
6   Remove the splash shield bolts and move the splash shield aside to access the air pump mounting bolts.
7   Remove the mounting bolts and remove the air pump assembly.
8   Installation is the reverse of removal.

Vacuum valve/solenoid
9   Disconnect the electrical connector from the valve/solenoid.
10   Label and disconnect the vacuum hoses from the valve/solenoid.
11   Remove the mounting nut and remove the vacuum valve/solenoid **(see Illustration)**.
12   Installation is the reverse of removal.

Check valve
13   Remove the vacuum valve/solenoid.
14   Remove the air hose from the check valve.
15   Disconnect the air tube from the exhaust pipe.
16   Remove the mounting bolts and remove the check valve and tube.
17   Installation is the reverse of removal.

19   Catalytic converter

Note: *Because of a Federally mandated warranty which covers emissions-related components such as the catalytic converter, check with a dealer service department before replacing the converter at your own expense.*
1   The catalytic converter is an emission control device added to the exhaust system to reduce pollutants from the exhaust gas stream. A three-way (reduction) catalyst design is used. The catalytic coating on the three-way catalyst contains platinum and rhodium, which lowers the levels of oxides of nitrogen (NOx) as well as hydrocarbons (HC) and carbon monoxide (CO).
2   The test equipment for a catalytic converter is expensive and highly sophisticated. If you suspect that the converter on your vehicle is malfunctioning, take it to a dealer or authorized emissions inspection facility for diagnosis and repair.

Check
3   Whenever the vehicle is raised for servicing of underbody components, check the converter for leaks, corrosion, dents and other damage. Check the welds/flange bolts that attach the front and rear ends of the converter to the exhaust system. If damage is discovered, the converter should be replaced.
4   A catalytic converter may become plugged. The easiest way to check for a restricted converter is to use a vacuum gauge to diagnose the effect of a blocked exhaust on intake vacuum.

a) Connect a vacuum gauge to an intake manifold vacuum source.
b) Warm the engine to operating temperature, place the transmission in Park and apply the parking brake.
c) Note and record the vacuum reading at idle.
d) Open the throttle until the engine speed is about 2000 rpm.
e) Release the throttle quickly and record the vacuum reading.
f) Perform the test three more times, recording the reading after each test.
g) If the reading after the fourth test is more than one in-Hg lower than the reading recorded at idle, the catalytic converter, muffler or exhaust pipes may be plugged or restricted.

Replacement
Refer to illustration 19.7
Note: *Refer to the exhaust system servicing section in Chapter 4 for additional information.*
5   Raise the vehicle and support it securely on jackstands.
6   Disconnect the electrical connector from the downstream oxygen sensor.
7   Remove the catalytic converter-to-exhaust pipe flange bolts and separate the exhaust pipe from the catalytic converter **(see Illustration)**. Support the exhaust pipe.
8   Remove the bolts and detach the catalytic converter header pipe from the exhaust manifold (see Chapter 2A or 2B). Remove the catalytic converter and pipe assembly.
9   Clean the carbon deposits from the mounting flanges and install new gaskets.
10   Installation is the reverse of removal.

6

Notes

Chapter 7 Part A
Manual transaxle

Contents

Specifications

7A

General

Lubricant type .... See Chapter 1

Torque specifications — Ft-lbs

	Ft-lbs
Back-up light switch	24
Transaxle-to-engine bolts	66

1 General information

The vehicles covered by this manual are equipped with either a 5-speed manual or an automatic transaxle. Information on the manual transaxle is included in this Part of Chapter 7. Information on the automatic transaxle is in Part B.

The 5-speed Getrag transaxle consists of a transmission and differential housed in a single all-aluminum assembly. Because of its complexity, the special tools needed to overhaul it, and the difficulty of obtaining replace-

ment parts, overhauling this unit is beyond the scope of the average home mechanic. The information in this chapter is limited to general diagnosis, external adjustments and removal and installation.

Depending on the cost, it may be a good idea to consider replacing the old unit with either a rebuilt or used transaxle instead of a new one. Your local dealer or transmission shop should be able to supply information concerning cost, availability and exchange policy. Regardless of how you decide to remedy a transaxle problem, however, you will save money by removing and installing it yourself.

2 Shift control assembly - removal and installation

1   Remove the shift lever knob, console (Chapter 11) and shift boot.
2   Disconnect the shift and select cables from the shift control assembly.
3   Detach the cables from their bracket on the shift control assembly base.
4   Remove the shift control assembly retaining nuts and remove the shift control assembly.
5   Installation is the reverse of removal.

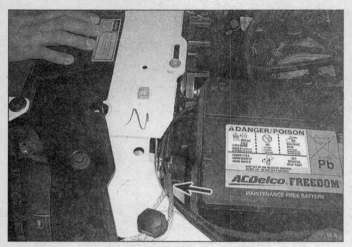

5.7a Run a length of rope under the radiator hose (as close to the radiator as possible) on the left side and tie it around the hood bumper

5.7b Also support the radiator on the right side with a piece of rope, run under the hose from the expansion tank

3 Shift cables - removal and installation

1   Disconnect the shift and select cables from the levers on the transaxle.
2   Remove the center console (see Chapter 11).
3   Disconnect the shift and select cables from the shift control assembly. The cables can be pried from the pins on the shift control assembly with a screwdriver.
4   Pry loose the grommet, pull the cables through into the passenger compartment and remove them from the vehicle.
5   Installation is the reverse of removal.

4 Back-up light switch - check and replacement

Check

1   Turn the ignition key to the On position and move the shift lever to the Reverse position. The switch should close the back-up light circuit and turn on the back-up lights.
2   If it doesn't, check the back-up light fuse (see Chapter 12).
3   If the fuse is okay, verify that there's voltage available on the battery side of the switch (with the ignition turned to On).
4   If there's no voltage on the battery side of the switch, check the wire between the fuse and the switch; if there is voltage, put the shift lever in reverse and see if there's voltage on the other side of the switch.
5   If there's no voltage on the other side of the switch, replace the switch; if there is voltage, note whether one or both back-up lights are out.
6   If only one bulb is out, replace it; if they're both out, the bulbs could be the problem, but it's more likely that the wire between the switch and the bulbs has an open somewhere.

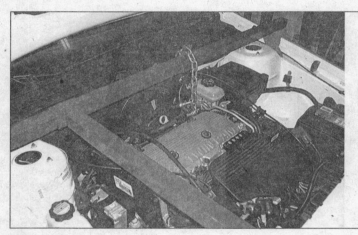

5.9 During transaxle removal, the preferred way to support the engine is with a support fixture designed for this purpose; they are often available from rental yards

Replacement

7   The back-up light switch is located on back of the left end (driver's side) of the transaxle.
8   Unplug the electrical connector from the switch.
9   Unscrew the switch.
10   Installation is the reverse of removal. Tighten the switch securely.

5 Manual transaxle - removal and installation

Refer to illustrations 5.7a, 5.7b and 5.9

Removal

1   Disconnect the negative battery cable. **Caution:** *On models equipped with the Theftlock audio system, be sure the lockout feature is turned off before performing any procedure which requires disconnecting the battery (see the information at the front of this manual).*
2   Remove the air filter housing (see Chapter 4). Also remove the mounting bracket.
3   Disconnect the shift cables from the levers on the transaxle, then remove the cable bracket.
4   Detach the hydraulic line from the clutch release cylinder. Unbolt the bracket securing the hydraulic line and reposition the line and bracket out of the way.
5   Disconnect the electrical connectors from the Vehicle Speed Sensor and the back-up light switch.
6   Remove the starter motor (see Chapter 5).
7   Support the radiator and condenser from above with two lengths of rope **(see illustrations)**.
8   Detach any wiring harnesses from the brackets along the top of the transaxle.
9   Attach an engine support fixture (which is recommended) or an engine hoist or to the engine and raise it sufficiently to just support the weight of the engine **(see illustration)**. **Note 1:** *The engine must remain supported while the transaxle is out of the vehicle.* **Note 2:** *If you use an engine hoist, position the hoist with its legs inserted under the vehicle from the right (passenger's) side. This will give you room to maneuver the transaxle out with a jack.*
10   Remove the upper transaxle-to-engine bolts.

11 Loosen the driveaxle/hub nuts and the front wheel lug nuts, then raise the vehicle and support it securely on jackstands placed underneath the rocker panel flanges, right behind the wheel openings on each side of the vehicle. Remove the front wheels.

12 Remove both inner fender splash shields.

13 Remove the driveaxles (see Chapter 8). **Note:** *Instead of separating the struts from the steering knuckles when removing the driveaxles, separate the control arm balljoints from the steering knuckles (see Chapter 10). This will allow you to pull out on the steering knuckles far enough to remove the driveaxles, and has to be done anyway.*

14 Detach both tie-rod ends from the steering knuckle arms (see Chapter 10).

15 Disconnect the electrical connectors from both front ABS wheel speed sensors harnesses. Detach the harnesses from their clips on the control arms and subframe (also known as the "cradle").

16 Unbolt the Anti-lock Brake System (ABS) hydraulic modulator from the subframe.

17 Disconnect all ground wires and any other electrical connectors accessible from underneath the vehicle.

18 Detach the brake line from the clips along the front of the subframe.

19 Remove the pinch bolt that secures the intermediate steering shaft to the steering gear (see Chapter 10).

20 Unscrew the power steering line fittings from the power steering gear (see Chapter 10).

21 Remove the bolt that secures the power steering line bracket at the right side of the subframe.

22 Remove the radiator/condenser support panel (see Chapter 3).

23 Unbolt all three mounts from the transaxle (see Chapter 2A or 2B).

24 Position two floor jacks under the subframe and remove the subframe mounting bolts. Slowly lower the jacks, making sure nothing is still connected to the subframe. Remove the subframe out from under the vehicle.

25 Support the transaxle with a jack, preferably a jack made for this purpose. Transmission jacks are commonly available at most equipment rental yards. These jacks are equipped with safety chains; use these chains to secure the transaxle to the jack.

26 Remove the transaxle mount brackets from the transaxle.

27 Remove the remaining transaxle-to-engine bolts.

28 Remove the transaxle from the engine by sliding it toward the left side of the vehicle. It may be necessary to lower the engine to provide clearance for the transaxle to pass below the left side inner body panel.

Installation

29 Installation is the reverse of removal, with attention paid to the following points:

a) *Apply a film of high-temperature grease to the transaxle input shaft splines.*

b) *When mating the transaxle to the engine, make sure the transaxle seats against the engine completely before tightening the bolts; if it doesn't, figure out why. Don't use the bolts to draw the transaxle into place, as you could break something.*

c) *Tighten all transaxle-to-engine bolts to the torque listed in this Chapter's Specifications.*

d) *Tighten the subframe mounting bolts to the torque listed in the Chapter 10 Specifications. Also tighten the intermediate shaft pinch bolt, the control arm balljoint-to-steering knuckle nuts and the tie-rod end-to-steering knuckle nuts to the torque values listed in the Chapter 10 Specifications.*

e) *Tighten the transmission (powertrain) mounting fasteners to the torque values listed in the Chapter 2A or 2B Specifications.*

f) *Tighten the wheel lug nuts to the torque listed in the Chapter 1 Specifications.*

g) *The wheel alignment should be checked and, if necessary, adjusted.*

h) *Check the transaxle lubricant level and add, as necessary, to bring it to the appropriate level (see Chapter 1).*

6   Manual transaxle overhaul - general information

Overhauling a manual transaxle is a difficult job for the do-it-yourselfer. It involves the disassembly and reassembly of many small parts. Numerous clearances must be precisely measured and, if necessary, changed with select fit spacers and snap-rings. As a result, if transaxle problems arise, it can be removed and installed by a competent do-it-yourselfer, but overhaul should be left to a transmission repair shop. Rebuilt transaxles may be available - check with your dealer parts department and auto parts stores. At any rate, the time and money involved in an overhaul is almost sure to exceed the cost of a rebuilt unit.

Nevertheless, it's not impossible for an inexperienced mechanic to rebuild a transaxle if the special tools are available and the job is done in a deliberate, step-by-step manner so nothing is overlooked.

The tools necessary for an overhaul include internal and external snap-ring pliers, a bearing puller, a slide hammer, a set of pin punches, a dial indicator and possibly a hydraulic press. In addition, a large, sturdy workbench and a vise or transaxle stand will be required.

During disassembly of the transaxle, make careful notes of how each piece comes off, where it fits in relation to other pieces and what holds it in place. Note how the parts are installed when you remove them; this will make it much easier to get the transaxle back together.

Before taking the transaxle apart for repair, it will help if you have some idea what area of the transaxle is malfunctioning. Certain problems can be closely tied to specific areas in the transaxle, which can make component examination and replacement easier. Refer to the Troubleshooting section at the front of this manual for information regarding possible sources of trouble.

7A

Notes

Chapter 7 Part B
Automatic transaxle

Contents

Specifications

Torque specifications

	Ft-lbs
Park/Neutral Position switch retaining bolts	18
Transaxle-to-engine bolts	66
Transaxle-to-engine brace bolts	32
Manual lever-to-shaft nut	15
Torque converter-to-driveplate bolts	46
Fluid pan bolts	See Chapter 1

1 General information

The vehicles covered by this manual are equipped with either a four-speed automatic transaxle or a five-speed manual transaxle. Two models of Hydra-matic automatic transaxles are used on these vehicles; the 4T40E and the 4T45E, both of which are electronically controlled four-speed transaxles.

Due to the complexity of the clutches and the hydraulic control system, and because of the special tools and expertise required to perform an automatic transmission overhaul, it should not be undertaken by the home mechanic. Therefore, the procedures in this Chapter are limited to general diagnosis, adjustment and transmission removal and installation.

If the transmission requires major repair work it should be left to a dealer service department or an automotive or transmission repair shop. You can, however, remove and install the transmission yourself and save the expense, even if the repair work is done by a transmission specialist.

Adjustments that the home mechanic may perform include those involving the shift linkage and the Park/Neutral Position (PNP) switch. **Caution:** *Never tow a disabled vehicle equipped with an automatic transaxle at speeds greater than 30 mph or distances over 50 miles unless the front wheels are off the ground. Failure to observe this precaution may result in severe transmission damage caused by lack of lubrication.*

2 Diagnosis - general

Note: *Automatic transaxle malfunctions may be caused by five general conditions: poor engine performance, improper adjustments, hydraulic malfunctions, mechanical malfunctions or malfunctions in the Powertrain Control Module or its signal network. Diagnosis of these problems should always begin with a check of the easily repaired items: fluid level and condition (see Chapter 1), and shift cable adjustment (see Section 4). Next, perform a road test to determine if the problem has been corrected or if more diagnosis is necessary. Because the transaxle relies on many sensors in the engine control system, and since the transaxle shift points are controlled* *by the Powertrain Control Module, you'll also want to check to see if any trouble codes have been stored in the PCM (see Chapter 6 for a list of trouble codes and how to extract them). If the problem persists after the preliminary tests and corrections are completed, additional diagnosis should be done by a dealer service department or transmission repair shop. Refer to the Troubleshooting section at the front of this manual for transaxle problem diagnosis.*

Preliminary checks

1   Drive the vehicle to warm the transaxle to normal operating temperature.

2   Check the fluid level as described in Chapter 1:

   a) *If the fluid level is unusually low, add enough fluid to bring the level within the designated area of the dipstick, then check for external leaks.*

   b) *If the fluid level is abnormally high, drain off the excess, then check the drained fluid for contamination by coolant. The presence of engine coolant in the automatic transmission fluid indicates that a failure has occurred in the internal radiator walls that separate the coolant from the transmission fluid (see Chapter 3).*

   c) *If the fluid is foaming, drain it and refill the transaxle, then check for coolant in the fluid or a high fluid level.*

3   Check the engine idle speed. **Note:** *If the engine is malfunctioning, do not proceed with the preliminary checks until it has been repaired and runs normally.*

4   Inspect the shift control cable (see Section 4). Make sure that it's properly adjusted and that it operates smoothly.

5   Check the Park/Neutral Position (PNP) switch adjustment (see Section 6).

Fluid leak diagnosis

6   Most fluid leaks are easy to locate visually. Repair usually consists of replacing a seal or gasket. If a leak is difficult to find, the following procedure may help.

7   Identify the fluid. Make sure it's transmission fluid and not engine oil or brake fluid (automatic transmission fluid is a deep red color).

8   Try to pinpoint the source of the leak. Drive the vehicle several miles, then park it over a large sheet of cardboard. After a minute or two, you should be able to locate the leak by determining the source of the fluid dripping onto the cardboard.

9   Make a careful visual inspection of the suspected component and the area immediately around it. Pay particular attention to gasket mating surfaces. A mirror is often helpful for finding leaks in areas that are hard to see.

10   If the leak still cannot be found, clean the suspected area thoroughly with a degreaser or solvent, then dry it.

11   Drive the vehicle for several miles at normal operating temperature and varying speeds. After driving the vehicle, visually

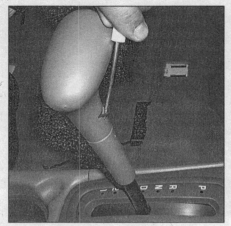

3.2a To remove the shift lever knob, pry out this locking clip . . .

inspect the suspected component again.

12   Once the leak has been located, the cause must be determined before it can be properly repaired. If a gasket is replaced but the sealing flange is bent, the new gasket will not stop the leak. The bent flange must be straightened.

13   Before attempting to repair a leak, check to make sure that the following conditions are corrected or they may cause another leak. **Note:** *Some of the following conditions cannot be fixed without highly specialized tools and expertise. Such problems must be referred to a transmission shop or a dealer service department.*

Gasket leaks

14   Check the pan periodically. Make sure the bolts are tight, no bolts are missing, the gasket is in good condition and the pan is flat (dents in the pan may indicate damage to the valve body inside).

15   If the pan gasket is leaking, the fluid level or the fluid pressure may be too high, the vent may be plugged, the pan bolts may be too tight, the pan sealing flange may be warped, the sealing surface of the transaxle housing may be damaged, the gasket may be damaged or the transaxle casting may be cracked or porous. If sealant instead of gasket material has been used to form a seal between the pan and the transaxle housing, it may be the wrong sealant.

Seal leaks

16   If a transaxle seal is leaking, the fluid level or pressure may be too high, the vent may be plugged, the seal bore may be damaged, the seal itself may be damaged or improperly installed, the surface of the shaft protruding through the seal may be damaged or a loose bearing may be causing excessive shaft movement.

17   Make sure the dipstick tube seal is in good condition and the tube is properly seated. Periodically check the area around the speedometer gear or sensor for leakage. If transmission fluid is evident, check the O-

3.2b . . . then pull the knob straight up

ring for damage. Also inspect the driveaxle oil seals for leakage.

Case leaks

18   If the case itself appears to be leaking, the casting is porous and will have to be repaired or replaced.

19   Make sure the oil cooler hose fittings are tight and in good condition.

Fluid comes out vent pipe or fill tube

20   If this condition occurs, the transaxle is overfilled, there is coolant in the fluid, the case is porous, the dipstick is incorrect, the vent is plugged or the drain back holes are plugged.

---

3   Shift lever - removal and installation

Removal

Refer to illustrations 3.2a, 3.2b and 3.6

1   Disconnect the negative battery cable. **Caution:** *On models equipped with the Theftlock audio system, be sure the lockout feature is turned off before performing any procedure which requires disconnecting the battery (see the information at the front of this manual).*

2   Remove the shift lever knob **(see illustrations)**.

3   Remove the console (see Chapter 11).

4   Disconnect the shift cable from the shift lever and base (see Section 4).

5   Unplug the electrical connector from the brake/transmission shift interlock solenoid, then pry the solenoid from its mounting pins. Also detach the Park/Lock cable from the shifter and base (see Section 5).

6   Remove the shift lever assembly retaining nuts **(see illustration)** and remove the shift lever assembly.

Installation

7   Place the shift lever in position on the mounting studs and install the nuts. Tighten

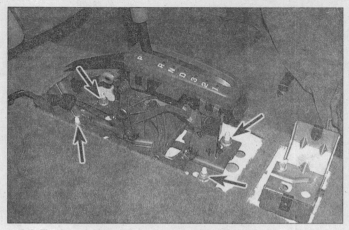

3.6 To detach the shift lever assembly from the floor, remove these four nuts (arrows)

4.2 To disconnect the shift cable from the transaxle manual lever, pry it off the pin on the lever with a prying tool or screwdriver

4.3a To disengage the shift cable from the cable bracket on the transaxle, remove this locking clip . . .

4.3b . . . then squeeze the two locking tangs and pull the cable housing from the bracket

4.5 To disconnect the shift cable from the pin on the shift lever, simply pry it loose

the nuts securely.

8   The remainder of installation is the reverse of removal. Adjust the shift cable (see Section 4) and the Park/Lock cable (see Section 5) when you're done.

---

4   Shift cable - replacement and adjustment

Replacement

Refer to illustrations 4.2, 4.3a, 4.3b, 4.5, 4.6a and 4.6b

1   Disconnect the negative battery cable. **Caution:** *On models equipped with the Theft-lock audio system, be sure the lockout feature is turned off before performing any procedure which requires disconnecting the battery (see the information at the front of this manual).* Remove the air filter housing (see Chapter 4).

2   Disconnect the shift cable from the transaxle manual lever **(see illustration).**

3   Disengage the shift cable from the cable bracket on the transaxle **(see illustrations).**

4   Remove the shift lever knob (see Sec-

4.6a To disengage the shift cable from the bracket on the shift lever base, remove this locking clip . . .

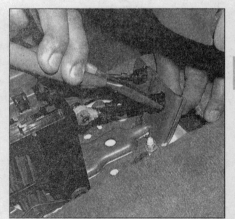

4.6b . . . then squeeze the locking tangs together and disengage the cable from the bracket

tion 3) and the console (see Chapter 11).

5   Disconnect the shift cable from the pin on the shift lever **(see illustration).**

6   Detach the cable from the bracket at the front of the shift lever base **(see illustrations).**

7   Trace the cable to the cable grommet (the point at which it goes through the firewall). Pry out the grommet and pull the cable through the hole and remove it.

8   Installation is the reverse of removal.

7B

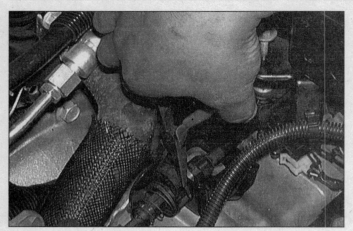

4.11 With the shift levers both in Neutral, lift up on the adjuster tab and connect the cable to the manual lever on the transaxle, then push the tab back into place

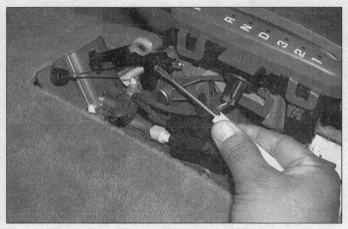

5.5 To release the Park/Lock cable, pry up the center portion on the cable connector (the cable will then automatically adjust); to lock the cable into place, press the center portion back down

When you're done installing the new cable, be sure to adjust it.

Adjustment

Refer to illustration 4.11

9   Detach the cable from the manual lever on the transaxle. Place the manual lever on the transaxle in the Neutral position; this is accomplished by rotating the lever clockwise from the Park position, through Reverse and into Neutral.

10   Place the shift lever inside the car in Neutral.

11   Pull up on the cable adjuster tab at the cable bracket on the transaxle **(see illustration)**, then connect the cable to the manual lever on the transaxle. The cable will automatically adjust itself. Push the adjuster tab back into place.

12   Make sure the engine will start in the Park and Neutral positions only.

13   If the engine can be started in any position other than Park or Neutral, check the adjustment of the Park/Neutral Position Switch (see Section 6), then adjust and check the shift cable again.

---

5   Park/Lock system - description, adjustment and component replacement

Description

1   The Park/Lock system prevents the shift lever from being moved out of Park unless the brake pedal is depressed simultaneously. It also prevents the ignition key from being removed from the ignition switch unless the shift lever is in the Park position. When the car is started, a solenoid is energized, locking the shift lever in Park; when the brake pedal is depressed, the solenoid is deenergized, unlocking the shift lever so that it can be moved into some other gear.

5.12 Pull the cable end to the rear to disengage it from the pin on the Park/Lock lever

Cable check and adjustment

Refer to illustration 5.5

2   Apply the parking brake. With the ignition key turned to the On position, depress the brake pedal and move the shift lever out of Park. Verify that the shift lever moves through all gear positions.

3   While moving the shift lever through all gear positions, verify that the ignition key can't be turned to the Lock position.

4   Verify that the ignition key can be removed when it's in the Lock position and the shift lever is in the Park position.

5   If the Park/Lock cable fails any of the above tests, adjust it as follows.

a) *Put the shift lever in Park.*
b) *Turn the ignition key to Lock.*
c) *Remove the shift lever knob (see Section 3).*
d) *Remove the console (see Chapter 11).*
e) *Pry up the locking tab on the cable connector at the Park/Lock lever **(see illustration)**. This releases the Park/Lock cable. The adjuster within the connector is spring loaded, so it will automatically remove the slack from the cable.*

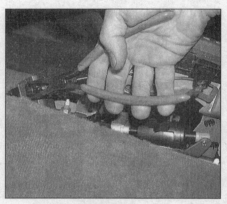

5.13 To disengage the housing from the cable bracket, depress the tangs on the sides of the cable housing

f) *Push in on the locking tab until it clicks back into place.*

6   Following Steps 2 through 4 above, check the cable as described.

Cable replacement

Refer to illustrations 5.12 and 5.13

7   Remove the shift lever knob (see Section 3).

8   Remove the console, left-side underdash panel and steering column knee bolster panel (see Chapter 11).

9   Put the shift lever in the Park position.

10   Turn the ignition key to the Run position.

11   Insert a screwdriver blade into the slot in the ignition switch inhibitor, depress the cable latch and detach the cable from the inhibitor.

12   Snap the rear end of the cable loose from the pin on the Park/Lock lever **(see illustration)**.

13   Depress the tabs on the sides of the cable housing **(see illustration)** and disengage the housing from the cable bracket.

14   Remove any cable clips and remove the Park/Lock cable.

15   Make sure the cable locking tab is in the

5.29 Pry the ends of the BTSI solenoid off its mounting pins

6.7 The Park/Neutral Position switch is located on top of the transaxle

up position and the shift lever is in Park.

16   With the ignition key in the Run position (this is very important), snap the cable into the inhibitor housing.

17   Snap the rear end of the cable onto the pin on the Park/Lock lever.

18   Turn the ignition key to the Lock position.

19   Push in on the locking tab of the cable connector until it clicks back into place.

20   Check the operation of the park/lock cable as described in Steps 2, 3 and 4.

21   If it operates as described above, the park/lock cable system is properly adjusted.

22   If the park/lock system does not operate as described, repeat the adjustment procedure. Push the cable connector down and recheck the operation.

23   Install the steering column knee bolster, the left-side under-dash panel and the console (see Chapter 11).

24   Install the shift lever knob (see Section 3).

Brake/Transmission Shift Interlock (BTSI) solenoid check

25   Turn the ignition key to the Run position and, without depressing the brake pedal, attempt to move the shift lever out of Park; you shouldn't be able to depress the button on the shift lever. If you can depress the button, either the solenoid is not receiving voltage, is not grounded, or it is defective.

a) *Check the NSBU fuse in the underhood fuse box.*

b) *Check the adjustment of the shift cable and the Park/Neutral Position switch.*

c) *Check the BTSI portion of the brake light switch; with the brake pedal at rest, there should be voltage available at the light green wire terminal (input) and the dark green/white wire (output). With the brake pedal depressed, there should only be voltage available at the light green wire terminal.*

d) *If the BTSI portion of the brake light switch is working properly, check for power to the BTSI solenoid on the dark*

green/white wire. *If voltage is not available, repair the circuit between the solenoid and the brake light switch. If voltage is available, check the ground circuit for continuity. If continuity is present, replace the BTSI solenoid.*

e) *If there is no voltage to the BTSI portion of the brake light switch on the light green wire terminal, remove the air filter housing and check for voltage to the Park/Neutral Position switch on the pink wire. If voltage is not present, repair the circuit between the PNP switch and the underhood fuse box. If voltage is available, check for voltage to the upper light green wire terminal at the Park/Neutral Position switch (with the shifter in Park). If no voltage is present, replace the switch.* **Note:** *Make sure you check the upper light green wire; the other light green wire, in the center of the lower row of the connector, is for the back-up light circuit.*

26   Now depress the brake pedal - you should be able to push the button on the shifter in and move the lever out of Park. If you can't, the solenoid isn't de-activating.

a) *Make sure the shift cable and Park/Neutral Position switch are properly adjusted.*

b) *Check the operation of the BTSI portion of the brake light switch as described in Step 25.*

c) *If voltage to the solenoid is cut when the brake pedal is depressed, but the solenoid still doesn't release, it is stuck; replace it.*

Brake/Transmission Shift Interlock (BTSI) solenoid replacement

Refer to illustration 5.29

27   Remove the shift lever knob (see Section 3) and the center console (see Chapter 11).

28   Unplug the electrical connector from the solenoid.

29   Pry off each end of the solenoid from its mounting pins **(see illustration)**.

30   Installation is the reverse of removal.

6   Park/Neutral Position (PNP) switch/back-up light switch - check and replacement

Check

1   Make sure the shift cable is properly adjusted (see Section 4). Remove the air filter housing (see Chapter 4).

2   Verify that there's voltage available at the Park/Neutral Position switch on the pink wire with the ignition key On **(see illustration 6.7 for switch location)**. If there isn't, check the NSBU fuse in the underhood fuse box. If the fuse is good, check the circuit to the switch.

3   If voltage is available, check for voltage to the upper light green wire terminal at the Park/Neutral Position switch (with the shifter in Park). If no voltage is present, try adjusting the switch. If that doesn't work, replace the switch.

4   Place the shift lever into reverse and check for voltage to the other light green wire in the connector (in the center of the lower row of wires). If no voltage is available, try adjusting the switch. If that doesn't work, replace the switch.

Replacement

Refer to illustrations 6.7, 6.14a and 6.14b

5   Turn the ignition Off and disconnect the cable from the negative terminal of the battery. **Caution:** *On models equipped with the Theftlock audio system, be sure the lockout feature is turned off before performing any procedure which requires disconnecting the battery (see the information at the front of this manual).*

6   Apply the parking brake and put the shift lever in Neutral.

7   Locate the Park/Neutral Position (PNP) switch **(see illustration)**, which is mounted on the transaxle at the manual lever.

8   Disconnect the shift cable from the manual lever.

9   Unplug the electrical connectors from the PNP switch **(see illustration 6.7)**.

7B

10   Remove the manual lever retaining nut and remove the manual lever. **Note:** *Be careful not to move the manual lever from the Neutral position while doing this.*

11   Remove the switch retaining bolts and detach the switch. **Note:** *Some models have a wiring harness clip fastened to a stud on the rear bolt. If this is the case, remove the pressed-metal nut with a pair of pliers and remove the clip from the stud.*

12   Remove the Park/Neutral Position switch.

13   Put the shift shaft in the Neutral position.

14   If you're installing the old switch, pry out the old set pin **(see illustration)**, insert a 3/32-inch gauge pin or drill bit into the service adjustment hole **(see illustration)** and rotate the switch until the pin drops down. Remove the gauge pin or drill bit. If you're installing a new switch you don't need to perform this Step; the switch comes equipped with a set pin that holds the switch in the Neutral position until it has been installed and the shift lever has been moved through its range, at which point the set pin will shear off.

15   Align the flats of the shift shaft with the flats of the Park/Neutral Position switch and install the switch.

16   Install the switch mounting bolts and tighten them to the torque listed in this Chapter's Specifications.

17   Install the manual lever, tightening the nut securely.

18   Attach the shift cable and reconnect the electrical connectors.

19   Verify that the engine will start only in Park or Neutral. If it starts in any other gear, readjust the switch (turn it slightly one way or the other and see if the engine now only starts in Park or Neutral; it isn't necessary to remove the switch and adjust it with a gauge pin).

20   The remainder of installation is the reverse of removal.

---

7   Driveaxle oil seals - replacement

Refer to illustration 7.4

1   Oil leaks frequently occur due to wear of the driveaxle oil seals. Replacement of these seals is relatively easy, since the repairs can be performed without removing the transaxle from the vehicle.

2   The driveaxle oil seals are located in the sides of the transaxle, where the driveaxles are attached. If leakage at the seal is suspected, raise the vehicle and support it securely on jackstands. If the seal is leaking, fluid will be found on the sides of the transaxle.

3   Remove the driveaxles (see Chapter 8).

4   If you're removing the right (passenger's) side seal, the stub axle shaft will have to be removed first. To do this, remove the snap-ring on the end of the stub shaft, then pull out on the stub axle and rotate it until the inner snap-ring seats in the differential side gear taper. Now attach a slide hammer and puller adapter to the stub axle and pull it from

6.14a To adjust the old Park/Neutral Position switch, pry out the old set pin . . .

6.14b . . . insert a 3/32-inch gauge pin or drill bit into the service adjustment hole and rotate the switch until the pin drops down

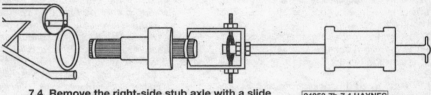

7.4 Remove the right-side stub axle with a slide hammer and adapter

24053-7b-7.4 HAYNES

the transaxle **(see illustration)**. Remove the inner snap-ring from the stub axle and discard both snap-rings (they aren't reusable).

5   Note how deep the seal is installed, then use a screwdriver or prybar to carefully pry the oil seal out of the transaxle bore. If the oil seal cannot be removed with a screwdriver or prybar, a special oil seal removal tool (available at most auto parts stores) will be required.

6   Compare the old seal to the new one to be sure it's the correct one.

7   Coat the outside and inside diameters of the new seal with a small amount of transmission fluid.

8   Using a seal installation tool, install the new oil seal. Drive it into the bore squarely and make sure it's seated to the original depth.

9   If you replaced the right side seal, install new snap-rings on the stub shaft, then carefully guide the stub shaft through the seal (don't let the shaft splines contact the seal lips). Tap the stub shaft into place until it is seated.

10   Install the driveaxle (see Chapter 8).

---

8   Transaxle - removal and Installation

Refer to illustrations 8.4a, 8.4b, 8.6, 8.7, 8.13, 8.16, 8.17, 8.31a and 8.31b

Removal

1   Disconnect the negative battery cable. **Caution:** *On models equipped with the Theft-lock audio system, be sure the lockout feature*

is turned off before performing any procedure which requires disconnecting the battery (see the information at the front of this manual).

2   Remove the air filter housing (see Chapter 4).

3   Disconnect the shift cable from the manual lever and the bracket on the transaxle (see Section 4).

4   Support the radiator and condenser from above with two lengths of rope **(see illustrations)**.

5   Clearly label, then unplug, all electrical connectors from the transaxle which are accessible from the top. Detach the harnesses from the brackets along the top of the transaxle.

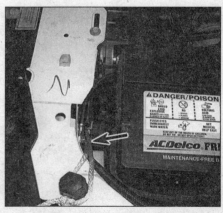

8.4a Run a length of rope under the radiator hose (as close to the radiator as possible) on the left side and tie it around the hood bumper

8.4b Also support the radiator on the right side with a piece of rope, run under the hose from the expansion tank

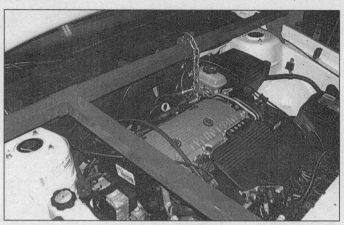

8.6 During transaxle removal, the preferred way to support the engine is with a support fixture designed for this purpose; they are often available from rental yards

8.7 Remove these three transaxle-to-engine bolts from above

8.13 Remove the engine-to-transaxle brace (arrow)

6   Attach an engine support fixture (which is recommended) or an engine hoist or to the engine and raise it sufficiently to just support the weight of the engine (see illustration). Note 1: *The engine must remain supported while the transaxle is out of the vehicle.* Note 2: *If you use an engine hoist, position the hoist with its legs inserted under the vehicle from the right (passenger's) side. This will give you room to maneuver the transaxle out with a jack.*

7   Remove the upper transaxle-to-engine bolts (see illustration).

8   Loosen the driveaxle/hub nuts and the front wheel lug nuts, then raise the vehicle and support it securely on jackstands placed underneath the rocker panel flanges, right behind the wheel openings on each side of the vehicle. Remove the front wheels.

9   Remove both inner fender splash shields.

10   Remove the driveaxles (see Chapter 8). Note: *Instead of separating the struts from the steering knuckles when removing the driveaxles, separate the control arm balljoints from the steering knuckles (see Chapter 10). This will allow you to pull out on the steering knuckles far enough to remove the driveaxles,*

and has to be done anyway.

11   Detach both tie-rod ends from the steering knuckle arms (see Chapter 10).

12   Disconnect the electrical connectors from both front ABS wheel speed sensors harnesses. Detach the harnesses from their clips on the control arms and subframe (also known as the "cradle").

13   Remove the engine-to-transaxle brace

8.16 There isn't much clearance between the transaxle and engine; use an offset wrench to remove the torque converter-to-driveplate bolts, and use a large screwdriver wedged in the ring gear teeth to prevent the driveplate from turning

(see illustration).

14   Remove the torque converter cover bolts and remove the cover.

15   Remove the starter motor (see Chapter 5).

16   Mark the relationship of the torque converter to the driveplate and remove the fly-wheel-to-torque converter bolts (see illustration). You can use a socket and breaker bar

7B

8.17 To detach the transaxle fluid cooler lines, remove the nut between the two lines

8.31a Remove these mounting nuts along the front side of the transaxle . . .

on the crankshaft pulley bolt to rotate the engine for access to the bolts, or you can turn the driveplate with a screwdriver by prying against the ring gear teeth.

17   Disconnect and plug the transaxle cooler lines **(see illustration)**.

18   Unbolt the Anti-lock Brake System (ABS) hydraulic modulator from the subframe.

19   Disconnect all ground wires and any other electrical connectors accessible from underneath the vehicle.

20   Detach the brake line from the clips along the front of the subframe.

21   Remove the transaxle cooler lines from the clip at the front of the subframe.

22   Remove the pinch bolt that secures the intermediate steering shaft to the steering gear (see Chapter 10).

23   Unscrew the power steering line fittings from the power steering gear (see Chapter 10).

24   Remove the bolt that secures the power steering line bracket at the right side of the subframe.

25   Remove the radiator/condenser support panel (see Chapter 3).

26   Unbolt all three mounts from the transaxle (see Chapter 2A or 2B).

27   Position two floor jacks under the subframe and remove the subframe mounting bolts. Slowly lower the jacks, making sure nothing is still connected to the subframe. Remove the subframe out from under the vehicle.

28   Support the transaxle with a jack, preferably a jack made for this purpose. Transmission jacks are commonly available at most equipment rental yards. These jacks are equipped with safety chains; use these

chains to secure the transaxle to the jack.

29   Remove the left (side) transaxle mount bracket-to-transaxle bolts and remove the mount and bracket assembly.

30   Remove the nut and bolt from the heater core hose pipe-to-transaxle bracket.

31   Remove the remaining transaxle-to-engine bolts **(see illustrations)**.

32   Remove the transaxle from the engine by sliding it toward the left side of the vehicle.

Installation

33   Installation is the reverse of removal, with attention paid to the following points:

a) *Before installing the transaxle, make sure the torque converter is completely seated. To do this, push in on the converter while turning it. If it wasn't seated, it will "clunk" into place (it may even "clunk" more than once).*

b) *Apply a film of multi-purpose grease to the nose of the converter.*

c) *When mating the transaxle to the engine, make sure the transaxle seats against the engine completely before tightening the bolts; if it doesn't, figure out why. Don't use the bolts to draw the transaxle into place, as you could break something.*

d) *Tighten all transaxle-to-engine bolts to the torque listed in this Chapter's Specifications.*

e) *Tighten the torque converter-to-driveplate bolts to the torque listed in this Chapter's Specifications.* **Note:** *Install all of the bolts before tightening any of them.*

f) *Tighten the subframe mounting bolts to the torque listed in the Chapter 10 Specifications. Also tighten the interme-*

8.31b . . . then remove this nut, unscrew the stud and remove the spacer at the lower rear side of the transaxle

diate shaft pinch bolt, the control arm balljoint-to-steering knuckle nuts and the tie-rod end-to-steering knuckle nuts to the torque values listed in the Chapter 10 Specifications.

g) *Tighten the transmission (powertrain) mounting fasteners to the torque values listed in the Chapter 2A or 2B Specifications.*

h) *Tighten the wheel lug nuts to the torque listed in the Chapter 1 Specifications.*

i) *The front end alignment should be checked and, if necessary, adjusted.*

j) *Adjust the shift cable (see Section 4).*

k) *Check the transaxle fluid level and add fluid, as necessary, to bring it to the appropriate level (see Chapter 1).*

Chapter 8
Clutch and driveaxles

Contents

Specifications

Clutch

Fluid type .... See Chapter 1

Inner CV joint boot length (see illustration 10.3s) .... 4-29/32 inches

Torque specifications

Ft-lbs (unless otherwise indicated)

Clutch master cylinder mounting nuts	180 in-lbs
Clutch pressure plate-to-flywheel bolts	180 in-lbs plus an additional 45-degrees rotation
Driveaxle/hub nut	
1997	30 plus an additional 235-degrees rotation
1998 on	284
Wheel lug nuts	See Chapter 1

1   General information

The information in this Chapter deals with the components from the rear of the engine to the front wheels, except for the transaxle, which is dealt with in Chapters 7A and 7B. For the purposes of this Chapter, these components are grouped into two categories: clutch and driveaxles. Separate Sections within this Chapter offer general descriptions and checking procedures for both groups.

Since nearly all the procedures covered in this Chapter involve working under the vehicle, make sure it's securely supported on sturdy jackstands or a hoist where the vehicle can be easily raised and lowered.

2   Clutch - description and check

1   All vehicles with a manual transaxle use a single dry plate, diaphragm spring type clutch. The clutch disc has a splined hub which allows it to slide along the splines of the transaxle input shaft. The clutch and pressure plate are held in contact by spring pressure exerted by the diaphragm in the pressure plate.

2   The clutch release system is operated by hydraulic pressure. The hydraulic release system consists of the clutch pedal, a master cylinder and fluid reservoir, the clutch fluid hydraulic line, and an integral clutch release cylinder and release bearing assembly.

3   When pressure is applied to the clutch pedal to release the clutch, hydraulic pressure is exerted against the release bearing, which pushes against the fingers of the diaphragm spring of the pressure plate assembly, which in turn releases the clutch disc.

4   Terminology can be a problem regarding the clutch components because common names have in some cases changed from that used by the manufacturer. For example, the driven plate is also called the clutch plate or disc, the pressure plate assembly is also referred to as the clutch cover, the clutch release bearing is also called a throw-out bearing, and the actuator cylinder is also known as a release or slave cylinder.

5   Unless you're replacing components that are obviously damaged, make the fol-

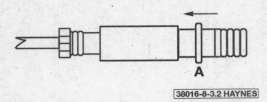

38016-8-3.2 HAYNES

3.5 To disconnect the clutch hydraulic line, push in the release slide (A), hold it there and pull the line and fitting apart

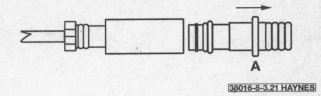

38016-8-3.21 HAYNES

3.7 To connect the clutch hydraulic line, pull back on the release slide (A), hold it there, and push the line into the fitting until you hear a click

lowing preliminary checks to determine the nature of the clutch system failure.

a) *Check the fluid level in the clutch master cylinder (see Chapter 1). If the fluid level is low, add fluid as necessary and inspect the hydraulic clutch system for leaks. If the master cylinder reservoir has run dry, bleed the system (see Section 5) and retest the clutch operation.*

b) *Check "clutch spin down time": run the engine at normal idle speed with the transaxle in Neutral (not with the clutch pedal depressed), depress the clutch pedal, wait several seconds and shift the transaxle into Reverse. You should not hear a grinding noise, the most likely cause of which is a defective pressure plate or clutch disc.*

c) *Check for complete clutch release: run the engine (with the parking brake applied) and hold the clutch pedal about 1/2-inch from the floor. Shift the transaxle between 1st gear and Reverse several times. If the shift is not smooth, component failure is indicated. Check the release cylinder pushrod travel. With the clutch pedal depressed completely the release cylinder pushrod should extend substantially. If it doesn't, check the fluid level in the clutch master cylinder.*

d) *Visually inspect the clutch pedal bushing at the top of the clutch pedal to make sure there is no sticking or excessive wear.*

3 Clutch master cylinder - removal and installation

Refer to illustrations 3.5 and 3.7

1   Remove the sound insulator from under the left side of the dashboard (see Chapter 11).

2   Using a flashlight, locate the clutch master cylinder pushrod. Disconnect the pushrod from the clutch pedal.

3   Remove the two nuts that attach the clutch master cylinder to the firewall. Detach the clutch master cylinder from the firewall.

4   Detach the clutch master cylinder remote reservoir from the firewall.

5   Disconnect the clutch hydraulic line from the clutch release cylinder **(see illustration)**. Have rags handy as some fluid will be

lost as the line is removed. **Caution:** *Don't allow brake fluid to come into contact with paint, as it will damage the finish.*

6   Remove the clutch master cylinder, remote reservoir and clutch hydraulic line as a single assembly.

7   Installation is the reverse of removal. Be sure to tighten the clutch master cylinder nuts to the torque listed in this Chapter's Specifications. Reconnect the clutch hydraulic line at the clutch release cylinder **(see illustration)**.

8   When you're done, bleed the clutch hydraulic system (see Section 5).

4 Clutch release cylinder and bearing - removal, inspection and installation

Warning: *Dust produced by clutch wear and deposited on clutch components may contain asbestos, which is hazardous to your health. DO NOT blow it out with compressed air and DO NOT inhale it. DO NOT use gasoline or petroleum-based solvents to remove the dust. Brake system cleaner should be used to flush it into a drain pan. After the clutch components are wiped clean with a rag, dispose of the contaminated rags and cleaner in a labeled, covered container.*

Removal

1   Remove the clutch master cylinder (see Section 3).

2   Remove the transaxle (see Chapter 7A).

3   Slide the clutch release cylinder and bearing assembly off the transaxle input shaft.

Inspection

4   Hold the bearing by the outer race and rotate the inner race while applying pressure. If the bearing doesn't turn smoothly or if it's noisy, replace the bearing/hub assembly with a new one. Wipe the bearing with a clean rag and inspect it for damage, wear and cracks. Don't immerse the bearing in solvent - it's sealed for life and to do so would ruin it. Also check the release cylinder for leaks. A thin coating of hydraulic fluid near the seal is acceptable, but a liberal amount of fluid indicates a damaged seal. If the release cylinder and bearing assembly is noisy or leaking, replace it.

Installation

5   Fill the inner groove of the release bearing with high-temperature grease. Also apply a light coat of the same grease to the transaxle input shaft splines.

6   Slide the release cylinder and bearing onto the input shaft.

7   Apply a light coat of high-temperature grease to the face of the release bearing where it contacts the pressure plate diaphragm fingers.

8   Install the transaxle (see Chapter 7A).

9   Install the clutch master cylinder (see Section 3).

10   Bleed the clutch hydraulic system (see Section 5).

5 Clutch hydraulic system - bleeding

1   The hydraulic system should be bled of all air whenever any part of the system has been removed or if the fluid level has been allowed to fall so low that air has been drawn into the master cylinder. The procedure is similar to bleeding a brake system.

2   Fill the master cylinder with new brake fluid conforming to DOT 3 specifications. **Caution:** *Do not reuse any of the fluid coming from the system during the bleeding operation or use fluid which has been inside an open container for an extended period of time.*

3   Raise the vehicle and place it securely on jackstands to gain access to the release cylinder, which is located on the left side of the clutch housing.

4   Locate the bleeder valve on the clutch release cylinder (right above the fitting for the hydraulic fluid line). Remove the dust cap which fits over the bleeder valve and push a length of clear hose over the valve. Place the other end of the hose into a clear container with about two inches of brake fluid in it. The hose end must be submerged in the fluid.

5   Have an assistant depress the clutch pedal and hold it. Open the bleeder valve on the release cylinder, allowing fluid to flow through the hose. Close the bleeder valve when fluid stops flowing from the hose. Once closed, have your assistant release the pedal.

6   Continue this process until all air is evacuated from the system, indicated by a

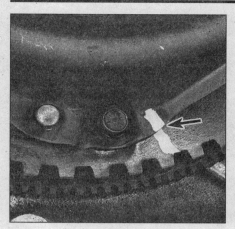

6.6 If you're going to re-use the same pressure plate, mark the relationship of the pressure plate to the flywheel

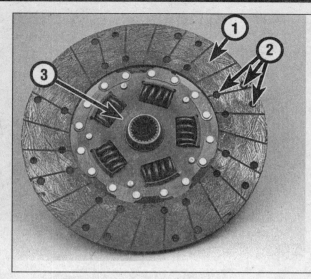

6.10 Examine the clutch disc for evidence of excessive wear, overheated friction material, loose rivets, worn hub splines and distorted damper cushions or springs

1 *Lining* - wears down in use
2 *Rivets* - secure the lining and can damage the flywheel or pressure plate if allowed to contact the surfaces
3 *Markings* - usually says something like "Flywheel side"

full, solid stream of fluid being ejected from the bleeder valve each time and no air bubbles in the hose or container. Keep a close watch on the fluid level inside the clutch master cylinder reservoir; if the level drops too low, air will be sucked back into the system and the process will have to be started all over again.

7 Install the dust cap and lower the vehicle. Check carefully for proper operation before placing the vehicle in normal service.

6 Clutch components - removal, inspection and installation

Warning: *Dust produced by clutch wear and deposited on clutch components may contain asbestos, which is hazardous to your health. DO NOT blow it out with compressed air and DO NOT inhale it. DO NOT use gasoline or petroleum based solvents to remove the dust. Brake system cleaner should be used to flush the dust into a drain pan. After the clutch components are wiped clean with a rag, dispose of the contaminated rags and cleaner in a labeled, covered container.*

Removal

Refer to illustration 6.6

1 Access to the clutch components is normally accomplished by removing the transaxle, leaving the engine in the vehicle. If, of course, the engine is being removed for major overhaul, then the opportunity should always be taken to check the clutch for wear and replace worn components as necessary. However, the relatively low cost of the clutch components compared to the time and labor involved in gaining access to them warrants their replacement any time the engine or transaxle is removed, unless they are new or in near-perfect condition. The following procedures assume that the engine will stay in place.

2 Disconnect the clutch hydraulic line from the clutch release cylinder **(see illustra-**

tion 3.5)**. Have rags handy as some fluid will be lost as the line is disconnected. **Caution:** *Don't allow brake fluid to come into contact with paint, as it will damage the finish.*

3 Remove the clutch master cylinder assembly (see Section 3).

4 Remove the transaxle from the vehicle (see Chapter 7A).

5 To support the clutch disc during removal, install a clutch alignment tool through the clutch disc hub **(see illustration 6.14)**.

6 Carefully inspect the flywheel and pressure plate for indexing marks. The marks are usually an X, an O or a white letter. If they cannot be found, scribe marks yourself so the pressure plate and the flywheel will be in the same alignment during installation **(see illustration)**.

7 Slowly loosen the pressure plate-to-flywheel bolts. Work in a diagonal pattern and loosen each bolt a little at a time until all spring pressure is relieved. Then hold the pressure plate securely and completely remove the bolts, followed by the pressure plate and clutch disc.

Inspection

Refer to illustrations 6.10, 6.12a and 6.12b

8 Ordinarily, when a problem occurs in the clutch, it can be attributed to wear of the clutch driven plate assembly (clutch disc). However, all components should be inspected at this time.

9 Inspect the flywheel for cracks, heat checking, score marks and other damage. If the imperfections are slight, a machine shop can resurface it to make it flat and smooth. Refer to Chapter 2A or 2B for the flywheel removal procedure.

10 Inspect the lining on the clutch disc. There should be at least 1/16-inch of lining above the rivet heads. Check for loose rivets, distortion, cracks, broken springs and other obvious damage **(see illustration)**. As mentioned above, ordinarily the clutch disc is replaced as a matter of course, so if in doubt about the condition, replace it with a new one.

11 Inspect the clutch release cylinder and bearing (see Section 4).

12 Check the machined surface and the diaphragm spring fingers of the pressure plate **(see illustrations)**. If the surface is grooved or otherwise damaged, replace the pressure plate assembly. Also check for obvious damage, distortion, cracking, etc. Light glazing can be removed with emery cloth or sandpaper. If a new pressure plate is indicated, new or factory rebuilt units are available.

NORMAL FINGER WEAR

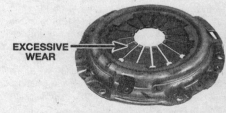

EXCESSIVE WEAR ←

EXCESSIVE FINGER WEAR

BROKEN OR BENT FINGERS

6.12a Replace the pressure plate if excessive wear or damage is noted

8

6.12b Examine the pressure plate friction surface for score marks, cracks and evidence of overheating (blue spots)

6.14 Center the clutch disc in the pressure plate with a clutch alignment tool

Installation

Refer to illustration 6.14

13 Before installation, wash the flywheel and pressure plate machined surfaces with brake system cleaner. It's important that no oil or grease is on these surfaces or the lining of the clutch disc. Handle these parts only with clean hands.

14 Position the clutch disc and pressure plate with the clutch held in place with an alignment tool **(see illustration)**. Make sure the disc is installed properly (most replacement clutch discs will be marked "flywheel side" or something similar - if not marked, install the clutch disc with the damper springs or cushion toward the transaxle).

15 Tighten the pressure plate-to-flywheel bolts only finger tight, working around the pressure plate.

16 Center the clutch disc by ensuring the alignment tool is through the splined hub and into the recess in the crankshaft. Wiggle the tool up, down or side-to-side as needed to bottom the tool. Tighten the pressure plate-to-flywheel bolts a little at a time, working in a crisscross pattern to prevent distortion of the cover. After all of the bolts are snug, tighten them to the torque listed in this Chapter's Specifications. Remove the alignment tool.

17 Using high-temperature grease, lubricate the inner groove of the release bearing. Also place grease on the transaxle input shaft bearing retainer.

18 Install the clutch release cylinder and bearing (see Section 4).

19 Install the transaxle (see Chapter 7A).

20 Install the clutch master cylinder (see Section 3).

21 Connect the clutch hydraulic line to the clutch release cylinder **(see illustration 3.7)**.

22 Bleed the clutch hydraulic system (see Section 5).

7   Clutch start switch - check and replacement

1 Remove the left side under-dash panel.

Check

2 Verify that the engine will not start when the clutch pedal is released. Now, depress the clutch pedal - the engine should start.

3 If the switch does not operate as described, locate the switch at the upper end of the clutch pedal and unplug the electrical connector.

4 Using an ohmmeter, verify that there is continuity between the terminals of the clutch start switch when the pedal is depressed. There should be no continuity when the pedal is released.

5 If the switch does not work as described, replace it.

Replacement

6 Unplug the electrical connector from the switch.

7 Detach the clutch start switch from the clutch pedal bracket.

8 Installation is the reverse of removal. The switch is self-adjusting, so there's no need for adjustment.

9 Verify that the engine doesn't start when the clutch pedal is released, and does start when the pedal is depressed.

8   Driveaxles - general information and inspection

1 Power is transmitted from the transaxle to the wheels through a pair of driveaxles. The inner ends of the driveaxles are splined onto the differential side gear shafts. The outer ends of the driveaxles are splined to the axle hubs and locked in place by a large nut.

2 The inner ends of the driveaxles are equipped with sliding constant velocity joints, which are capable of both angular and axial motion. These joint assemblies consist of a tripot bearing and a joint housing (outer race) in which the joint is free to slide in and out as the driveaxle moves up and down with the wheel. The inner joints can be disassembled, cleaned, inspected and repacked, but they cannot be overhauled. If any parts are dam-

aged, an inner joint must be replaced as a unit.

3 The outer CV joints are of the cross-groove, or "ball-and-cage" type. The outer joints are capable of angular but not axial movement. The outer joints can be disassembled, cleaned, inspected and repacked, but they cannot be overhauled. If any parts are damaged, an outer joint must be replaced as a unit.

4 The boots should be inspected periodically for damage and leaking lubricant. Torn CV joint boots must be replaced immediately or the joints can be damaged. Boot replacement involves removal of the driveaxle (see Section 9). **Note:** *Some auto parts stores carry "split" type replacement boots, which can be installed without removing the driveaxle from the vehicle. This is a convenient alternative; however, the driveaxle should be removed and the CV joint disassembled and cleaned to ensure the joint is free from contaminants such as moisture and dirt which will accelerate CV joint wear.* The most common symptom of worn or damaged CV joints, besides lubricant leaks, is a clicking noise in turns, a clunk when accelerating after coasting and vibration at highway speeds. To check for wear in the CV joints and driveaxle shafts, grasp each axle (one at a time) and rotate it in both directions while holding the CV joint housings, feeling for play indicating worn splines or sloppy CV joints. Also check the driveaxle shafts for cracks, dents and distortion.

9   Driveaxle - removal and installation

Removal

Refer to illustrations 9.3, 9.7, 9.9 and 9.10

1 Disconnect the cable from the negative terminal of the battery. **Caution:** *If the stereo in your vehicle is equipped with an anti-theft system, make sure you have the correct activation code before disconnecting the battery.*

2 Set the parking brake.

3 Remove the wheel cover, then break the

9.3 Loosen the driveaxle/hub nut with a long breaker bar

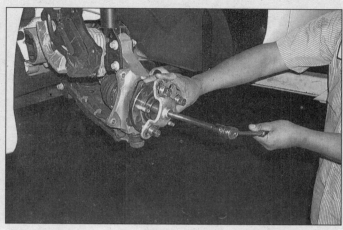

9.7 Attach a puller to the hub and tighten it just enough to break the hub splines loose

9.9 Pull the steering knuckle out and slide the end of the driveaxle out of the hub

9.10 To separate the inner end of the driveaxle from the transaxle, pry on the CV joint housing like this with a large screwdriver or prybar - you may need to give the prybar a sharp rap with a brass hammer

hub nut loose with a socket and large breaker bar **(see illustration)**. **Note:** *If the socket won't fit through the opening in the center of the wheel, remove the wheel and install the spare (the nut is very tight and is easier to loosen when the wheel is on the ground).*

4   Loosen the front wheel lug nuts, raise the vehicle and support it securely on jackstands. Remove the wheel.

5   Remove the driveaxle/hub nut and washer. To prevent the disc/hub from turning, insert a long punch into the brake disc cooling vanes and allow it to rest against the caliper mounting bracket.

6   Remove the brake caliper (see Chapter 9) and suspend it with a piece of wire from the strut coil spring.

7   Attach a puller to the hub flange and tighten it just to the point where the driveaxle moves in the hub **(see illustration)**. **Note:** *If you are using a jaw-type puller it will be necessary to remove the caliper mounting bracket and the brake disc.* **Caution:** *Don't attempt to push the end of the driveaxle through the hub yet. Applying force to the end of the driveaxle, beyond just breaking it*

loose from the hub, can damage the driveaxle or transaxle.

8   Separate the strut from the steering knuckle (see Chapter 10). Place a drain pan underneath the transaxle to catch any lubricant that may spill out when the driveaxle is removed.

9   Tighten the puller to push the end of the driveaxle through the hub, then pull out on the steering knuckle and detach the driveaxle from the hub **(see illustration)**. Don't let the driveaxle hang by the inner CV joint after the outer end has been detached from the steering knuckle, as the inner joint could become damaged. Support the outer end of the driveaxle with a piece of wire, if necessary.

10   Carefully pry the inner CV joint out of the transaxle **(see illustration)**.

11   Refer to Chapter 7 for the driveaxle oil seal replacement procedure, if necessary.

Installation

Refer to illustration 9.12

12   Installation is the reverse of the removal procedure, but with the following additional points:

a) *Replace the retaining ring on the differential side gear shaft* **(see illustration)**.

b) *Seat the inner CV joint in the differential side gear by positioning the end of a*

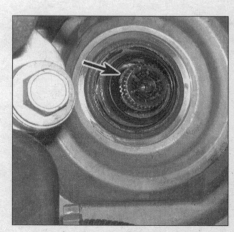

9.12 Replace the driveaxle retaining ring on the differential side gear shaft (left side shown)

8

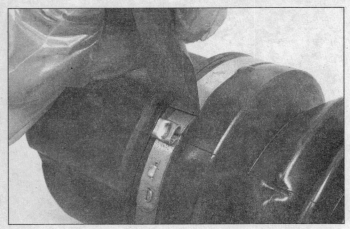

10.3a Cut off the boot retaining clamps, using wire cutters or a chisel and hammer

10.3b Slide the housing off the spider assembly

10.3c Slide the boot towards the center of the driveaxle

10.3d Spread the ends of the stop ring apart and slide it towards the center of the shaft

large screwdriver in the groove in the CV joint housing and tapping it into position with a hammer. Once this has been done pull out on the joint housing to make sure the retaining ring has seated.

c) *Tighten the strut-to-knuckle bolts/nuts to the torque listed in the Chapter 10 Specifications.*

d) *Install a new driveaxle/hub nut, but don't tighten it completely until the vehicle has been lowered (don't exceed 30 ft-lbs at this time).*

e) *If removed, install the caliper mounting bracket and tighten the bolts to the torque listed in the Chapter 9 Specifications.*

f) *Install the brake caliper and tighten the mounting bolts to the torque listed in the Chapter 9 Specifications.*

g) *Install the wheel and lug nuts, lower the vehicle and tighten the lug nuts to the torque listed in the Chapter 1 Specifications. Now tighten the driveaxle/hub nut to the torque listed in this Chapter's Specifications.*

h) *Check the transaxle lubricant and add, if necessary, to bring it to the proper level (see Chapter 1).*

10   Driveaxle boot - replacement

Note: *If the CV joint boots must be replaced, explore all options before beginning the job. Complete rebuilt driveaxles are available on an exchange basis, which eliminates much time and work. Whichever route you choose to take, check on the cost and availability of parts before disassembling the vehicle.*

1   Remove the driveaxle (see Section 9).
2   Place the driveaxle in a vise lined with rags to avoid damage to the axleshaft. Check the CV joint for excessive play in the radial direction, which indicates worn parts. Check for smooth operation throughout the full range of motion for each CV joint. If a boot is torn, disassemble the joint, clean the components and inspect for damage due to loss of lubrication and possible contamination by foreign matter.

Inner CV joint

Refer to illustrations 10.3a through 10.3t
3   To replace the inner boot, refer to the accompanying illustrations **(see illustrations 10.3a through 10.3t).**

Outer CV joint

Refer to illustrations 10.4a through 10.4r
4   Refer to the accompanying illustrations and perform the outer CV joint boot replacement procedure **(see illustrations 10.4a through 10.4r).**

10.3e Slide the spider assembly back to expose the retaining ring and pry off the ring

10.3f Carefully tap the spider off the axleshaft with a brass punch (but don't hit it so hard that it flies off, or you'll be picking up needle bearings!)

10.3g When you slide the spider off the driveaxle, hold the bearings in place with your hand; even better, use tape or a cloth wrapped around the spider bearing assembly to retain them

10.3h Slide the boot and the stop ring off the axleshaft

10.3i Clean all of the old grease out of the housing and spider assembly, then remove each bearing, one at time

10.3j Carefully disassemble each section of the spider assembly, clean the needle bearings with solvent and inspect the rollers, spider cross, bearings and housing for scoring, pitting and other signs of abnormal wear

10.3k Apply a coat of CV joint grease to the inner bearing surfaces to hold the needle bearings in place and slide the bearing over them

10.3l Wrap the axleshaft splines with tape to avoid damaging the boot, then slide the small clamp and boot onto the axleshaft

10.3m Slide the spider stop ring onto the axleshaft, past the groove in which it seats

8

10.3n Install the spider bearing with the recess in the counterbore facing the end of the driveaxle

10.3o Install the spider retaining ring, then slide the spider assembly against it and install the stop ring in its groove

10.3p Pack the housing with half of the grease furnished with the new boot and place the remainder in the boot

10.3q With the retaining clamps in place (but not tightened), install the tripot housing

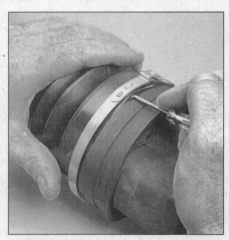

10.3r Seat the boot in the housing and axle seal grooves - a small screwdriver can make the job easier (make sure the boot isn't dimpled, stretched or out of shape)

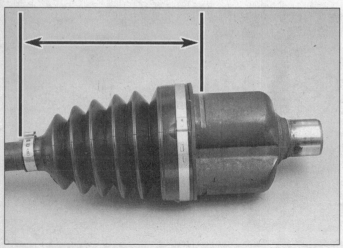

11.3s Adjust the length of the joint so the distance from the small end of the boot to the groove in the housing is the same as the dimension listed in this Chapter's Specifications

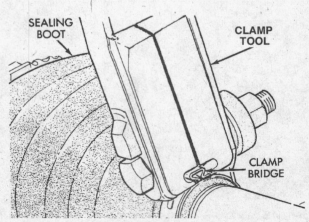

10.3t With the joint at the proper length, equalize the pressure in the boot by inserting a small screwdriver between the boot and the housing, then secure the boot clamps with a clamp crimping tool (available at auto parts stores)

10.4a Cut off the boot retaining clamps, using wire cutters or a chisel and hammer

10.4b Spread apart the ends of the internal snap-ring, then slide the CV joint off the shaft

10.4c Press down on the inner race far enough to allow a ball bearing to be removed - if it's difficult to tilt, gently tap the cage and inner race with a brass punch and hammer

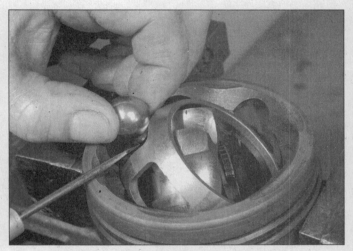

10.4d Pry the balls out of the cage, one at a time

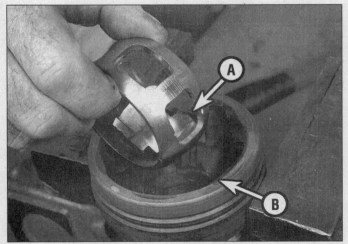

10.4e Tilt the inner race and cage 90-degrees, then align the windows in the cage (A) with the lands of the housing (B) and rotate the inner race up and out of the outer race

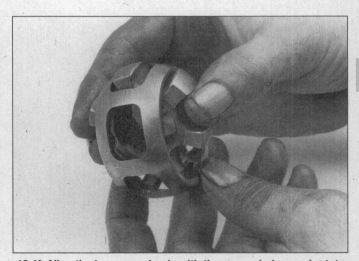

10.4f Align the inner race lands with the cage window and rotate the inner race out of the cage

8

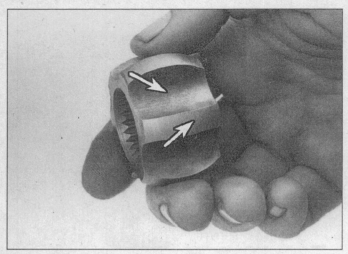

10.4g After cleaning the components with solvent, check the inner race lands and grooves for pitting and score marks

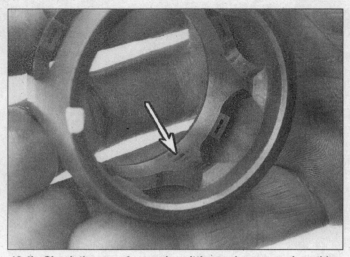

10.4h Check the cage for cracks, pitting and score marks - shiny spots are normal and don't affect operation

10.4i With the race and cage tilted at 90-degrees, lower the assembly into the housing

10.4j Rotate the assembly by gently tapping with a hammer and brass punch, then . . .

10.4k . . . press the balls into the cage windows, repeating until all of the balls are installed

10.4l Use needle-nose pliers to lower a new snap-ring into the groove . . .

10.4m . . . then seat it into the groove with snap-ring pliers

10.4n Apply grease through the splined hole, then insert a wooden dowel (with a diameter slightly less than that of the axle) through the splined hole and push down - the dowel will force the grease into the joint - repeat until the bearing is completely packed

10.4o Install the small clamp and the boot on the driveaxle and apply grease to the inside of the axle boot . . .

10.4p . . . until the level is up to the end of axle

10.4q Position the CV joint assembly on the driveaxle, aligning the splines, then use a soft-face hammer to drive the joint onto the driveaxle until the snap-ring is seated in the groove

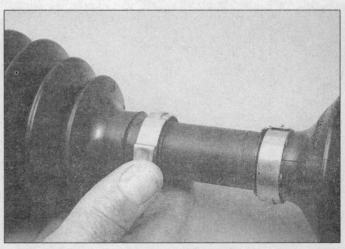

10.4r Seat the inner end of the boot in the groove and install the retaining clamp, then do the same on the other end of the boot - tighten boot clamps with the special tool (see illustration 10.3t)

8

Notes

Chapter 9 Brakes

Contents

Specifications

General
Brake fluid type	See Chapter 1

Disc brakes
Minimum pad thickness	See Chapter 1
Brake disc minimum thickness	Cast into disc
Maximum disc runout	0.0025 inch
Maximum disc thickness variation	0.0005 inch

Rear drum brakes
Shoe friction material minimum thickness	See Chapter 1
Maximum inside diameter	Cast into drum
Maximum out-of-round	0.006 inch

Brake pedal travel (maximum)
1997 models	2-1/2 inches
1998 and later models	2-29/32 inches

Brake light switch
Plunger-to-pedal stopper clearance	0 to 5/64-inch

9

Torque specifications

	Ft-lbs (unless otherwise indicated)
Brake booster mounting nuts	20
Brake caliper	
Caliper mounting bolts	
Malibu/Cutlass	
1997	38
1998 on	23
Grand Am/Alero	
Front	23
Rear	81
Caliper mounting bracket bolts (all)	85
Brake hose-to-caliper banjo bolt	
Malibu/Cutlass	37
Grand Am/Alero	40
Brake hose-to-wheel cylinder banjo bolt	17
Master cylinder-to-brake booster retaining nuts	20
Wheel cylinder retaining bolts	15
Wheel lug nuts	See Chapter 1

1   General information

The vehicles covered by this manual are equipped with hydraulically operated front and rear brake systems. The front brakes are disc type and the rear brakes are disc or drum type. Both the front and rear brakes are self adjusting. The disc brakes automatically compensate for pad wear, while the drum brakes incorporate an adjustment mechanism which is activated whenever the brakes are applied.

Hydraulic system

The hydraulic system consists of two separate circuits. The master cylinder has separate reservoirs for the two circuits, and, in the event of a leak or failure in one hydraulic circuit, the other circuit will remain operative. A proportioning valve provides brake balance between the front and rear brakes.

Power brake booster

The power brake booster, utilizing engine manifold vacuum and atmospheric pressure to provide assistance to the hydraulically operated brakes, is mounted on the firewall in the engine compartment.

Parking brake

The parking brake operates the rear brakes only, through cable actuation. It's activated by a pedal mounted under the left end of the instrument panel.

Service

After completing any operation involving disassembly of any part of the brake system, always test drive the vehicle to check for proper braking performance before resuming normal driving. When testing the brakes, perform the tests on a clean, dry, flat surface. Conditions other than these can lead to inaccurate test results.

Test the brakes at various speeds with both light and heavy pedal pressure. The vehicle should stop evenly without pulling to one side or the other.

Tires, vehicle load and wheel alignment are factors which also affect braking performance. **Caution:** *On models equipped with the "Theftlock" audio system, be sure the lockout feature is turned off before performing any procedure which requires disconnecting the battery (see the front of this manual).*

2   Anti-lock Brake System (ABS) - general information

Refer to illustrations 2.2 and 2.3

Anti-lock Brake Systems (ABS) maintain vehicle maneuverability, directional stability, and optimum deceleration under severe braking conditions on most road surfaces. It does so by monitoring the rotational speed of the wheels and controlling the brake line pressure to the wheels during braking. This prevents the wheels from locking up on slippery roads or during hard braking.

2.2 The ABS hydraulic control unit/motor pack is located along the left frame rail, underneath the battery tray

Hydraulic modulator/motor pack assembly

The hydraulic modulator/motor pack assembly, mounted on the frame underneath the battery, controls hydraulic pressure to the front calipers and rear wheel cylinders or calipers by modulating hydraulic pressure to prevent wheel lock-up **(see illustration)**. Basically, this unit bleeds off pressure in a brake line when the Electronic Brake Control Module (EBCM) detects an abnormal deceleration in the speed of a wheel. When the speed of the wheel is restored to normal, the modulator once again allows full pressure to the brake. This cycle is repeated as many times as necessary, which results in a pulsing of the brake pedal. **Note:** *The modulator/motor pack assembly can't increase brake line pressure above that which is generated by the master cylinder, and it can't apply the brakes by itself.*

Electronic Brake Control Module (EBCM)

The Electronic Brake Control Module

2.3 The Electronic Brake Control Module (EBCM) is located behind the left front wheel inner splash shield

3.5a Before disassembling the brake, wash it thoroughly with brake system cleaner and allow it to dry - position a drain pan under the brake to catch the residue - DO NOT use compressed air to blow off brake dust!

3.5b To make room for the new pads, use a C-clamp to depress the piston into the caliper before removing the caliper and pads - do this a little at a time, keeping an eye on the fluid level in the master cylinder to make sure it doesn't overflow

(EBCM) is located behind the inner splash shield of the left front wheel **(see illustration)**. The EBCM monitors the ABS system and controls the anti-lock valve solenoids. It accepts and processes information received from the brake switch and wheel speed sensors to control the hydraulic line pressure and avoid wheel lock up. It also monitors the system and stores fault codes which indicate specific problems.

Wheel speed sensors

Each wheel is equipped with a speed sensor, which is self-contained in each wheel bearing. The sensors are neither adjustable nor rebuildable. If a sensor malfunctions, the wheel bearing assembly must be replaced.

A wheel speed sensor measures wheel speed by monitoring the rotation of a toothed ring. As the teeth of the ring move through the magnetic field of the sensor, an AC voltage signal is generated. This signal frequency increases or decreases in proportion to the speed of the wheel. The EBCM monitors these signals for changes in wheel speed; if it detects the sudden deceleration of a wheel, i.e. wheel lockup, the EBCM activates the ABS system.

Warning lights

The ABS system has self-diagnostic capabilities. Each time the vehicle is started, the EBCM runs a self-test. There are two warning lights on the instrument panel, a red BRAKE light and an amber ABS light, each with their own functions. During starting, these lights should come on briefly then go out. If the red BRAKE light stays on, it indicates a problem with the main braking system, such as low fluid level detected or the parking brake is still on. If the light stays on after the parking brake is released, check the brake fluid level in the master cylinder reservoir (see Chapter 1).

The amber ABS light indicates a problem with the ABS system, not the main or basic brake system. If the light stays on, it indicates that there is a problem with the ABS system, but the main system is still working. Take the vehicle to a dealer service department or other qualified repair shop for diagnosis and repair.

Checks

Although a special electronic tester is necessary to properly diagnose the system, the home mechanic can perform a few preliminary checks before taking the vehicle to a dealer service department or other repair shop which is equipped with this tester:

a) *Check the fuses.*
b) *Check the electrical connectors at the EBCM and the hydraulic modulator/motor pack.*
c) *Follow the wiring harness to the speed sensors and brake light switch and make sure all connections are secure and the wiring isn't damaged.*
d) *Make sure the brake lines, calipers and wheel cylinders are in good condition.*

If the above preliminary checks don't rectify the problem, the vehicle should be diagnosed by a dealer service department or other qualified repair shop.

---

3 Disc brake pads - replacement

Refer to illustrations 3.5a through 3.5m
Warning: *Disc brake pads must be replaced on both front or both rear wheels at the same time - never replace the pads on only one wheel. Also, the dust created by the brake system is harmful to your health. Never blow it out with compressed air and don't inhale any of it. An approved filtering mask should be worn when working on the brakes. Do not, under any circumstances, use petroleum-based solvents to clean brake parts. Use brake system cleaner only!*
Note: *This procedure applies to the front and rear brake pads.*

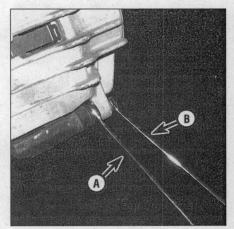

3.5c Hold the caliper slide pin with an open-end wrench (A) (front caliper only) and loosen the mounting bolt with another wrench (B); if you're replacing the front pads, remove the lower bolt - if you're replacing the rear pads, remove the upper bolt

1 Remove the cap from the brake fluid reservoir. Remove about two-thirds of the fluid from the reservoir. **Caution:** *Brake fluid will damage paint. If any fluid is spilled, wash it off immediately with plenty of clean, cold water.*
2 Loosen the front or rear wheel lug nuts, raise the front or rear of the vehicle and support it securely on jackstands. Block the wheels at the opposite end.
3 Remove the wheels. Work on one brake assembly at a time, using the assembled brake for reference if necessary.
4 Inspect the brake disc carefully as outlined in Section 5. If machining is necessary, follow the information in that Section to remove the disc.
5 Follow the accompanying photo sequence for the actual pad replacement procedure **(see illustrations 3.5a through**

3.5d Pivot the caliper out of its bracket and support it in this position for access to the brake pads

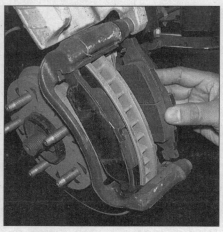

3.5e Remove the inner brake pad

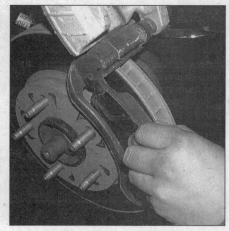

3.5f Remove the outer brake pad

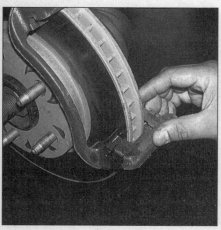

3.5g Remove the upper and lower pad retainers from the caliper mounting bracket

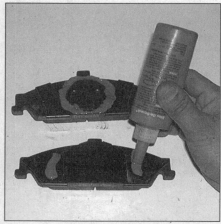

3.5h Apply anti-squeal compound to the back of both pads (let the compound "set up" a few minutes before installing them)

3.5i Install the upper and lower pad retainers

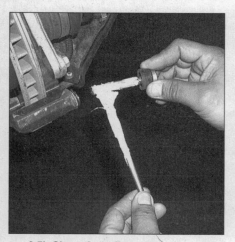

3.5j Clean the caliper slide pin and inspect it for scoring and corrosion; coat the pin with high-temperature grease

3.5k Install the inner brake pad; if you're replacing the front pads, it's the one with the wear indicator on it, which must be positioned at the top (on the rear brakes, the outer pad has the wear indicator)

3.5l Install the outer brake pad; if you're replacing the rear pads, it's the one with the wear indicator on it, which must be positioned at the bottom

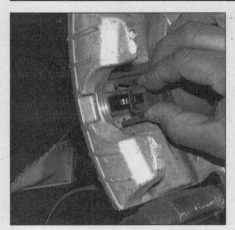

3.5m Check the condition of the anti-rattle spring in the center of the caliper, replacing it if necessary. Swing the caliper down over the pads and install the mounting bolt, tightening it to the torque listed in this Chapter's Specifications. *Note: If the caliper won't fit over the pads, use a C-clamp to push the piston into the caliper a little further*

3.5m). Be sure to stay in order and read the caption under each illustration.

6   When reinstalling the caliper, be sure to tighten the mounting bolts to the torque listed in this Chapter's Specifications. Tighten the wheel lug nuts to the torque listed in the Chapter 1 Specifications.

7   After the job has been completed, firmly depress the brake pedal a few times to bring the pads into contact with the disc. Check the level of the brake fluid, adding some if necessary (see Chapter 1). Check the operation of the brakes carefully before placing the vehicle into normal service.

4   Disc brake caliper - removal and installation

Refer to illustration 4.2

Warning: *The dust created by the brake system may contain asbestos, which is harmful to your health. Never blow it out with compressed air and don't inhale any of it. An approved filtering mask should be worn when working on the brakes. Do not, under any circumstances, use petroleum-based solvents to clean brake parts. Use brake system cleaner only!*

Removal

1   Loosen the front or rear wheel lug nuts, raise the front or rear of the vehicle and place it securely on jackstands. Block the wheels at the opposite end. Remove the front or rear wheel.

2   Remove the banjo bolt and disconnect the brake hose from the caliper **(see illustration)**. Discard the old sealing washers. Plug the brake hose immediately to keep contaminants and air out of the brake system and to

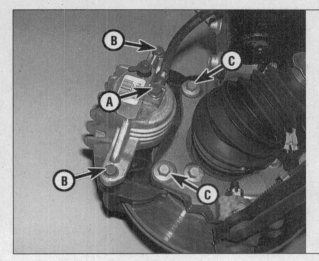

4.2 Brake caliper mounting details

A   *Brake hose banjo fitting bolt*
B   *Caliper mounting bolts*
C   *Caliper mount-to-steering knuckle bolts (don't remove these unless the brake disc must be removed)*

prevent losing any more brake fluid than is necessary. **Note:** *If you are simply removing the caliper for access to other components, leave the brake hose connected and suspend the caliper with a length of wire don't let it hang by the hose).*

3   Remove the caliper mounting bolts and detach the caliper from the mounting bracket. To prevent the caliper slide pins from turning when the mounting bolts are unscrewed, hold the slide pins with an open-end wrench **(see illustration 3.5c)**.

Installation

4   Installation is the reverse of removal. Don't forget to use new sealing washers on each side of the brake hose banjo fitting and be sure to tighten the banjo fitting bolt and the caliper mounting bolts to the torque listed in this Chapter's Specifications.

5   Bleed the brake system (see Section 11). **Note:** *If the brake hose was not disconnected, bleeding won't be required.* Make sure there are no leaks from the hose connections. Test the brakes carefully before returning the vehicle to normal service.

5.2 Hang the caliper out of the way with a piece of wire - don't let it hang by the brake hose!

5.3 The brake pads on this vehicle were obviously neglected, as they wore down completely and cut deep grooves into the disc - wear this severe means the disc must be replaced

5   Brake disc - inspection, removal and installation

Inspection

Refer to illustrations 5.2, 5.3, 5.4a, 5.4b and 5.5

1   Loosen the wheel lug nuts, raise the vehicle and support it securely on jackstands. Remove the wheel and install the lug nuts to hold the disc in place. **Note:** *If the lug nuts don't contact the disc when screwed on all the way, install washers under them.*

2   Remove the brake caliper. It isn't necessary to disconnect the brake hose. After removing the caliper bolts, suspend the caliper out of the way with a piece of wire **(see illustration)**.

3   Visually inspect the disc surface for score marks and other damage. Light scratches and shallow grooves are normal after use and may not always be detrimental to brake operation, but deep scoring requires disc removal and refinishing by an automo-

9

5.4a To check disc runout, mount a dial indicator as shown and rotate the disc

5.4b Using a swirling motion, remove the glaze from the disc with sandpaper or emery cloth

5.5 Use a micrometer to measure disc thickness

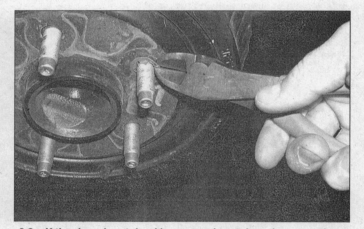

6.2a If the drum is retained by pressed-metal washers, cut them off and discard them (there is no need to reinstall them)

tive machine shop. Be sure to check both sides of the disc **(see illustration)**. If pulsating has been noticed during application of the brakes, suspect disc runout.

4   To check disc runout, place a dial indicator at a point about 1/2-inch from the outer edge of the disc **(see illustration)**. Set the indicator to zero and turn the disc. The indicator reading should not exceed the specified allowable runout limit. If it does, the disc should be refinished by an automotive machine shop. **Note:** *When replacing the brake pads, it's a good idea to resurface the discs regardless of the dial indicator reading, as this will impart a smooth finish and ensure a perfectly flat surface, eliminating any brake pedal pulsation or other undesirable symptoms related to questionable discs. At the very least, if you elect not to have the discs resurfaced, remove the glaze from the surface with emery cloth or sandpaper, using a swirling motion* **(see illustration)**.

5   It's absolutely critical that the disc not be machined to a thickness under the specified minimum thickness. The minimum wear (or discard) thickness is cast into the disc. The disc thickness can be checked with a micrometer **(see illustration)**.

Removal

6   Remove the two caliper mounting bracket bolts and detach the mounting bracket **(see illustration 4.2)**.
7   Remove the lug nuts which you installed to hold the disc in place and slide the disc off the hub.

Installation

8   Place the disc in position over the threaded studs.
9   Install the mounting bracket and tighten the bolts to the torque listed in this Chapter's Specifications. Install the brake pads.
10   Install the caliper onto the mounting bracket, tightening the bolts to the torque listed in this Chapter's Specifications.
11   Install the wheel and lug nuts. Lower the vehicle and tighten the lug nuts to the torque listed in the Chapter 1 Specifications. Depress the brake pedal a few times to bring the brake pads into contact with the disc. Bleeding won't be necessary unless the brake hose was disconnected from the caliper. Check the operation of the brakes carefully before driving the vehicle.

6   Drum brake shoes - replacement

Refer to illustrations 6.2a, 6.2b, 6.3, 6.4a through 6.4q, 6.5a and 6.5b

Warning: *Drum brake shoes must be replaced on both wheels at the same time - never replace the shoes on only one wheel. Also, the dust created by the brake system is harmful to your health. Never blow it out with compressed air and don't inhale any of it. An approved filtering mask should be worn when working on the brakes. Do not, under any circumstances, use petroleum-based solvents to clean brake parts. Use brake system cleaner only!*

Caution: *Whenever the brake shoes are replaced, the return and hold-down springs should be replaced. Due to the continuous heating/cooling cycle the springs are subjected to, they lose tension over a period of time and may allow the shoes to drag on the drum and wear at a much faster rate than normal.*

1   Loosen the wheel lug nuts, raise the rear of the vehicle and support it securely on jackstands. Block the front wheels to keep the

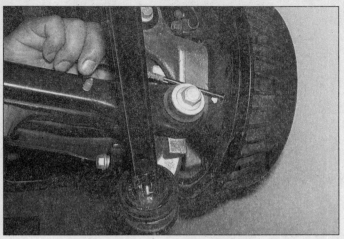

6.2b If the shoes have worn into the drum, preventing its removal, remove the plug from the backing plate and push the parking brake lever off its stop (this will retract the brake shoes slightly)

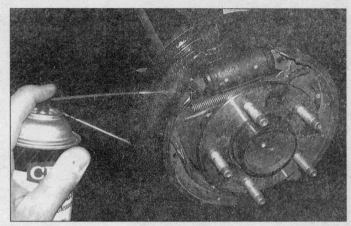

6.3 Before removing any internal drum brake components, wash them off with brake system cleaner and allow them to dry - position a drain pan under the brake to catch the residue - DO NOT USE COMPRESSED AIR TO BLOW THE DUST FROM THE PARTS

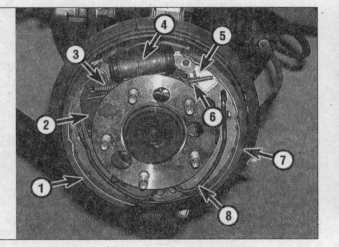

6.4a Drum brake components (left side shown)

1   *Leading shoe*
2   *Parking brake lever*
3   *Actuator spring*
4   *Wheel cylinder*
5   *Adjuster actuator*
6   *Adjuster screw assembly*
7   *Trailing shoe*
8   *Retractor spring*

6.4b Detach the spring from the adjuster actuator

vehicle from rolling. Remove the wheels.

2   Release the parking brake and remove the brake drums **(see illustration)**. If the shoes have worn into the drum, preventing drum removal, remove the access plug from the backing plate, insert a small screwdriver

through the hole and pry the parking brake lever from its stop **(see illustration)**.

3   Before disassembling anything, clean off the brake assembly with brake system cleaner **(see illustration)**.

4   Follow the accompanying illustrations

for the actual shoe replacement procedure **(see illustrations 6.4a through 6.4q)**. Be sure to stay in order and read the caption under each illustration. **Note:** *All four rear brake shoes must be replaced at the same*

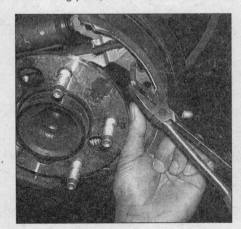

6.4c Pull the retractor spring out of its hole in the trailing shoe . . .

6.4d . . . and also from the leading shoe

6.4e Unscrew the bolt that retains the guide for the parking brake cable

9

6.4f Remove the trailing shoe and adjuster actuator, ~~en~~ the adjuster screw assembly (arrow)

6.4g Pull the retractor spring out of the way, then remove the leading shoe and parking brake lever

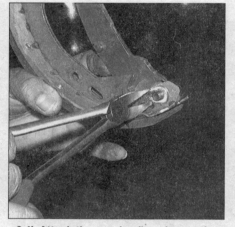

6.4h Force the C-clip off the post and detach the leading shoe from the parking brake lever. **Note:** *It isn't necessary to detach the parking brake cable from the lever*

6.4i Attach the new leading shoe to the parking brake lever and secure it with a new C-clip

6.4j Clean the backing plate, then lubricate the shoe contact areas with a thin film of high-temperature grease

time, but to avoid mixing up parts, work on only one brake assembly at a time.

5   Before reinstalling the drum it should be checked for cracks, score marks, deep scratches and hard spots, which will appear as small, discolored areas. If the hard spots cannot be removed with emery cloth or if any of the other conditions listed above exist, the drum must be taken to an automotive machine shop to have it resurfaced. **Note:** *Professionals recommend resurfacing the*

6.4k Position the leading shoe on the backing plate and install the end of the retractor spring in its hole

6.4l Clean the adjuster screw assembly, then lubricate the threads and socket end with high-temperature grease

6.4m Install the adjuster screw assembly, making sure it engages properly with the leading shoe and the parking brake lever

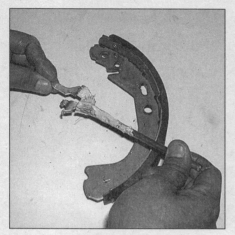

6.4n Lubricate the adjuster actuator . . .

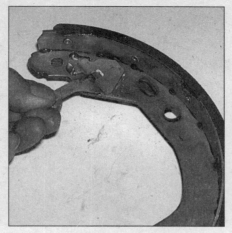

6.4o . . . and install it on the trailing shoe

6.4p Position the trailing shoe on the backing plate, making sure it (and the adjuster actuator) engage properly with the adjuster screw assembly, then insert the retractor spring into its hole in the shoe

6.4q Insert the actuator spring into its hole in the leading shoe, then stretch it across and connect it to the adjuster actuator. Install the parking brake cable guide to the anchor at the bottom of the shoes and tighten the bolt securely

the shoes. When turning the drum, the shoes should not rub; if they do, remove the drum and back off the star wheel a little bit so they don't. This adjustment is just to get the shoes close to the drum; the brake shoes will self-adjust after you depress the pedal a few times).

7 Wheel cylinder - removal and installation

Refer to illustration 7.4

Removal

1   Raise the rear of the vehicle and support it securely on jackstands. Block the front wheels to keep the vehicle from rolling.
2   Remove the brake shoe assembly (see Section 6).
3   Remove all dirt and foreign material from around the wheel cylinder.

drums whenever a brake job is done. Resurfacing will eliminate the possibility of out-of-round or tapered drums. If the drums are worn so much that they can't be resurfaced without exceeding the maximum allowable diameter (stamped into the drum) **(see illustration)**, *then new ones will be required. At*

the very least, if you elect not to have the drums resurfaced, remove the glazing from the surface with emery cloth or sandpaper using a swirling motion **(see illustration)**.
6   When installing the drum, adjust the brake shoes by turning the star wheel on the adjuster screw until the drum just slips over

6.5a The maximum permissible diameter is cast into the drum

6.5b Remove the glaze from the drum surface with sandpaper or emery cloth

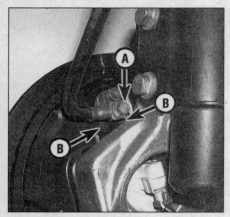

7.4 Remove the brake hose banjo fitting bolt (A), then remove the wheel cylinder mounting bolts (B)

8.2 Master cylinder mounting details

1. *Electrical connector for fluid level sensor*
2. *Brake line fittings*
3. *Mounting nuts*

4   Disconnect the brake hose by removing the banjo fitting bolt **(see illustration)**. Discard the sealing washers on either side of the fitting - new ones should be used during installation. Immediately plug the brake hose to prevent fluid loss and the entry of air and contaminants.

5   Remove the wheel cylinder mounting bolts.

6   Detach the wheel cylinder from the brake backing plate.

Installation

7   Place the wheel cylinder in position and install the bolts, tightening them to the torque listed in this Chapter's Specifications.

8   Connect the brake hose, using new sealing washers on either side of the fitting. Tighten the banjo fitting bolt to the torque listed in this Chapter's Specifications.

9   Install the brake shoes (see Section 6).

10   Bleed the brakes (see Section 11).

11   Check the operation of the brakes carefully before driving the vehicle.

8   Master cylinder - removal and installation

Removal

Refer to illustration 8.2

1   Disconnect the cable from the negative battery terminal. **Caution:** *On models equipped with the Theftlock audio system, be sure the lockout feature is turned off before performing any procedure which requires disconnecting the battery (see the front of this manual).*

2   Unplug the electrical connector for the fluid level warning switch **(see illustration)**.

3   Remove as much fluid as possible from the reservoir with a syringe.

4   Place rags under the fittings and prepare caps or plastic bags to cover the ends of the lines once they're disconnected. **Caution:** *Brake fluid will damage paint. Cover all body parts and be careful not to spill fluid*

8.8 The best way to bleed air from the master cylinder before installing it on the vehicle is with a pair of bleeder tubes that direct brake fluid into the reservoir during bleeding

during this procedure. Loosen the fittings at the ends of the brake lines where they enter the master cylinder. To prevent rounding off the flats, use a flare-nut wrench, which wraps around the fitting hex.

5   Pull the brake lines away from the master cylinder and plug the ends to prevent contamination.

6   Remove the nuts attaching the master cylinder to the power booster **(see illustration 8.2)**. Pull the master cylinder off the studs to remove it. Again, be careful not to spill the fluid as this is done. Remove and discard the old gasket between the master cylinder and the power brake booster.

Installation

Refer to illustrations 8.8 and 8.16

7   Bench bleed the new master cylinder before installing it. Mount the master cylinder in a vise, with the jaws of the vise clamping on the mounting flange.

8   Attach a pair of master cylinder bleeder tubes to the outlet ports of the master cylinder **(see illustration)**.

9   Fill the reservoir with brake fluid of the recommended type (see Chapter 1).

10   Slowly push the pistons into the master

8.16 Have an assistant depress the brake pedal and hold it down, then loosen the fitting nut, allowing air and fluid to escape; repeat this procedure on both fittings until the fluid is clear of air bubbles

cylinder (a large Phillips screwdriver can be used for this) - air will be expelled from the pressure chambers and into the reservoir. Because the tubes are submerged in fluid, air can't be drawn back into the master cylinder when you release the pistons.

11   Repeat the procedure until no more air bubbles are present.

12   Remove the bleed tubes, one at a time, and install plugs in the open ports to prevent fluid leakage and air from entering. Install the reservoir cap.·

13   Install the master cylinder over the studs on the power brake booster and tighten the attaching nuts only finger tight at this time. Don't forget to use a new gasket.

14   Thread the brake line fittings into the master cylinder. Since the master cylinder is still a bit loose, it can be moved slightly so the fittings thread in easily. Don't strip the threads as the fittings are tightened.

15   Tighten the mounting nuts to the torque listed in this Chapter's Specifications. Tighten the brake line fittings securely.

16   Fill the master cylinder reservoir with fluid, then bleed the lines at the master cylinder, followed by bleeding the remainder of the brake system (see Section 11). To bleed

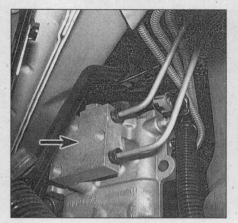

9.1 The proportioning valve (arrow) is mounted to the side of the ABS hydraulic modulator

10.3 Using a flare-nut wrench, unscrew the threaded fitting on the brake line (1), then pry the U-clip (2) off the end of the hose and separate the hose from the bracket

the lines at the master cylinder, have an assistant depress the brake pedal and hold it down. Loosen the fitting to allow air and fluid to escape **(see illustration)**. Tighten the fitting, then allow your assistant to return the pedal to its rest position. Repeat this procedure on both fittings until the fluid is free of air bubbles, then bleed the rest of the system. **Caution:** *Brake fluid will damage paint. Wash it off immediately with plenty of clean, cold water.* Check the operation of the brake system carefully before driving the vehicle. **Warning:** *If you do not have a firm brake pedal at the end of the bleeding procedure, or have any doubts as to the effectiveness of the brake system, DO NOT drive the vehicle. Have it towed to a dealer service department or other qualified repair shop for diagnosis.*

9 Proportioning valve - replacement

Refer to illustration 9.1

The proportioning valve is mounted on the side of the ABS hydraulic modulator/motor pack assembly **(see illustration)**. Proportioning valves rarely fail, but if replacement is required, the ABS hydraulic modulator/motor pack assembly must first be removed from the vehicle. The proportioning valve can then be separated from the ABS unit; however, due to the sensitive nature of the ABS system, this job is best left to a qualified repair shop.

10 Brake hoses and lines - inspection and replacement

Inspection

1   About every six months, with the vehicle raised and supported securely on jackstands, the rubber hoses which connect the steel brake lines with the front and rear brake assemblies should be inspected for cracks,

chafing of the outer cover, leaks, blisters and other damage. These are important and vulnerable parts of the brake system and inspection should be complete. A light and mirror will be helpful for a thorough check. If a hose exhibits any of the above conditions, replace it with a new one.

Replacement
Flexible brake hose

Refer to illustration 10.3

2   Loosen the wheel lug nuts, raise the vehicle and support it securely on jackstands. Remove the wheel.

3   At the bracket, unscrew the brake line fitting from the hose **(see illustration)**. Use a flare-nut wrench to prevent rounding off the corners.

4   Remove the U-clip from the female fitting at the bracket with a pair of pliers, then pass the hose through the bracket.

5   At the caliper end of the hose, remove the banjo bolt, then separate the hose from the caliper. Note that there are two copper sealing washers on either side of the banjo fitting - they should be replaced with new ones during installation.

6   To install the hose, connect the fitting to the caliper with the banjo bolt and new copper washers.

7   Route the hose into the frame bracket, making sure it isn't twisted, then connect the brake line fitting, starting the threads by hand. Install the U-clip, then tighten the fitting securely.

8   Bleed the caliper (see Section 11).

9   Install the wheel and lug nuts, lower the vehicle and tighten the lug nuts to the torque listed in the Chapter 1 Specifications.

Metal brake lines

10   When replacing brake lines, be sure to use the correct parts. Don't use copper tubing for any brake system components. Purchase steel brake lines from a dealer or auto parts store.

11   Prefabricated brake line, with the tube ends already flared and fittings installed, is available at auto parts stores and dealer parts departments. These lines must be bent to the proper shapes using a tubing bender.

12   When installing the new line, make sure it's securely supported in the brackets and has plenty of clearance between moving or hot components.

13   After installation, check the master cylinder fluid level and add fluid as necessary. Bleed the brake system (see Section 11) and test the brakes carefully before driving the vehicle in traffic.

11 Brake hydraulic system - bleeding

Refer to illustrations 11.10, 11.11 and 11.19

Warning: *Wear eye protection when bleeding the brake system. If the fluid comes in contact with your eyes, immediately rinse them with water and seek medical attention.*

Note: *Bleeding the hydraulic system is necessary to remove any air that manages to find its way into the system when it's been opened during removal and installation of a hose, line, caliper or master cylinder.*

1   You'll probably have to bleed the system at all four brakes if air has entered it due to low fluid level, or if the brake lines have been disconnected at the master cylinder.

2   If a brake line was disconnected only at a wheel, then only that caliper or wheel cylinder must be bled.

3   If a brake line is disconnected at a fitting located between the master cylinder and any of the brakes, the entire system must be bled.

4   Remove any residual vacuum from the brake power booster by applying the brake several times with the engine off. If the entire system is to be bled, remove the battery and the battery tray (see Chapter 5).

5   Remove the master cylinder reservoir cap and fill the reservoir with brake fluid. Reinstall the cap. **Note:** *Check the fluid level often during the bleeding operation and add fluid as necessary to prevent the fluid level from falling low enough to allow air bubbles into the master cylinder.*

6   Have an assistant on hand, as well as a supply of new brake fluid, a clear plastic container partially filled with clean brake fluid, a length of clear tubing to fit over the bleeder valves and a wrench to open and close the bleeder valves.

7   If it is necessary to bleed the entire system, begin the bleeding procedure with the next Step. If it is only necessary to bleed one caliper or wheel cylinder, skip to Step 16.

8   Before bleeding the entire system, sit in the driver's seat and:

a)   *Take your foot off the brake pedal.*
b)   *Start the engine and let it run for a minimum of 10 seconds. Watch the amber ABS light on the dash.*

9

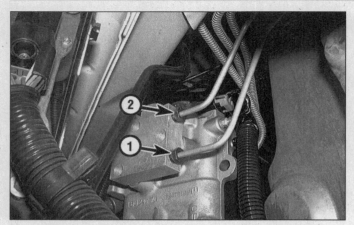

11.10 When bleeding the brake line fittings at the proportioning valve, start with the lower fitting (1), then bleed the upper fitting (2)

11.11 When bleeding the hydraulic modulator, begin with the rear bleeder valve (1), then move to the front bleeder valve (2)

c) *If the light comes on and does not turn off after 10 seconds, have the vehicle towed to a dealer service department or other qualified repair shop. A scan tool will have to be used to diagnose the ABS system.*

d) *If the ABS light goes off after three seconds or so, turn off the ignition.*

e) *Repeat paragraphs a) through d) one more time, then proceed with the bleeding operation.*

9   Begin the bleeding procedure by bleeding the master cylinder (see Section 8, Step 16).

10   Using the same technique as for bleeding the master cylinder, bleed the lower brake line fitting at the proportioning valve, followed by the upper brake line fitting **(see illustration)**. Recheck the fluid level in the master cylinder reservoir, adding fluid as necessary.

11   Next, prime the ABS hydraulic modulator. Connect the bleeder hose to the rear bleeder valve on the modulator **(see illustration)** and place the other end into a container partially filled with clean brake fluid. Make sure the end of the hose is submerged.

12   Open the bleeder valve slowly (approximately 1/2 to 3/4 turn), then have an assistant depress the brake pedal and hold it in the depressed position. When the flow of fluid ceases, close the bleeder valve.

13   Repeat Step 12 until no air bubbles are present in the fluid. Tighten the bleeder valve securely.

14   Move the bleeder hose to the front bleeder valve and perform Steps 12 and 13.

15   Recheck the fluid level in the master cylinder reservoir, adding fluid as necessary.

16   Raise the vehicle and support it securely on jackstands.

17   Bleed the brakes at the wheels in the following sequence:

1997 through 1999 Malibu and Cutlass models

    Right rear
    Left rear
    Right front
    Left front

2000 Malibu and Cutlass models, all Grand Am and Alero models

    Right rear
    Left front
    Left rear
    Right front

18   Beginning at the first wheel in the bleeding sequence, loosen the bleeder valve slightly, then tighten it to a point where it's snug but can still be loosened quickly and easily.

19   Place one end of the tubing over the bleeder valve and submerge the other end in brake fluid in the container **(see illustration)**.

20   Have the assistant depress the brake pedal slowly, holding it in the depressed position.

21   While the pedal is held down, open the bleeder valve just enough to allow a flow of fluid to leave the valve. Watch for air bubbles to exit the submerged end of the tube. When the fluid flow slows after a couple of seconds, close the valve and have your assistant release the pedal.

22   Repeat Steps 20 and 21 until no more air is seen leaving the tube, then tighten the bleeder valve.

23   Perform Steps 19 through 22 at the remaining wheels, following the bleeding sequence. Be sure to check the fluid in the master cylinder reservoir frequently.

24   Refill the master cylinder with fluid at the end of the operation. If you're working on a model with ABS, be sure to reconnect the electrical connectors to the ABS actuator. **Warning:** *Never use old brake fluid. It contains moisture which can cause the fluid to boil, rendering the brake system inoperative.*

25   If removed, install the battery tray and the battery.

26   Check the operation of the brakes. Turn the ignition key to the On position, then Off (don't start the engine). The pedal should feel solid when depressed, with no sponginess. Now start the engine and recheck the pedal travel and feel.

27   If the pedal travel or feel is not satisfactory, take your foot off the brake pedal and

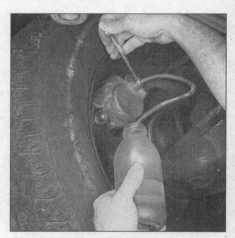

11.19 When bleeding the brakes, a hose is connected to the bleed screw at the caliper or wheel cylinder and then submerged in brake fluid - air will be seen as bubbles in the tube and container (all air must be expelled before moving to the next wheel)

start the engine. Let the engine run for a minimum of 10 seconds, then turn it off (don't depress the pedal during this time). Repeat this Step 5 times, which should dislodge any air trapped in the ABS hydraulic modulator, then repeat the entire bleeding procedure.

28   Road test the vehicle in a safe area before returning it to normal service. **Warning:** *Do not operate the vehicle if you're in doubt about the effectiveness of the brake system.*

12   Power brake booster - check, removal and installation

Refer to illustrations 12.10 and 12.11

Operating check

1   Depress the pedal and start the engine. If the pedal goes down slightly, operation is normal.

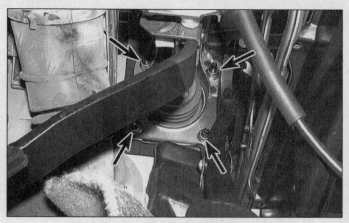

12.10 Pry off the clip retaining the booster pushrod to the pin on the brake pedal

12.11 Unscrew the booster mounting nuts

2   Depress the brake pedal several times with the engine running and make sure that there is no change in the pedal reserve distance.

Airtightness check

3   Start the engine and turn it off after one or two minutes. Depress the brake pedal several times slowly. If the pedal goes down farther the first time but gradually rises after the second or third depression, the booster is airtight.
4   Depress the brake pedal while the engine is running, then stop the engine with the pedal depressed. If there is no change in the pedal reserve travel after holding the pedal for 30 seconds, the booster is airtight.

Removal

5   Remove the air filter housing (see Chapter 4).
6   Pull the underhood fuse box forward, straight out of its bracket. Remove the bracket.
7   Remove the master cylinder without detaching the brake lines. Pull it forward and position it aside.
8   Detach the vacuum hose from the booster.
9   Remove the left side under-dash panel.

10   Remove the pushrod retaining clip **(see illustration)** and slip the pushrod off the pin.
11   Peel the insulation on the firewall back to gain access to the booster mounting nuts. Remove the four nuts holding the brake booster to the firewall **(see illustration)**.
12   Slide the booster straight out from the firewall until the studs clear the holes and pull the booster and gasket from the engine compartment.

Installation

13   Installation is the reverse of removal. Be sure to tighten the booster mounting nuts and the master cylinder mounting nuts to the torque values listed in this Chapter's Specifications.

13   Brake pedal travel - check

1   The brake pedal is not adjustable, but the travel should be checked if the pedal seems low. You'll need a tape measure, yardstick or ruler for this procedure.
2   Depress the pedal a few times to deplete the vacuum reserve in the power brake booster.
3   Measure the position of the pedal at rest. You can either measure from the floor to

the pedal or from the pedal to the steering wheel. Record your reading.
4   Now, depress the pedal (exerting approximately 100 lbs. of force) and measure how far the pedal has traveled. Compare your findings with the measurement listed in this Chapter's Specifications.
5   If the pedal travel is excessive, check for air in the system (bleed the brakes - see Section 11). If that doesn't cure the problem, remove the rear brake shoes and make sure the adjuster screw assemblies are operating properly (see Section 6). Also check the condition of the brake shoes, replacing them if they're worn excessively.

14   Parking brake - adjustment

The parking brake is self-adjusting. If the parking brake pedal travel is excessive or won't hold the vehicle on an incline, the rear brake shoes may need to be adjusted or replaced (see Section 6).

15   Parking brake cables - replacement

Front cable

Malibu and Cutlass models
Refer to illustrations 15.5, 15.7 and 15.8
Note: *This procedure is much easier if the instrument panel is removed (see Chapter 11).*
1   Release the parking brake. Remove the left-side under-dash panel.
2   Remove the front and rear sill plates and peel back the carpet.
3   Remove the Powertrain Control Module (PCM) (see Chapter 6). Also remove the PCM bracket.
4   Raise the vehicle and support it securely on jackstands.
5   Detach the end of the front cable from the connector that attaches it to the right rear cable **(see illustration)**. Detach the cable

15.5 Pull on the front parking brake cable to create some slack and hold it in that position with a pair of locking pliers (1), then grip the cable with a pair of pliers (2), push the cable into the connector (3) and detach it

9

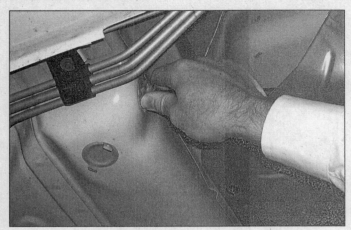

15.7 Push the cable grommet through the floor pan

15.8 Prevent the cable adjuster reel from turning, then detach the cable from the reel (instrument panel removed for clarity)

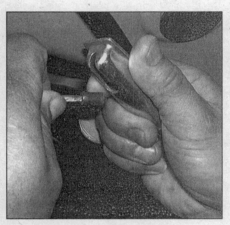

15.18 Pass the end of the left rear cable through the large portion of the slot in the equalizer

15.22 Retract the spring and pass the cable through the slot in the parking brake lever

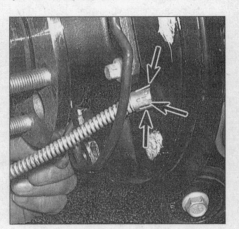

15.23 Depress the tangs on the cable retainer, then pass the cable through the backing plate

from the equalizer.

6   Unbolt the cable bracket from the floor-pan.

7   Push the cable grommet through the hole in floorpan (see illustration). Don't try to pass the cable into the interior of the vehicle yet.

8   Working up at the parking brake pedal and adjuster assembly, pull on the cable and hold the adjuster reel from rotating back, then remove the cable from the groove in the reel, align the cable with the slot and detach the cable end from the reel (see illustration).

9   Depress the tangs on the cable retainer and detach the cable casing from the parking brake pedal bracket.

10   Pull the cable into the interior of the vehicle and remove it.

11   Begin installation of the new cable by routing the cable through the parking brake pedal bracket, making sure the retainer tangs pass through the hole, then attaching the cable end to the adjuster reel. Make sure the pedal is in the released position; it may be necessary to carefully pry the cable over the reel and into place.

12   The remainder of installation is the reverse of removal. Make sure the grommet

seats properly in the floorpan.

13   Actuate the parking brake several times to adjust the cable.

Grand Am and Alero models

14   The procedure for removing the front cable on these models is essentially the same as for the Malibu and Cutlass, but on the Grand Am and Alero the parking brake is actuated by a lever in the center console instead of a pedal. Follow the above proce-dure, but ignore Steps 1, 2 and 3. Instead, remove the center console (see Chapter 11) for access to the adjuster reel on the parking brake pedal.

Rear cables

Refer to illustrations 15.18, 15.22 and 15.23

15   Release the parking brake.

16   Loosen the rear wheel lug nuts, raise the rear of the vehicle and support it securely on jackstands.

17   Disconnect the front parking brake cable from its connector at the rear of the vehicle (see illustration 15.5).

18   If you're removing the left rear cable, disconnect the cable end from the equalizer

(see illustration).

19   Depress the tangs on the cable retainer and detach the cable from the bracket.

Disc brake models

20   Remove the clip from the cable and detach the cable from the bracket. Detach the cable end from the actuator on the back-ing plate and remove the cable.

Drum brake models

21   If you're working on a model with drum brakes, remove the brake shoes (see Sec-tion 6).

22   Pull back the spring and detach the cable end from the parking brake lever (see illustration).

23   Depress the tangs of the cable retainer, then pass the cable through the backing plate (see illustration).

All models

24   Installation is the reverse of the removal procedure. Actuate the parking brake several times to adjust the cable.

25   Tighten the wheel lug nuts to the torque listed in the Chapter 1 Specifications.

16 Parking brake shoes - replacement

Note: *This procedure applies only to models with rear disc brakes.*

1   Loosen the rear wheel lug nuts, raise the rear of the vehicle and support it securely on jackstands. Remove the wheels.

2   Remove the brake caliper (see Section 4) and the brake disc (see Section 5).

3   Disconnect the parking brake cable from the bracket and the actuator lever.

4   Unbolt the parking brake cable bracket.

5   Remove the rear hub and bearing assembly (see Chapter 10).

6   Remove the parking brake actuator and the parking brake shoe.

7   When installing the new shoe and lining assembly, turn the adjuster screw until the shoe lining just drags on the braking surface inside the disc. Then remove the disc and back-off the adjuster screw until the shoe lining doesn't drag when the disc is installed and turned.

8   Installation is otherwise the reverse of the removal procedure. Be sure to tighten the hub and bearing assembly bolts to the torque listed in the Chapter 10 Specifications, the caliper mounting bolts to the torque listed in the Chapter 9 Specifications, and the wheel lug nuts to the torque listed in the Chapter 1 Specifications.

17 Brake light switch - check, adjustment and replacement

Refer to illustration 17.1

Check

1   The brake light switch **(see illustration)** is located on a bracket at the top of the brake pedal (it's the switch on the left side; the one on the right is for the cruise control system). The switch activates the brake lights at the rear of the vehicle when the pedal is depressed.

2   If the brake lights are inoperative, check

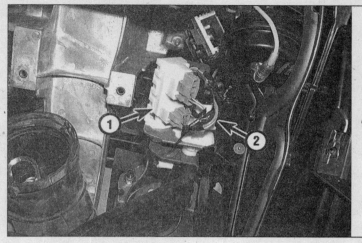

17.1 Location of the brake light switch (1) and the cruise control switch (2)

the fuse first (see Chapter 12).

3   If the fuse is good, try adjusting the switch (see Steps 9 through 12).

4   If the brake lights still don't work, check for voltage to the switch on the orange wire. If no voltage is present, repair the wire between the switch and the fuse box.

5   If voltage is present, depress the brake pedal and check for voltage at the light blue wire terminal. If no voltage is present, replace the switch.

6   If voltage is present, check for power on the light blue wire at the taillight housings (with the brake pedal depressed). If voltage is not present, repair the circuit between the switch and the brake lights (refer to the wiring diagrams at the end of Chapter 12).

7   If voltage is present, check for a bad ground; using a jumper wire connected to a good ground, probe the black wire terminal at the taillight connector. If the brake lights go on, repair the ground circuit (follow the black wire from the taillight housing).

8   Keep in mind that the brake light bulbs *could* be burned out, but the likelihood of all the bulbs being burned out is very slim.

Adjustment

9   Remove left-side under-dash panel.

10   Turn the cruise control switch counter-

clockwise to unlock it, then pull it out of the bracket.

11   Turn the brake light switch counter-clockwise to unlock it, then depress the brake pedal and push in on the switch. Slowly release the brake pedal while lightly pushing on the switch. When the brake pedal comes to rest, turn the switch clockwise to lock it. The switch plunger-to-switch body clearance should be within the range listed in this Chapter's Specifications.

12   Install the cruise control switch and adjust it using the same method.

Replacement

13   Remove left-side under-dash panel, if not already done.

14   Turn the cruise control switch counter-clockwise to unlock it, then pull it out of the bracket.

15   Turn the brake light switch counter-clockwise to unlock it, then remove it from the bracket.

16   Unplug the electrical connectors from the switch.

17   To install the new switch, reverse the removal procedure, then adjust the switch as described in Step 11.

18   Install and adjust the cruise control switch using the same method.

Notes

Chapter 10
Suspension and steering systems

Contents

Specifications

Torque specifications

Ft-lbs (unless otherwise indicated)

Front suspension

Balljoint-to-steering knuckle nut	41 (up to 50, to install cotter pin)
Control arm	
Front pivot bolt/nut	
1997	89
1998 and 1999	79
2000	84
Rear vertical bushing bolt	
1997	125
1998 and 1999	81
2000	180
Hub and bearing assembly-to-steering knuckle bolts	70
Stabilizer bar	
Link nuts	156 in-lbs
Bushing clamp bolts	49
Strut assembly	
Strut-to-steering knuckle nuts	133
Strut-to-body nuts	18
Strut-to-body bolt	18
Strut damper shaft nut	52
Subframe bolts (except subframe-to-body bolts)	
1997 models	71, plus an additional 90-degrees rotation
1998 and later models	81
Subframe-to-body bolts	61

10

Torque specifications

Rear suspension   **Ft-lbs** (unless otherwise indicated)

Hub and bearing assembly retaining bolts	
1999 and earlier	70
2000	62
Strut assembly	
Strut-to-knuckle nuts	89
Strut-to-body nuts	18
Strut damper shaft nut	52
Trailing arm-to-body bolt	48, plus an additional 120-degrees rotation
Trailing arm-to-knuckle bolt	51
Lateral link-to-knuckle bolt	89
Lateral link-to-suspension support bolt	89
Stabilizer bar	
Link-to-knuckle bolt	
1999 and earlier	51
2000	37
Bushing clamp bolts	39
Suspension support assembly bolts	89

Steering

Intermediate shaft pinch bolts	16
Steering column mounting bolts	19
Steering gear mounting bolts	89
Power steering pump mounting bolts	
Four-cylinder engine	19
V6 engines	25
Steering wheel nut	30
Tie-rod end-to-steering knuckle nuts	
1997	168 in-lbs, plus an additional 210-degrees
1998 and 1999	33
2000	15, plus an additional 180-degrees rotation
Wheel lug nuts	See Chapter 1

1 General information

Refer to illustrations 1.1 and 1.2

The front suspension **(see illustration)** is a MacPherson strut design. The upper ends of the struts are attached to the body; the lower ends of the struts are bolted to the steering knuckles. The lower ends of the steering knuckles are attached to the control arms by balljoints. The inner ends of the control arms are attached to the subframe. A stabilizer bar reduces body lean during cornering. The stabilizer bar is attached to the subframe by a pair of bushing clamps and to the control arms by link bolts.

The rear suspension is independent, with a coil spring/strut assembly on each side bolted to knuckle assemblies that are located by trailing arms, parallel lateral links, and a stabilizer bar **(see illustration)**.

The rack-and-pinion steering gear, which is located behind the engine/transaxle

assembly, is bolted to the suspension subframe. The steering gear turns the steering knuckles via a pair of tie-rod assemblies, each of which consists of an inner tie-rod and a tie-rod end. The inner tie-rods are attached to the steering gear; the outer tie-rods, or tie-rod ends, are attached to the steering knuckles. All models are equipped with power steering.

Warning: *Whenever any of the suspension or steering fasteners are loosened or removed, they must be inspected and, if necessary, replaced with new ones of the same part number or of original equipment quality and design. Torque specifications must be followed for proper reassembly and component retention. Never attempt to heat or straighten any suspension or steering components. Instead, replace any bent or damaged part with a new one.*

Note 1: *These vehicles have a combination of standard and metric fasteners on the various suspension and steering components, so it*

would be a good idea to have both types of tools available when beginning work.

Note 2: *On models equipped with the Theft-lock audio system, be sure the lockout feature is turned off and make sure you have the correct activation code before performing any procedure which requires disconnecting the battery.*

2 Strut/coil spring assembly (front) - removal, inspection and installation

Removal

Refer to illustrations 2.2, 2.3 and 2.5

1 Loosen the wheel lug nuts, raise the front of the vehicle and support it securely on jackstands. Apply the parking brake and block the rear wheels to keep the vehicle from rolling off the jackstands. Remove the wheel.

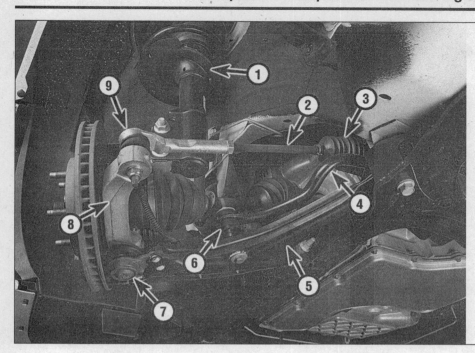

1.1 Front suspension and steering components

1   Strut and spring assembly
2   Tie-rod
3   Steering gear boot
4   Stabilizer bar
5   Control arm
6   Stabilizer link
7   Balljoint
8   Steering knuckle
9   Tie-rod end

1.2 Rear suspension components

1 Toe-adjuster cam bolt	4 Rear knuckle	7 Lateral link (front)
2 Lateral link (rear)	5 Trailing arm	8 Stabilizer bar
3 Strut and spring assembly	6 Stabilizer link	

10

2.2 To mark the relationship of the strut to the steering knuckle, paint or scribe around the strut-to-knuckle bolt heads

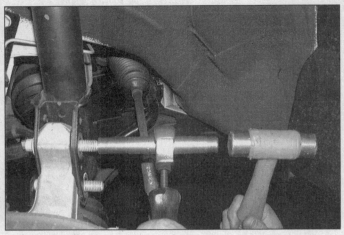

2.3 To remove the strut-to-knuckle bolts, drive them out with a hammer and brass punch

2   Mark the strut-to-steering knuckle relationship by making a line around the strut-to-steering knuckle bolt heads **(see illustration)**.

3   Remove the nuts from the strut-to-knuckle bolts and knock the bolts out with a brass punch and a hammer **(see illustration)**.

4   Separate the strut from the steering knuckle. Be careful not to overextend the inner CV joint or stretch the brake hose. If necessary, support the control arm with a jack.

5   Support the strut and spring assembly with one hand and remove the upper strut mounting nuts and bolt **(see illustration)**. Remove the strut and spring assembly.

Inspection

6   Check the strut body for leaking fluid, dents, cracks and other obvious damage which would warrant repair or replacement.

7   Check the coil spring for chips and cracks in the spring coating (this will cause premature spring failure due to corrosion). Inspect the spring seat for hardening, cracks and general deterioration.

8   If wear or damage is evident, replace the strut and/or coil spring as necessary (see Section 3).

Installation

9   Guide the strut assembly up into the fenderwell and insert the upper mounting studs through the holes in the shock tower. Once the studs protrude from the shock tower, install the nuts and bolt so the strut won't fall back through. This may require an assistant, since the strut is quite heavy and awkward.

10   Slide the steering knuckle into the strut flange and insert the two bolts. Install the nuts, align the marks you made prior to disassembly and tighten the nuts to the torque listed in this Chapter's Specifications.

11   Install the wheel, lower the vehicle and tighten the wheel lug nuts to the torque listed in the Chapter 1 Specifications.

12   Tighten the upper mounting nuts and bolt to the torque listed in this Chapter's Specifications.

13   Drive the vehicle to a dealer service department or an alignment shop to have the front wheel alignment checked and, if necessary, adjusted (this is only necessary if the strut has been modified for camber adjustment).

3   Strut/coil spring - replacement

1   If the struts or coil springs exhibit the telltale signs of wear (leaking fluid, loss of damping capability, chipped, sagging or cracked coil springs) explore all options before beginning any work. The strut assemblies are not serviceable and must be replaced if a problem develops. However, strut assemblies complete with springs may be available on an exchange basis, which eliminates much time and work. Whichever route you choose to take, check on the cost and availability of parts before disassembling your vehicle. **Warning:** *Disassembling a strut is potentially dangerous and utmost attention must be directed to the job, or serious injury may result. Use only a high-quality spring compressor and carefully follow the manufacturer's instructions furnished with the tool. After removing the coil spring from the strut assembly, set it aside in a safe, isolated area.*

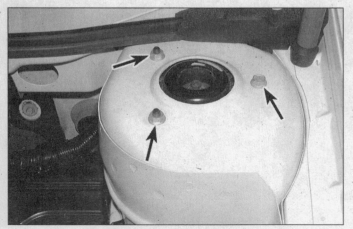

2.5 Remove the strut upper mounting nuts and bolt while supporting the strut assembly

3.3a Mark the relationship of the coil spring to the upper spring seat and insulator and to the strut mount . . .

3.3b . . . and to the lower seat of the strut

Disassembly

Refer to illustrations 3.3a, 3.3b, 3.4, 3.5, 3.6a, 3.6b, 3.7, 3.8 and 3.9

2   Remove the strut and spring assembly (see Section 2 [front] or Section 9 [rear]). Mount the strut assembly in a vise. Line the vise jaws with wood or rags to prevent damage to the unit and don't tighten the vise excessively.

3   Mark the relationship of the coil spring to the upper insulator and mount and to the lower seat **(see illustrations)**.

4   Following the tool manufacturer's instructions, install the spring compressor (which can be obtained at most auto parts stores or at equipment rental yards) on the spring and compress it sufficiently to relieve all pressure from the upper spring seat **(see illustration)**. This can be verified by wiggling the spring.

5   Loosen the damper shaft nut **(see illustration)**.

6   Remove the washer and strut mount **(see illustrations)**. Inspect the bearing in the strut mount for smooth operation. If it doesn't turn smoothly, replace the strut mount. Check the rubber portion of the strut mount for cracking and general deterioration. If there is any separation of the rubber, replace it.

7   Remove the upper spring seat and insu-

3.4 Following the tool manufacturer's instructions, install the spring compressor on the spring and compress it sufficiently to relieve all pressure from the upper spring seat

3.6a Remove the washer . . .

lator from the damper shaft **(see illustration)**. Check the insulator for cracking and hardness; replace it if necessary.

8   Remove the rubber jounce bumper and dust shield from the damper shaft **(see illustration)**.

3.5 Using a wrench on the damper shaft to prevent it from turning, loosen the damper shaft nut

3.6b . . . and the strut mount; inspect the bearing in the mount for smooth operation and the rubber portion of the mount for cracking and general deterioration, if the bearing doesn't turn smoothly, or if there's any separation of the rubber, replace the mount

3.7 Remove the upper spring seat and insulator from the damper shaft; inspect the insulator for cracking and hardness and, if necessary, replace it

3.8 Remove the rubber jounce bumper and dust shield from the damper shaft

10

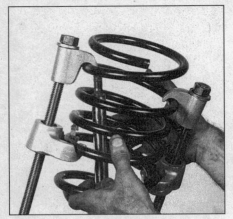

3.9 Carefully lift the compressed spring from the assembly and set it in a safe place; do NOT *place your head near the end of the spring!*

3.12 Place the coil spring onto the lower insulator, with the end of the spring butted against the spring stop on the insulator

3.15a Install the strut mount . . .

9   Carefully lift the compressed spring from the assembly **(see illustration)** and set it in a safe place. **Warning:** *Never place your head near the end of the spring!*
10   Check the lower insulator for wear, cracking and hardness and replace it if necessary.

Reassembly

Refer to illustrations 3.12, 3.15a, 3.15b and 3.16

11   If the lower insulator is being replaced, set it into position with the dropped portion seated in the lowest part of the seat. Extend the damper rod to its full length and install the rubber bumper.
12   Place the coil spring onto the lower insulator, with the end of the spring butted against the spring stop on the insulator **(see illustration)**.
13   Install the dust shield and rubber jounce bumper.
14   Install the upper insulator and spring seat.
15   Install the strut mount and washer **(see illustrations)**. Align the marks made previously.
16   Install the nut and tighten it securely **(see illustration)**.
17   Install the strut/spring assembly (see Section 2 [front] or Section 9 [rear]).
18   Repeat this entire procedure for the other strut or shock absorber/coil spring assembly.
19   After the vehicle has been lowered to the ground, tighten the damper shaft nuts to the torque listed in this Chapter's Specifications.

4   Steering knuckle and hub - removal and installation

Warning: *Dust created by the brake system is harmful to your health. Never blow it out with compressed air and don't inhale any of it. Do*

3.15b . . . and washer

not, under any circumstances, use petroleum-based solvents to clean brake parts. Use brake system cleaner only.

Removal

1   Remove the wheel cover, then break the driveaxle/hub nut loose with a socket and large breaker bar. **Note:** *If the socket won't fit through the opening in the center of the wheel, remove the wheel and install the spare (the nut is very tight and is easier to loosen when the wheel is on the ground).*
2   Loosen the front wheel lug nuts, raise the vehicle and support it securely on jackstands. Remove the wheel.
3   Remove the driveaxle/hub nut and washer. To prevent the disc/hub from turning, insert a long punch into the brake disc cooling vanes and allow it to rest against the caliper mounting bracket.
4   Remove the caliper and suspend it out of the way with a piece of wire. Remove the caliper mounting bracket, then lift the disc off the hub (see Chapter 9).
5   Disconnect the electrical connector for the wheel speed sensor **(see illustration 5.5)**.

3.16 Hold the damper shaft from turning, then tighten the nut securely (unless you have a special socket with a "window" in it for the back-up wrench, you won't be able to torque the nut until the strut assembly is installed in the vehicle and the vehicle is on the ground)

6   Mark the position of the two strut-to-knuckle bolts heads **(see illustration 2.2)**, then remove the nuts. Don't drive out the bolts at this time.
7   Separate the control arm balljoint from the steering knuckle (see Section 7).
8   Attach a puller to the hub flange and push the driveaxle out of the hub (see Chapter 8). Hang the driveaxle with a piece of wire to prevent damage to the inner CV joint. **Caution:** *Be careful not to pull outward on the driveaxle, as this could separate the inner CV joint components.*
9   Support the knuckle and drive out the two strut-to-knuckle bolts with a hammer and brass punch. Separate the steering knuckle from the strut. If necessary, remove the hub and bearing assembly from the steering knuckle (see Section 5).

Installation

10   Position the knuckle in the strut and insert the two splined bolts. Tap the bolts into

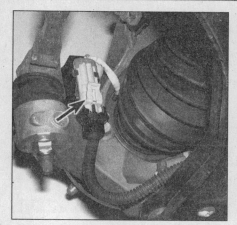

5.5 Unplug the electrical connector for the wheel speed sensor, then unclip the harness connector from the disc shield

5.6 Remove the hub retaining bolts (arrows); using a swivel socket and a short extension it is possible to remove the bolts without removing the driveaxle (if you don't have a swivel socket or U-joint attachment, you'll have to remove the driveaxle)

place and install the nuts, but don't tighten them at this time.

11   Insert the end of the driveaxle into the hub.

12   Connect the control arm balljoint to the steering knuckle and tighten the nut to the torque listed in this Chapter's Specifications. Install a new cotter pin. If necessary, tighten the nut a little more to align the slots in the nut with the hole in the balljoint stud; don't loosen the nut in order to insert the cotter pin.

13   Align the strut-to-knuckle bolt heads with the previously applied marks and tighten the nuts to the torque listed in this Chapter's Specifications.

14   Reconnect the electrical connector for the wheel speed sensor.

15   Install the brake disc, caliper mounting bracket and caliper. Tighten the fasteners to the torque values listed in the Chapter 9 Specifications.

16   Tighten the driveaxle/hub nut securely to seat the driveaxle in the hub.

17   Install the wheel, lower the vehicle and

tighten the lug nuts to the torque listed in the Chapter 1 Specifications.

18   Tighten the driveaxle/hub nut to the torque listed in the Chapter 8 Specifications.

5   Hub and bearing assembly (front) - removal and installation

Refer to illustrations 5.5, 5.6, 5.7a and 5.7b

Warning: *Dust created by the brake system is harmful to your health. Never blow it out with compressed air and don't inhale any of it. Do not, under any circumstances, use petroleum-based solvents to clean brake parts. Use brake cleaner or denatured alcohol only.*

Note: *The hub and bearing assembly is sealed-for-life. If worn or damaged, it must be replaced as a unit.*

1   Remove the wheel cover, then break the

driveaxle/hub nut loose with a socket and large breaker bar. **Note:** *If the socket won't fit through the opening in the center of the wheel, remove the wheel and install the spare (the nut is very tight and is easier to loosen when the wheel is on the ground).*

2   Loosen the front wheel lug nuts, raise the vehicle and support it securely on jackstands. Remove the wheel.

3   Remove the driveaxle/hub nut and washer. To prevent the disc/hub from turning, insert a long punch into the brake disc cooling vanes and allow it to rest against the caliper mounting bracket.

4   Remove the caliper and hang it out of the way with a piece of wire, then remove the caliper mounting bracket (see Chapter 9). Pull the disc off the hub.

5   Unplug the electrical connector for the wheel speed sensor **(see illustration)**.

6   Working from the back side of the steering knuckle, remove the hub retaining bolts from the steering knuckle **(see illustration)**. Remove the disc shield.

7   Attach a puller to the hub flange and draw it off the driveaxle **(see illustration)**. The hub assembly should come right out of the steering knuckle, but if it doesn't, tap it from side-to-side to free it. Carefully guide the wiring harness and electrical connector for the wheel speed sensor through the opening between the driveaxle outer CV joint and the steering knuckle **(see illustration)**. **Caution:** *Be careful not to pull outward on the driveaxle, as this could separate the inner CV joint components.*

8   Clean the mating surfaces on the steering knuckle, bearing flange and knuckle bore.

9   Insert the hub and bearing assembly into the steering knuckle and onto the end of the driveaxle. Position the disc shield and install the three bolts, tightening them to the torque listed in this Chapter's Specifications.

10   Install the brake disc, caliper mounting bracket and caliper (see Chapter 9).

11   Install the hub nut and tighten it securely to seat the driveaxle in the hub. Prevent the

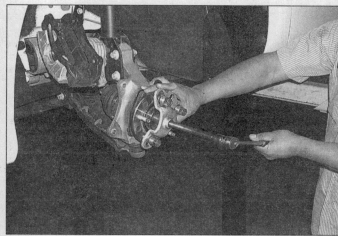

5.7a Use a puller to draw the hub and bearing assembly off the end of the driveaxle

5.7b While removing the hub and bearing assembly, carefully guide the wiring harness and electrical connector for the wheel speed sensor through the opening

10

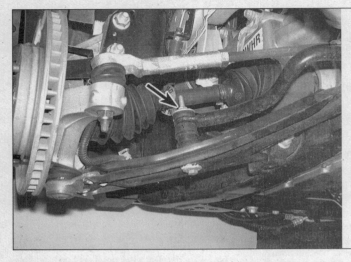

6.3 To disconnect the stabilizer bar from the link bolts that connect it to the control arms, remove the link bolt nut (arrow) from each end; make sure that you note the order in which the bushings, spacers and washers on the link are installed

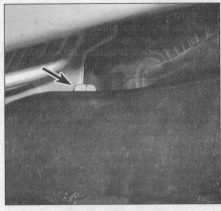

6.4 To detach the stabilizer bar from the subframe, remove the bushing clamp bolt (arrow) and clamp from each bushing

6.5 To detach the rear engine mount bracket from the subframe, remove these fasteners

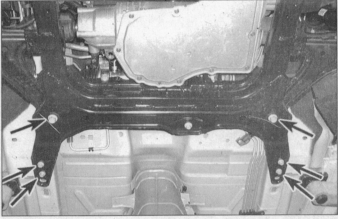

6.6 To lower the rear part of the subframe, remove these six bolts (arrows), then loosen the bolts at the front of the subframe; be sure to put a sturdy floor jack under the rear of the subframe before unbolting it

axle from turning by inserting a screwdriver through the caliper and into a disc cooling vane.

12 Install the wheel, lower the vehicle and tighten the lug nuts to the torque listed in the Chapter 1 Specifications.

13 Tighten the driveaxle/hub nut to the torque listed in the Chapter 8 Specifications.

6 Stabilizer bar and bushings (front) - removal and installation

Removal

Refer to illustrations 6.3, 6.4, 6.5 and 6.6

1 Loosen the lug nuts on both front wheels, raise the front of the vehicle and support it securely on jackstands. Apply the parking brake and block the rear wheels to keep the vehicle from rolling off the jackstands. Remove the front wheels.

2 Detach the tie-rod ends from the steering knuckles (see Section 16).

3 Remove the nuts from the upper ends of

the link bolts that connect the stabilizer bar to the control arms **(see illustration)**. Note the order in which the bushings, spacers and washers are installed on the links; they must be installed in exactly the same order in which they are removed.

4 Remove the stabilizer bar bushing clamp bolts and bushing clamps from the upper side of the subframe **(see illustration)**.

5 Unbolt the rear engine mount bracket from the subframe **(see illustration)**.

6 Support the rear of the subframe with a jack, then remove the rear subframe retaining bolts **(see illustration)**. Loosen the subframe front mounting bolts, then lower the subframe about three inches.

7 Remove the stabilizer bar and bushings through the wheel well.

8 Inspect the stabilizer bushings for wear and damage and replace them if necessary. To ease installation of the new bushings, spray the inside and outside of the bushings with a silicone-based lubricant. Do not use petroleum-based lubricants on rubber parts. Inspect the link bushings, spacers and washers for wear and replace as necessary.

Installation

9 Install the stabilizer bar bushings and clamps, guide the stabilizer through the wheel well, over the subframe and into position.

10 Center the stabilizer bar, install the bushing clamp bolts and tighten them to the torque listed in this Chapter's Specifications.

11 Raise the subframe, install the subframe retaining bolts and tighten them to the torque listed in this Chapter's Specifications (don't forget to tighten the front bolts).

12 Install the rear engine mounting bracket bolts, tightening them to the torque listed in the Chapter 2A or 2B Specifications.

13 Install the spacers, bushings and washers on the links in the same order in which they were removed. Tighten the link nuts to the torque listed in this Chapter's Specifications.

14 Connect the tie-rod ends to the steering knuckle arms, tightening the nuts to the torque listed in this Chapter's Specifications.

15 Install the wheels and lower the vehicle. Tighten the lug nuts to the torque listed in the Chapter 1 Specifications.

7.3 Remove the cotter pin and loosen - but don't remove - the nut (arrow) from the balljoint stud

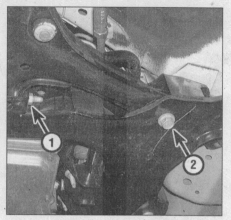

7.6 The front end of the control arm is secured to the subframe with a pivot bolt and nut (1); at the rear end it's secured by a vertical bushing bolt (2)

7 Control arm (front) - removal and installation

Removal

Refer to illustrations 7.3 and 7.6

1   Loosen the wheel lug nuts, raise the front of the vehicle and support it securely on jackstands. Apply the parking brake and block the rear wheels to keep the vehicle from rolling off the jackstands. Remove the wheel(s).

2   Disconnect the stabilizer bar from the control arm being removed (see Section 6). (If only one control arm is being removed, disconnect only that end of the stabilizer bar; if both control arms are being removed, disconnect both ends.)

3   Remove the cotter pin and loosen the balljoint stud-to-steering knuckle nut a few turns **(see illustration)**.

4   Using a balljoint separator, detach the balljoint from the steering knuckle. A "picklefork" -type tool will also work, but keep in mind that a picklefork will damage or destroy the boot, so it should be used only as a last resort.

5   Remove the nut from the ballstud. Using

a large prybar positioned between the control arm and steering knuckle, separate the ballstud from the knuckle. **Caution:** *When removing the balljoint from the knuckle, be careful not to overextend the inner CV joint or it may be damaged.*

6   Remove the front control arm pivot bolt/nut and the rear vertical bushing bolt **(see illustration)**. Detach the control arm.

7   The control arm bushings are replaceable, but special tools and expertise are necessary to do the job. Carefully inspect the bushings for hardening, excessive wear and cracks. If they appear to be worn or deteriorated, take the control arm to an automotive machine shop or other repair facility.

Installation

8   Position the control arm in the subframe and install the front pivot bolt and the rear vertical bushing bolt. Do not tighten them completely at this time.

9   Insert the balljoint stud into the steering knuckle boss, install the nut and tighten it to the torque listed in this Chapter's Specifications. If necessary, tighten the nut a little

more (up to, but not beyond, the specified maximum listed in the Specifications) if the cotter pin hole doesn't line up with an opening on the nut. Install a new cotter pin.

10   Install the stabilizer bar-to-control arm link bolt, bushings, spacers and washers (see Section 6) and tighten the link nut to the torque listed in this Chapter's Specifications.

11   Install the wheel and lower the vehicle. Tighten the lug nuts to the torque listed in the Chapter 1 Specifications.

12   With the weight of the vehicle on the suspension, tighten the control arm pivot bolt and the rear vertical bushing bolt to the torque listed in this Chapter's Specifications. **Note:** *You can raise the suspension with a floor jack positioned under the balljoint to simulate normal ride height.* **Caution:** *If the bolts aren't tightened with the weight of the vehicle on the suspension, control arm bushing damage may occur.*

8 Balljoints - check and replacement

Check

Refer to illustrations 8.3a and 8.3b

1   Raise the front of the vehicle and support it securely on jackstands. Apply the parking brake and block the rear wheels to keep the vehicle from rolling off the jackstands.

2   Visually inspect the rubber dust boot for damage, deterioration and leaking grease. If the boot is damaged, deteriorated or leaking, replace the balljoint.

3   Place a large prybar under the balljoint and resting on the wheel, then try to pry the balljoint up while feeling for movement between the balljoint and steering knuckle **(see illustration)**. Now, pry between the control arm and the steering knuckle and try to lever the control arm down while feeling for movement between the balljoint and steering knuckle **(see illustration)**. If any movement is evident in either check, the balljoint is worn.

10

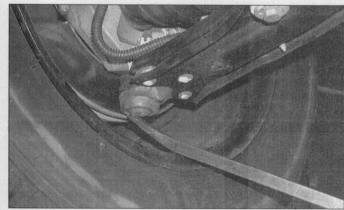

8.3a Check for movement between the balljoint and steering knuckle when prying up

8.3b With the prybar positioned between the steering knuckle boss and the balljoint, pry down and check for play in the balljoint - if there's any play, replace the balljoint

9.2 Peel back the carpet in the trunk and remove this nut (arrow) from the strut upper mount

9.4 To detach the rear strut from the knuckle, remove these two nuts and drive out the bolts with a hammer and brass punch

9.5 The remaining mounting fasteners for the rear strut are accessed through the fenderwell

4   Have an assistant grasp the tire at the top and bottom and move the top of the tire in-and-out. Touch the balljoint stud nut. If any looseness is felt, suspect a worn balljoint stud or a widened hole in the steering knuckle boss. If the latter problem exists, the steering knuckle should be replaced as well as the balljoint.

5   Separate the control arm from the steering knuckle (Section 7). Using your fingers (don't use pliers), try to twist the stud in the socket. If the stud turns, replace the balljoint.

Replacement

6   Loosen the wheel lug nuts, raise the front of the vehicle and support it securely on jackstands. Apply the parking brake and block the rear wheels to keep the vehicle from rolling off the jackstands. Remove the wheel.

7   Separate the control arm from the steering knuckle (see Section 7). Temporarily insert the balljoint stud back into the steering knuckle (loosely). This will ease balljoint removal after Step 9 has been performed, as well as hold the assembly stationary while drilling out the rivets.

8   Using a 1/8-inch drill bit, drill a pilot hole into the center of each balljoint-to-control arm rivet. Be careful not to damage the CV joint boot in the process.

9   Using a 1/2-inch drill bit, drill the head off each rivet. Work slowly and carefully to avoid deforming the holes in the control arm.

10   Loosen (but don't remove) the stabilizer bar-to-control arm link nut. Pull the control arm and balljoint down to remove the balljoint stud from the steering knuckle, then dislodge the balljoint from the control arm.

11   Position the new balljoint on the control arm and install the bolts (supplied in the balljoint kit) from the top of the control arm. Tighten the bolts to the torque specified in the new balljoint instruction sheet.

12   Insert the balljoint into the steering knuckle, install the castle nut, tighten it to the torque listed in this Chapter's Specifications

and install a new cotter pin. It may be necessary to tighten the nut a little more to align the cotter pin hole with an opening in the nut, which is acceptable. Never loosen the nut to allow cotter pin insertion.

13   Tighten the stabilizer bar-to-control arm link nut to the torque listed in this Chapter's Specifications.

14   Install the wheel, lower the vehicle and tighten the lug nuts to the torque listed in the Chapter 1 Specifications.

9   Strut/coil spring assembly (rear) - removal, inspection and installation

Refer to illustrations 9.2, 9.4 and 9.5

1   Open the trunk and peel back the side trim panel to expose the strut upper mount.

2   Remove the mount nut **(see illustration)**. **Note:** *The two bolts you see must be removed from the wheel well side, after the vehicle is raised and the wheel is removed.*

3   Loosen the wheel lug nuts, raise the rear of the vehicle and support it securely on jackstands. Block the front wheels to keep the vehicle from rolling off the jackstands. Remove the rear wheels.

4   Mark the strut-to-knuckle relationship by making a line around the strut-to-knuckle bolt heads **(see illustration 2.2)**. Support the rear knuckle with a jack and remove the strut-to-knuckle nuts **(see illustration)**, then drive out the bolts with a hammer and brass punch **(see illustration 2.3)**. Separate the knuckle from the strut.

5   Remove the two upper mount bolts **(see illustration)** and detach the strut assembly from the vehicle. Check the shock absorber for dents and leaking fluid. Check the coil spring for chips and cracks.

6   Inspect the upper mount for cracks, hardening, separation and other damage.

7   If any of these conditions are found,

replace the strut or coil spring as necessary (see Section 3 for the disassembly and reassembly procedure - it's similar to a front strut).

8   Installation is the reverse of the removal procedure. Be sure to tighten all fasteners to the torque values listed in this Chapter's Specifications.

10   Rear knuckle - removal and installation

1   Mark the strut-to-knuckle relationship by making a line around the strut-to-knuckle bolt heads **(see illustration 2.2)**.

2   Unbolt the stabilizer bar link from the knuckle (see Section 12).

3   Refer to Section 11 and remove the rear hub/bearing assembly.

4   On models with rear drum brakes, remove the brake shoe assembly and the backing plate.

5   Remove the bolts/nuts and detach the lateral links from the knuckle (see Section 13).

6   Remove the bolt that secures the trailing arm to the knuckle, then remove the washer and bushing.

7   Remove the strut-to-knuckle nuts and knock the bolts out with a brass or plastic hammer. Separate the knuckle from the strut.

8   Slide the knuckle off the trailing arm.

9   To install the knuckle, slide it into the strut flange and insert the two bolts. Install the nuts finger tight. Connect the lateral link rods and trailing arm to the knuckle, but do not tighten the fasteners at this time. Align the marks made in Step 1 and tighten the strut-to-knuckle bolts to the torque listed in this Chapter's Specifications. The remainder of installation is the reverse of removal.

10   Don't tighten the trailing arm or lateral link rod bolts/nuts to the torque listed in this Chapter's Specifications until the vehicle weight has been lowered onto the suspension (you can simulate normal ride height by raising the suspension with a floor jack).

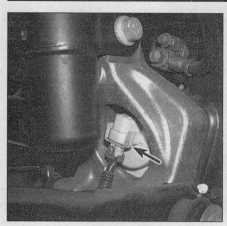

11.3 Disconnect the electrical connector from the rear wheel speed sensor

11.4 Removing the rear hub bolts

12.2 Remove the stabilizer bar link-to-rear knuckle bolt

11   Hub and bearing assembly (rear) - removal and installation

Note: *The rear hub and wheel bearing assembly is sealed-for-life and must be replaced as a unit.*

Removal

Refer to illustrations 11.3 and 11.4

1   Loosen the wheel lug nuts, raise the rear of the vehicle and support it securely on jackstands. Block the front wheels to keep the vehicle from rolling off the jackstands. Remove the wheel.

2   Remove the brake drum or the brake caliper, caliper mounting bracket and disc (see Chapter 9). Support the caliper with a piece of wire.

3   Unplug the electrical connector from the wheel speed sensor on the back side of the knuckle **(see illustration)**.

4   Remove the four hub-to-knuckle bolts **(see illustration)**.

5   Remove the hub and bearing assembly from the knuckle. Temporarily reinstall a couple of bolts through the brake backing plate to support the brake assembly rather than letting it hang by the brake hose (drum brake models only).

Installation

6   Position the hub and bearing assembly on the knuckle and align the holes. Install the bolts. After all four bolts have been installed, tighten them to the torque listed in this Chapter's Specifications.

7   Reconnect the electrical connector for the wheel speed sensor.

8   Install the brake drum or the brake caliper and disc, tightening the caliper mounting bracket bolts and the caliper bolts to the torque listed in the Chapter 9 Specifications. Install the wheel. Lower the vehicle and tighten the wheel lug nuts to the torque listed in the Chapter 1 Specifications.

12   Stabilizer bar and bushings (rear) - removal and installation

Refer to illustrations 12.2 and 12.3

1   Raise the rear of the vehicle and support it securely on jackstands.

2   Remove the bolt that connects the stabilizer bar link to the knuckle **(see illustration)**.

3   Remove the nut from each stabilizer bar bracket, then detach the brackets from the subframe **(see illustration)**.

4   Guide the stabilizer bar and links out from under the vehicle.

5   Inspect the stabilizer bar bushings for wear and damage and replace them if necessary. To ease installation of the new bushings, spray the inside and outside of the bushings with a silicone-based lubricant. Do not use petroleum-based lubricants on rubber parts. Inspect the link bushings for wear and replace as necessary.

6   Installation is the reverse of removal. Be sure to tighten the fasteners to the torque listed in this Chapter's Specifications.

13   Suspension arms (rear) - removal and installation

1   Loosen the wheel lug nuts, raise the rear of the vehicle and support it securely on jackstands. Block the front wheels to keep the vehicle from rolling off the jackstands. Remove the wheels.

Trailing arms

Refer to illustrations 13.2 and 13.3

2   Unscrew the bolt that secures the washer and bushing to the trailing arm, then remove the washer and bushing **(see illustration)**.

10

12.3 Each stabilizer bar bracket is secured by one bolt

13.2 Remove the bolt that secures the trailing arm rear bushing to the trailing arm, then remove the washer and bushing

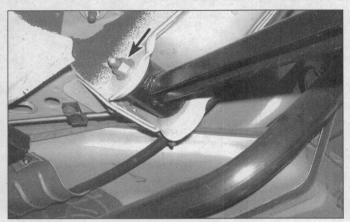

13.3 Remove the nut and pivot bolt from the forward end of the trailing arm

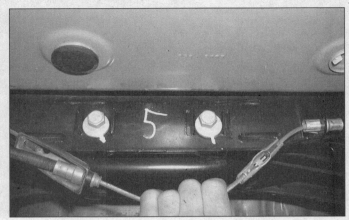

13.7 Before loosening the bolt at the inner end of the lateral link, mark the position of the toe adjuster cam to the crossmember

3   Remove the trailing arm-to-chassis nut and bolt **(see illustration)**.
4   Pull the front end of the trailing arm down, then remove the arm from the knuckle.
5   Check the rear bushings for wear and deterioration, replacing them if necessary.

13.8 To disconnect the lateral links from the rear knuckle, remove the nut and long through-bolt

6   Installation is the reverse of removal, but be sure to install the washers properly (the concave sides face away from the bushings). Do not tighten the nuts and bolts to the torque listed in this Chapters Specifications until the vehicle weight has been lowered onto the suspension (you can simulate normal ride height by raising the suspension with a floor jack).

Lateral links

Refer to illustrations 13.7 and 13.8

7   Mark the relationship of the toe adjuster cams to the crossmember **(see illustration)**.
8   Unclip the ABS wiring harness from the forward lateral link, then remove the nuts and bolts connecting the links to the knuckle **(see illustration)**.
9   Remove the link inner bolts at the crossmember.
10   Installation is the reverse of removal. Don't tighten the bolts to the torque listed in this Chapter's Specifications until the vehicle weight has been lowered onto the suspension (you can simulate normal ride height by raising the suspension with a floor jack). After installation have the wheel alignment checked and, if necessary, adjusted.

14   Steering wheel - removal and installation

Warning: *These models are equipped with airbags. Always turn the steering wheel to the straight ahead position, place the ignition switch in the Lock position and disable the airbag system (see Chapter 12) before working in the vicinity of any airbag system component to avoid the possibility of accidental deployment of the airbag, which could cause personal injury.*

Removal

Refer to illustrations 14.2, 14.3a, 14.3b, 14.3c, 14.3d, 14.5 and 14.6

1   Disconnect the cable from the negative terminal of the battery and disable the airbag system (see Chapter 12). **Note:** *On models equipped with the Theftlock audio system, be sure the lockout feature is turned off before performing any procedure which requires disconnecting the battery (see the front of this manual).* Remove the steering column covers (see Chapter 11).
2   Using a number 30 Torx bit, remove the two screws that secure the airbag module to

14.2 Remove the airbag module retaining screws using a no. 30 Torx bit

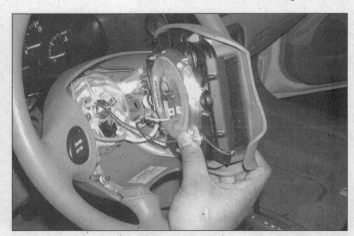

14.3a Lift the airbag module away from the steering wheel . . .

14.3b ... twist the connector counterclockwise and detach the horn wire ...

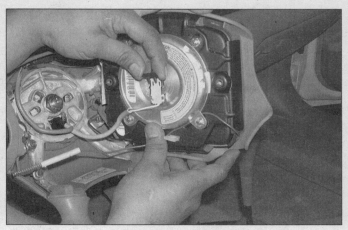

14.3c ... squeeze the tabs on the airbag module electrical connector ...

the steering wheel **(see illustration)**.

3   Lift the airbag module carefully away from the steering wheel and unplug the horn wire and the airbag electrical connector **(see illustrations)**. Remove the module. **Warning:** *When carrying the airbag module, keep the driver's (trim) side of it away from your body, and when you place it on the bench, have the driver's side facing up.*

4   Unplug the cruise control electrical connector.

5   Remove the steering wheel nut. If no marks are present, mark the relationship of the steering wheel to the shaft **(see illustration)**.

6   Install a puller and turn the center bolt until the wheel is free **(see illustration)**. **Note:** *A conventional steering wheel puller will not work on these vehicles, but a small two-jaw puller will.* **Warning:** *Do not hammer on the shaft or the puller in an attempt to loosen the wheel from the shaft.*

7   Lift the steering wheel from the shaft. **Warning:** *Don't allow the steering shaft to turn with the steering wheel removed. If the shaft turns, the airbag clockspring will become uncentered, which may cause the*

14.3d ... and unplug the connector from the module

wire inside to break when the vehicle is returned to service.

Installation

Refer to illustrations 14.8, 14.9 and 14.10

8   Before installing the steering wheel,

14.5 There should already be alignment marks on the steering wheel and steering shaft; if not, make your own

make sure the airbag clockspring is centered **(see illustration)**.

9   If the airbag system clockspring is not centered, remove the steering column covers (see Chapter 11) and the multi-function

14.6 Special puller attachments are available for steering wheel removal, but a small jaw-type puller of good quality can be used to remove the steering wheel from the shaft

14.8 When the clockspring is centered, the arrow on the housing will be aligned with the arrow on the hub

switch (see Chapter 12). Unplug the electrical connectors, remove the snap-ring **(see illustration)** and lift the clockspring off the steering column. You may have to cut a plastic wire-tie securing the clockspring harness to the steering column. **Note:** *The clockspring has a locking device to prevent the hub from turning when the steering wheel is off, but it is a good idea to place a piece of tape across the body of the clockspring and the hub, in case the lock is inadvertently depressed.*

10   To center the clockspring, turn the clockspring over, depress the lock lever and turn the hub in the direction of the arrow until it stops (don't apply too much force) **(see illustration)**. Then, turn the hub in the opposite direction approximately 2-3/4 turns (1997 models) or 2-1/2 turns (1998 and later models), aligning the arrows on the front. Release the lock lever and install the clockspring and snap-ring. Secure the wiring harness with a new wire tie, making sure the harness isn't kinked. Also install the steering column covers.

11   Pull the electrical leads for the airbag module and cruise control switches through the steering wheel and install the wheel on the steering shaft, aligning the marks.

12   Install the steering wheel nut and tighten it to the torque listed in this Chapter's Specifications.

13   Connect the airbag connector to the back of the airbag module and the horn wire to the horn ring.

14   Position the airbag module on the steering wheel and install the bolts, tightening them to the torque listed in this Chapter's Specifications.

15   Refer to Chapter 12 for the procedure to enable the airbag system.

15   Steering column - removal and installation

Refer to illustrations 15.7 and 15.10
Warning: *These models have airbags. Always disable the airbag system before working in*

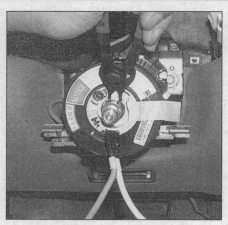

14.9 The clockspring is retained to the steering shaft with a snap-ring

the vicinity of any airbag system components to avoid the possibility of accidental deployment of the airbag(s), which could cause personal injury (see Chapter 12).

Removal

1   Park the vehicle with the front wheels pointing straight ahead.

2   Disconnect the cable from the negative battery terminal and disable the airbag system (see Chapter 12).**Caution:** *On models equipped with the Theftlock audio system, be sure the lockout feature is turned off before performing any procedure which requires disconnecting the battery (see the front of this manual).*

3   Remove the steering wheel, multi-function switch and airbag system clockspring (see Section 14).

4   Remove the left-side under-dash panel, knee bolster and reinforcement (see Chapter 11).

5   Remove the instrument cluster (see Chapter 12).

6   Follow the airbag clockspring wiring harness along the steering column and cut any wire ties that may be securing it to the column.

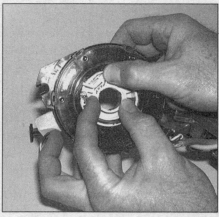

14.10 To center the clockspring, hold it with its underside facing up, depress the spring lock and rotate the hub in the direction of the arrow until it stops, then turn it in the opposite direction the number of turns indicated in the text

7   Pull back the steering column boot and remove the steering shaft-to-intermediate shaft pinch bolt **(see illustration)**.

8   Remove the steering column lower mounting bolts **(see illustration 15.7)**.

9   Insert a large screwdriver into the gap in the intermediate shaft joint and spread it apart slightly to loosen it.

10   Remove the steering column upper mounting bolts **(see illustration)**.

11   Lower the column slightly, turn it 45-degrees counterclockwise, then pull it out from under the instrument panel.

Installation

12   Installation is the reverse of the removal procedure, with the following points:

a) *Install all of the steering column fasteners before tightening any of them, then tighten them to the torque listed in this Chapter's Specifications in the following order:*

15.7 Steering column lower mounting details

   A   Intermediate shaft pinch bolt
   B   Column lower mounting bolts

15.10 Steering column upper mounting bolts

16.2a Before removing the tie-rod end, loosen the jam nut . . .

16.2b . . . and mark the position of the tie-rod end
on the inner tie-rod

16.3 To separate the tie-rod end from the steering knuckle, loosen - but don't remove - the ballstud nut, then install a balljoint removal tool (shown) or a small puller to pop the ballstud out of the knuckle (DO NOT pound on the stud!)

17.4a Squeeze the outer boot clamp with
a pair of pliers and slide it down
the tie-rod

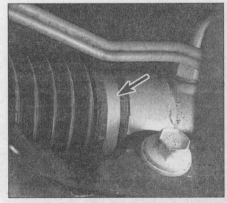

17.4b Cut off the inner boot clamps
(arrow) with diagonal cutters and slide off
the old boots

1) Upper mounting bolts
2) Lower mounting bolts
3) Intermediate shaft pinch bolt
b) Center and install the airbag system clockspring as described in Section 14.
c) Secure the clockspring wiring harness with new wire ties.
d) Install the steering wheel as described in Section 14.
e) Refer to Chapter 12 for the procedure to enable the airbag system.

16 Tie-rod ends - removal and installation

Refer to illustrations 16.2a, 16.2b and 16.3
1   Loosen the wheel lug nuts, raise the front of the vehicle and support it securely on jackstands. Remove the wheel.
2   Loosen the tie-rod end jam nut **(see illustration)** and mark the position of the tie-rod end on the threaded portion of the tie-rod **(see illustration)**.
3   Remove the cotter pin and loosen (but do not remove) the castle nut from the tie-rod end balljoint stud, then install a small puller **(see illustration)** and break loose the tie-rod end from the steering knuckle. Remove the nut and detach the tie-rod end.

4   Unscrew the old tie-rod end and install the new one. Make sure the new tie-rod end is aligned with the mark you made on the threads of the tie-rod.
5   Installation is the reverse of removal. Be sure to tighten the tie-rod end balljoint nut to the torque listed in this Chapter's Specifications. Tighten the jam nut securely.
6   Have the wheel alignment checked and if necessary, adjusted.

17 Steering gear boots - replacement

Refer to illustrations 17.4a and 17.4b
1   If a steering gear boot is torn, dirt and moisture can damage the steering gear. Replace it.
2   Loosen the wheel lug nuts, raise the vehicle and place it securely on jackstands. Remove the front wheels.
3   Disconnect the tie-rod ends from the steering knuckles and remove them from the tie-rods (see Section 16). Also remove the jam nuts.
4   Remove the boot clamps **(see illustrations)** and slide the boots off the tie-rods.
5   Installation is the reverse of removal. Be sure to use new clamps on the boots.

18 Steering gear - removal and installation

Refer to illustrations 18.2, 18.4 and 18.5
Warning: These models are equipped with airbags. Make sure the steering column shaft is not turned while the steering gear is removed or you could damage the airbag sys-

10

18.2 Push up on the rubber boot for access to the intermediate shaft-to-steering gear pinch bolt

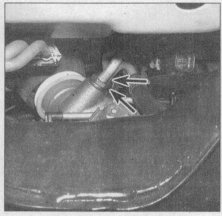

18.4 Unscrew the power steering line fittings

18.5 The head of the right (passenger's) side steering gear mounting bolt faces toward the front of the vehicle; the left side bolt head faces the toward the rear

tem. To prevent the shaft from turning, turn the ignition key to the lock position before beginning work or run the seat belt through the steering wheel and clip the seat belt into place. Due to the possible damage to the airbag system, we recommend only experienced mechanics attempt this procedure.

Removal

1   Refer to Section 6 and perform Steps 1 through 4.
2   Push the rubber boot up and remove the intermediate shaft-to-steering gear pinch bolt **(see illustration)**.
3   Refer to Section 6 again and perform Steps 5 through 7 to remove the stabilizer bar.
4   Unscrew the power steering pressure and return line fittings from the steering gear **(see illustration)**. Cap the ends to prevent fluid loss and the entry of contaminants.
5   Unscrew the steering gear mounting

bolts **(see illustration)**. The right bolt is accessed from the front side of the steering gear; the left bolt is accessed from the rear of the steering gear.
6   Carefully guide the steering gear through the left (driver's) side wheel opening.

Installation

7   Installation is the reverse of the removal procedure, with the following points:

  a) Tighten all fasteners to the torque values listed in this Chapter's Specifications.
  b) Tighten the power steering pressure and return line fittings securely.
  c) Add power steering fluid to the pump reservoir to bring it up to the desired level (see Chapter 1).
  d) Lower the vehicle and tighten the lug nuts to the torque listed in the Chapter 1 Specifications.
  e) Bleed the power steering system (see Section 20).
  f) Have the wheel alignment checked and if necessary, adjusted.

19   Power steering pump - removal and installation

Refer to illustrations 19.4a, 19.4b, 19.6, 19.8a and 19.8b

Removal

1   Disconnect the cable from the negative battery terminal. **Caution:** On models equipped with the Theftlock audio system, be sure the lockout feature is turned off before performing any procedure which requires disconnecting the battery (see the front of this manual).
2   If you're working on a model with a V6 engine, remove the drivebelt (see Chapter 1).
3   If you're working on a model with a four-cylinder engine, remove the air intake duct (see Chapter 4).
4   Using a suction gun or large syringe, remove as much fluid as possible from the

19.4a Power steering pump mounting details - four-cylinder engine

A   Fluid return hose   C   Pump mounting bolts
B   Fluid pressure line

19.4b Power steering pump mounting details - V6 engine

A   Fluid return hose   C   Variable Assist Steering
B   Fluid pressure line   Actuator electrical
                                       connector

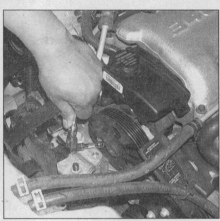

19.6 The power steering pump mounting bolts on V6 models are accessed through the holes in the pulley. You'll need a flex socket or a U-joint attachment to unscrew the lower bolt

19.8a A typical power steering pump pulley removal tool

19.8b A typical power steering pump pulley installation tool

power steering pump. Disconnect the pressure line and return hose from the power steering pump **(see illustrations)**. Plug the line and hose to prevent fluid spillage and the entry of contaminants. Also, if equipped, unplug the electrical connector for the Variable Assist Steering Actuator.

5   If you're working on a model with a V6 engine, remove the cruise control actuator (see Chapter 12).

6   Remove the power steering pump mounting bolts **(see accompanying illustration [V6 engine] or illustration 19.4a [four-cylinder engine])**.

7   Remove the pump.

8   If you're installing a new pump on a V6 model, remove the pulley from the pump with a suitable pulley removal tool and install the pulley on the new pump using a special installation tool **(see illustrations)**. Pulley removal and installation tools are available at most auto parts stores. **Caution:** *Do not use a press to install the pulley.* The pulley should be installed so the face of the pulley is flush with the end of the pump shaft.

9   The remainder of installation is the reverse of the removal procedure. Tighten the pump bolts and the fittings securely.

10   Install the drivebelt (see Chapter 1).

11   Bleed the power steering system (see Section 20).

20   Power steering system - bleeding

1   This is not a routine operation and normally will only be required when the system has been dismantled and reassembled.

2   Fill the reservoir to the correct level with fluid of the recommended type and allow it to remain undisturbed for at least two minutes.

3   Start the engine and run it for two or three seconds only. Check the reservoir and add more fluid as necessary.

4   Repeat the operations described in the preceding paragraph until the fluid level

remains constant.

5   Raise the front of the vehicle until the wheels are clear of the ground.

6   Start the engine and increase the speed to about 1500 rpm. Now turn the steering wheel gently from stop-to-stop. Check the reservoir fluid level.

7   Lower the vehicle to the ground and, with the engine still running, move the vehicle forward sufficiently to obtain full right lock followed by full left lock (but *don't* hold the steering wheel firmly against the stops). Recheck the fluid level. If the fluid in the reservoir is extremely foamy, allow the vehicle to stand for a few minutes with the engine switched off and then repeat the previous operations. At the same time, check the belt tightness and check for a bent or loose pulley. Check also to make sure the power steering hoses are not touching any other

part of the vehicle, especially sheet metal or the exhaust manifold.

8   The procedures above will normally remedy an extreme foam condition and/or an objectionably noisy pump (low fluid level and/or air in the power steering fluid are the leading causes of this condition). If, however, either or both conditions persist after a few trials, the power steering system will have to be thoroughly checked. Do not drive the vehicle until the condition(s) have been remedied.

21   Wheels and tires - general information

Refer to illustration 21.1

All vehicles covered by this manual are equipped with metric-sized steel-belted radial

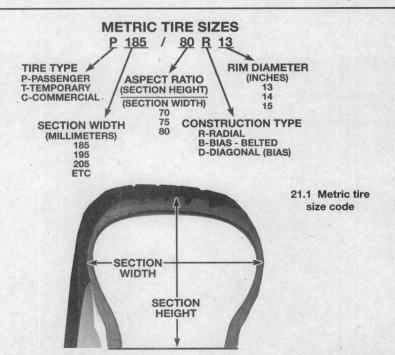

METRIC TIRE SIZES

P 185 / 80 R 13

TIRE TYPE
P-PASSENGER
T-TEMPORARY
C-COMMERCIAL

ASPECT RATIO
(SECTION HEIGHT)
(SECTION WIDTH)
70
75
80

RIM DIAMETER
(INCHES)
13
14
15

SECTION WIDTH
(MILLIMETERS)
185
195
205
ETC

CONSTRUCTION TYPE
R-RADIAL
B-BIAS - BELTED
D-DIAGONAL (BIAS)

SECTION WIDTH

SECTION HEIGHT

21.1 Metric tire size code

tires **(see illustration).** Use of other size or type of tires may affect the ride and handling of the vehicle. Don't mix different types of tires, such as radials and bias belted, on the same vehicle as handling may be seriously affected. It's recommended that tires be replaced in pairs on the same axle, but if only one tire is being replaced, be sure it's the same size, structure and tread design as the other. Because tire pressure has a substantial effect on handling and wear, the pressure on all tires should be checked at least once a month or before any extended trips (see Chapter 1).

Wheels must be replaced if they are bent, dented, leak air, have elongated bolt holes, are heavily rusted, out of vertical symmetry or if the lug nuts won't stay tight. Wheel repairs that use welding or peening are not recommended.

Tire and wheel balance is important to the overall handling, braking and performance of the vehicle. Unbalanced wheels can adversely affect handling and ride characteristics as well as tire life. Whenever a tire is installed on a wheel, the tire and wheel should be balanced by a shop with the proper equipment.

22   Wheel alignment - general information

Refer to illustration 22.1

A wheel alignment refers to the adjustments made to the wheels so they are in proper angular relationship to the suspension and the ground **(see illustration).** Wheels that are out of proper alignment not only affect steering control, but also increase tire wear. Camber and toe-in are the only angles that can be adjusted on the vehicles covered by this manual, but caster should also be measured to determine if any suspension parts are bent.

Getting the proper wheel alignment is a very exacting process, one in which complicated and expensive machines are necessary to perform the job properly. Because of this, you should have a technician with the proper

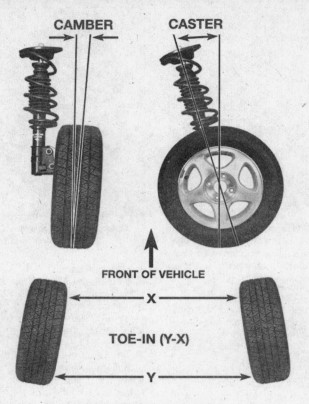

22.1 Camber, caster and toe-in angles

equipment perform these tasks. We will, however, use this space to give you a basic idea of what is involved with wheel alignment so you can better understand the process and deal intelligently with the shop that does the work.

Camber is the tilting of the wheels from vertical when viewed from the end of the vehicle. On the vehicles covered in this manual, camber can only be adjusted by elongating the lower strut-to-knuckle hole.

Caster is the tilting of the top of the front steering axis from the vertical: a tilt toward the rear is positive caster and a tilt toward the front is negative caster. Caster is not adjustable on these vehicles.

Toe-in is the turning in of the front or rear wheels. The purpose of a toe specification is to ensure parallel rolling of the front or rear wheels. In a vehicle with zero toe-in, the distance between the front edges of the wheels will be the same as the distance between the rear edges of the wheels. The actual amount of toe-in is normally only a fraction of an inch. On the front end, toe-in adjustment is controlled by the position of the tie-rod end on the tie-rod. On the rear end, it's adjusted by turning the inner mounting (cam) bolt for the rear lateral link. Incorrect toe-in will cause the tires to wear improperly by making them scrub against the road surface.

Chapter 11 Body

Contents

1 General information

The Chevrolet Malibu, Oldsmobile Cutlass, Oldsmobile Alero and Pontiac Grand Am models feature "unibody" construction, in which the major body components, floor pan and front and rear frame side rails are welded together to create a rigid structure which supports the remaining body components, drivetrain, front and rear suspension and other components.

Certain components are particularly vulnerable to accident damage and can be unbolted and repaired or replaced. Among these parts are the body moldings, bumpers, front fenders, doors, the hood and trunk lid. Only general body maintenance practices and body panel repair procedures within the scope of the do-it-yourselfer are included in this chapter.

Although all models are very similar, some procedures may differ somewhat from one body to another.

2 Body - maintenance

1 The condition of your vehicle's body is very important, because the resale value depends a great deal on it. It's much more difficult to repair a neglected or damaged body than it is to repair mechanical components. The hidden areas of the body, such as the wheel wells, the frame and the engine compartment, are equally important, although they don't require as frequent atten-

11

tion as the rest of the body.

2   Once a year, or every 12,000 miles, it's a good idea to have the underside of the body steam cleaned. All traces of dirt and oil will be removed and the area can then be inspected carefully for rust, damaged brake lines, frayed electrical wires, damaged cables and other problems.

3   At the same time, clean the engine and the engine compartment with a steam cleaner or water-soluble degreaser.

4   The wheel wells should be given close attention, since undercoating can peel away and stones and dirt thrown up by the tires can cause the paint to chip and flake, allowing rust to set in. If rust is found, clean down to the bare metal and apply an anti-rust paint.

5   The body should be washed about once a week. Wet the vehicle thoroughly to soften the dirt, then wash it down with a soft sponge and plenty of clean soapy water. If the surplus dirt is not washed off very carefully, it can wear down the paint.

6   Spots of tar or asphalt thrown up from the road should be removed with a cloth soaked in solvent.

7   Once every six months, wax the body and chrome trim. If a chrome cleaner is used to remove rust from any of the vehicle's plated parts, remember that the cleaner also removes part of the chrome, so use it sparingly. After cleaning chrome trim, apply paste wax to preserve it.

3   Vinyl trim - maintenance

Don't clean vinyl trim with detergents, caustic soap or petroleum-based cleaners. Plain soap and water works just fine, with a soft brush to clean dirt that may be ingrained. Wash the vinyl as frequently as the rest of the vehicle. After cleaning, application of a high-quality rubber and vinyl protectant will help prevent oxidation and cracks. The protectant can also be applied to weatherstripping, vacuum lines and rubber hoses, which often fail as a result of chemical degradation, and to the tires.

4   Upholstery and carpets - maintenance

1   Every three months, remove the floor-mats and clean the interior of the vehicle (more frequently if necessary). Use a stiff whiskbroom to brush the carpeting and loosen dirt and dust, then vacuum the upholstery and carpets thoroughly, especially along seams and crevices.

2   Dirt and stains can be removed from carpeting with basic household or automotive carpet shampoos available in spray cans. Follow the directions and vacuum again, then use a stiff brush to bring back the "nap" of the carpet.

3   Most interiors have cloth or vinyl upholstery, either of which can be cleaned and

maintained with a number of material-specific cleaners or shampoos available in auto supply stores. Follow the directions on the product for usage, and always spot-test any upholstery cleaner on an inconspicuous area (bottom edge of a backseat cushion) to ensure that it doesn't cause a color shift in the material.

4   After cleaning, vinyl upholstery should be treated with a protectant. **Note:** *Make sure the protectant container indicates the product can be used on seats - some products may make a seat too slippery.* **Caution:** *Do not use protectant on vinyl-covered steering wheels.*

5   Leather upholstery requires special care. It should be cleaned regularly with saddlesoap or leather cleaner. Never use alcohol, gasoline, water, nail polish remover or thinner to clean leather upholstery.

6   After cleaning, regularly treat leather upholstery with a leather conditioner, rubbed in with a soft cotton cloth. Never use car wax on leather upholstery.

7   In areas where the interior of the vehicle is subject to bright sunlight, cover leather seating areas of the seats with a sheet if the vehicle is to be left out for any length of time.

5   Body repair - minor damage

Plastic body panels (front and rear bumper fascia)

The following repair procedures are for minor scratches and gouges. Repair of more serious damage should be left to a dealer service department or qualified auto body shop. Below is a list of the equipment and materials necessary to perform the following repair procedures on plastic body panels. Although a specific brand of material may be mentioned, it should be noted that equivalent products from other manufacturers may be used instead.

Wax, grease and silicone removing solvent
Cloth-backed body tape
Sanding discs
Drill motor with three-inch disc holder
Hand sanding block
Rubber squeegees
Sandpaper
Non-porous mixing palette
Wood paddle or putty knife
Curved-tooth body file
Flexible parts repair material

1   Remove the damaged panel, if necessary or desirable. In most cases, repairs can be carried out with the panel installed.

2   Clean the area(s) to be repaired with a wax, grease and silicone removing solvent applied with a water-dampened cloth.

3   If the damage is structural, that is, if it extends through the panel, clean the backside of the panel area to be repaired as well. Wipe dry.

4   Sand the rear surface about 1-1/2 inches beyond the break.

5   Cut two pieces of fiberglass cloth large enough to overlap the break by about 1-1/2 inches. Cut only to the required length.

6   Mix the adhesive from the repair kit according to the instructions included with the kit, and apply a layer of the mixture approximately 1/8-inch thick on the backside of the panel. Overlap the break by at least 1-1/2 inches.

7   Apply one piece of fiberglass cloth to the adhesive and cover the cloth with additional adhesive. Apply a second piece of fiberglass cloth to the adhesive and immediately cover the cloth with additional adhesive in sufficient quantity to fill the weave.

8   Allow the repair to cure for 20 to 30 minutes at 60-degrees to 80-degrees F.

9   If necessary, trim the excess repair material at the edge.

10   Remove all of the paint film over and around the area(s) to be repaired. The repair material should not overlap the painted surface.

11   With a drill motor and a sanding disc (or a rotary file), cut a "V" along the break line approximately 1/2-inch wide. Remove all dust and loose particles from the repair area.

12   Mix and apply the repair material. Apply a light coat first over the damaged area; then continue applying material until it reaches a level slightly higher than the surrounding finish.

13   Cure the mixture for 20 to 30 minutes at 60-degrees to 80-degrees F.

14   Roughly establish the contour of the area being repaired with a body file. If low areas or pits remain, mix and apply additional adhesive.

15   Block sand the damaged area with sandpaper to establish the actual contour of the surrounding surface.

16   If desired, the repaired area can be temporarily protected with several light coats of primer. Because of the special paints and techniques required for flexible body panels, it is recommended that the vehicle be taken to a paint shop for completion of the body repair.

Steel body panels

See photo sequence

Repair of minor scratches

17   If the scratch is superficial and does not penetrate to the metal of the body, repair is very simple. Lightly rub the scratched area with a fine rubbing compound to remove loose paint and built up wax. Rinse the area with clean water.

18   Apply touch-up paint to the scratch, using a small brush. Continue to apply thin layers of paint until the surface of the paint in the scratch is level with the surrounding paint. Allow the new paint at least two weeks to harden, then blend it into the surrounding paint by rubbing with a very fine rubbing compound. Finally, apply a coat of wax to the scratch area.

19   If the scratch has penetrated the paint and exposed the metal of the body, causing the metal to rust, a different repair technique is required. Remove all loose rust from the bottom of the scratch with a pocketknife, then apply rust inhibiting paint to prevent the formation of rust in the future. Using a rubber or nylon applicator, coat the scratched area with glaze-type filler. If required, the filler can be mixed with thinner to provide a very thin paste, which is ideal for filling narrow scratches. Before the glaze filler in the scratch hardens, wrap a piece of smooth cotton cloth around the tip of a finger. Dip the cloth in thinner and then quickly wipe it along the surface of the scratch. This will ensure that the surface of the filler is slightly hollow. The scratch can now be painted over as described earlier in this Section.

Repair of dents

20   When repairing dents, the first job is to pull the dent out until the affected area is as close as possible to its original shape. There is no point in trying to restore the original shape completely as the metal in the damaged area will have stretched on impact and cannot be restored to its original contours. It is better to bring the level of the dent up to a point that is about 1/8-inch below the level of the surrounding metal. In cases where the dent is very shallow, it is not worth trying to pull it out at all.

21   If the backside of the dent is accessible, it can be hammered out gently from behind using a soft-face hammer. While doing this, hold a block of wood firmly against the opposite side of the metal to absorb the hammer blows and prevent the metal from being stretched.

22   If the dent is in a section of the body which has double layers, or some other factor makes it inaccessible from behind, a different technique is required. Drill several small holes through the metal inside the damaged area, particularly in the deeper sections. Screw long, self-tapping screws into the holes just enough for them to get a good grip in the metal. Now pulling on the protruding heads of the screws with locking pliers can pull out the dent.

23   The next stage of repair is the removal of paint from the damaged area and from an inch or so of the surrounding metal. This is easily done with a wire brush or sanding disk in a drill motor, although it can be done just as effectively by hand with sandpaper. To complete the preparation for filling, score the surface of the bare metal with a screwdriver or the tang of a file or drill small holes in the affected area. This will provide a good grip for the filler material. To complete the repair, see the Section on filling and painting.

Repair of rust holes or gashes

24   Remove all paint from the affected area and from an inch or so of the surrounding metal using a sanding disk or wire brush mounted in a drill motor. If these are not available, a few sheets of sandpaper will do

the job just as effectively.

25   With the paint removed, you will be able to determine the severity of the corrosion and decide whether to replace the whole panel, if possible, or repair the affected area. New body panels are not as expensive as most people think and it is often quicker to install a new panel than to repair large areas of rust.

26   Remove all trim pieces from the affected area except those which will act as a guide to the original shape of the damaged body, such as headlight shells, etc. Using metal snips or a hacksaw blade, remove all loose metal and any other metal that is badly affected by rust. Hammer the edges of the hole in to create a slight depression for the filler material.

27   Wire-brush the affected area to remove the powdery rust from the surface of the metal. If the back of the rusted area is accessible, treat it with rust inhibiting paint.

28   Before filling is done, block the hole in some way. This can be done with sheet metal riveted or screwed into place, or by stuffing the hole with wire mesh.

29   Once the hole is blocked off, the affected area can be filled and painted. See the following subsection on filling and painting.

Filling and painting

30   Many types of body fillers are available, but generally speaking, body repair kits which contain filler paste and a tube of resin hardener are best for this type of repair work. A wide, flexible plastic or nylon applicator will be necessary for imparting a smooth and contoured finish to the surface of the filler material. Mix up a small amount of filler on a clean piece of wood or cardboard (use the hardener sparingly). Follow the manufacturer's instructions on the package, otherwise the filler will set incorrectly.

31   Using the applicator, apply the filler paste to the prepared area. Draw the applicator across the surface of the filler to achieve the desired contour and to level the filler surface. As soon as a contour that approximates the original one is achieved, stop working the paste. If you continue, the paste will begin to stick to the applicator. Continue to add thin layers of paste at 20-minute intervals until the level of the filler is just above the surrounding metal.

32   Once the filler has hardened, the excess can be removed with a body file. From then on, progressively finer grades of sandpaper should be used, starting with a 180-grit paper and finishing with 600-grit wet-or-dry paper. Always wrap the sandpaper around a flat rubber or wooden block, otherwise the surface of the filler will not be completely flat. During the sanding of the filler surface, the wet-or-dry paper should be periodically rinsed in water. This will ensure that a very smooth finish is produced in the final stage.

33   At this point, the repair area should be surrounded by a ring of bare metal, which in turn should be encircled by the finely feathered edge of good paint. Rinse the repair

area with clean water until all of the dust produced by the sanding operation is gone.

34   Spray the entire area with a light coat of primer. This will reveal any imperfections in the surface of the filler. Repair the imperfections with fresh filler paste or glaze filler and once more smooth the surface with sandpaper. Repeat this spray-and-repair procedure until you are satisfied that the surface of the filler and the feathered edge of the paint are perfect. Rinse the area with clean water and allow it to dry completely.

35   The repair area is now ready for painting. Spray painting must be carried out in a warm, dry, windless and dust free atmosphere. These conditions can be created if you have access to a large indoor work area, but if you are forced to work in the open, you will have to pick the day very carefully. If you are working indoors, dousing the floor in the work area with water will help settle the dust that would otherwise be in the air. If the repair area is confined to one body panel, mask off the surrounding panels. This will help minimize the effects of a slight mismatch in paint color. Trim pieces such as chrome strips, door handles, etc., will also need to be masked off or removed. Use masking tape and several thickness of newspaper for the masking operations.

36   Before spraying, shake the paint can thoroughly, then spray a test area until the spray painting technique is mastered. Cover the repair area with a thick coat of primer. The thickness should be built up using several thin layers of primer rather than one thick one. Using 600-grit wet-or-dry sandpaper, rub down the surface of the primer until it is very smooth. While doing this, the work area should be thoroughly rinsed with water and the wet-or-dry sandpaper periodically rinsed as well. Allow the primer to dry before spraying additional coats.

37   Spray on the top coat, again building up the thickness by using several thin layers of paint. Begin spraying in the center of the repair area and then, using a circular motion, work out until the whole repair area and about two inches of the surrounding original paint is covered. Remove all masking material 10 to 15 minutes after spraying on the final coat of paint. Allow the new paint at least two weeks to harden, then use a very fine rubbing compound to blend the edges of the new paint into the existing paint. Finally, apply a coat of wax.

6   Body repair - major damage

1   Major damage must be repaired by an auto body shop specifically equipped to perform unibody repairs. These shops have the specialized equipment required to do the job properly.

2   If the damage is extensive, the body must be checked for proper alignment or the vehicle's handling characteristics may be adversely affected and other components may wear at an accelerated rate.

11

These photos illustrate a method of repairing simple dents. They are intended to supplement *Body repair - minor damage* in this Chapter and should not be used as the sole instructions for body repair on these vehicles.

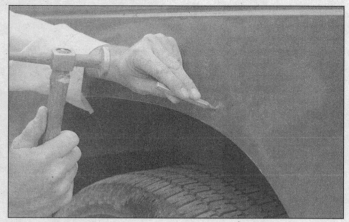

1   If you can't access the backside of the body panel to hammer out the dent, pull it out with a slide-hammer-type dent puller. In the deepest portion of the dent or along the crease line, drill or punch hole(s) at least one inch apart . . .

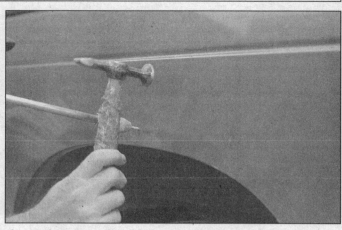

2   . . . then screw the slide-hammer into the hole and operate it. Tap with a hammer near the edge of the dent to help 'pop' the metal back to its original shape. When you're finished, the dent area should be close to its original contour and about 1/8-inch below the surface of the surrounding metal

3   Using coarse-grit sandpaper, remove the paint down to the bare metal. Hand sanding works fine, but the disc sander shown here makes the job faster. Use finer (about 320-grit) sandpaper to feather-edge the paint at least one inch around the dent area

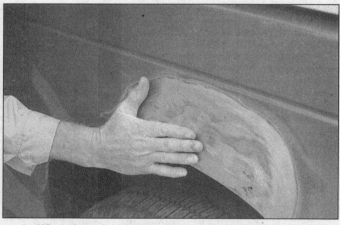

4   When the paint is removed, touch will probably be more helpful than sight for telling if the metal is straight. Hammer down the high spots or raise the low spots as necessary. Clean the repair area with wax/silicone remover

5   Following label instructions, mix up a batch of plastic filler and hardener. The ratio of filler to hardener is critical, and, if you mix it incorrectly, it will either not cure properly or cure too quickly (you won't have time to file and sand it into shape)

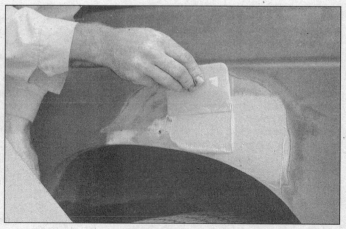

6   Working quickly so the filler doesn't harden, use a plastic applicator to press the body filler firmly into the metal, assuring it bonds completely. Work the filler until it matches the original contour and is slightly above the surrounding metal

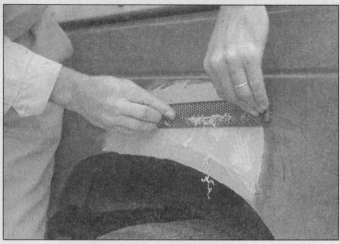

7   Let the filler harden until you can just dent it with your fingernail. Use a body file or Surform tool (shown here) to rough-shape the filler

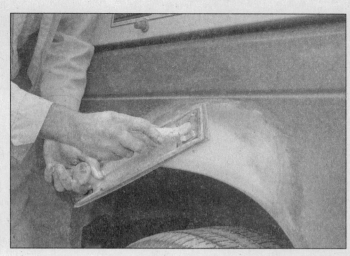

8   Use coarse-grit sandpaper and a sanding board or block to work the filler down until it's smooth and even. Work down to finer grits of sandpaper - always using a board or block - ending up with 360 or 400 grit

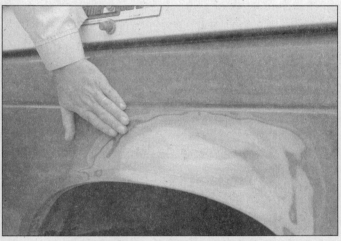

9   You shouldn't be able to feel any ridge at the transition from the filler to the bare metal or from the bare metal to the old paint. As soon as the repair is flat and uniform, remove the dust and mask off the adjacent panels or trim pieces

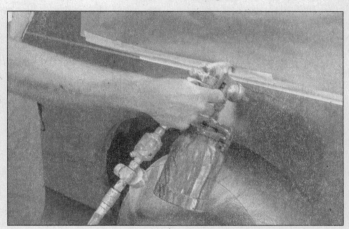

10   Apply several layers of primer to the area. Don't spray the primer on too heavy, so it sags or runs, and make sure each coat is dry before you spray on the next one. A professional-type spray gun is being used here, but aerosol spray primer is available inexpensively from auto parts stores

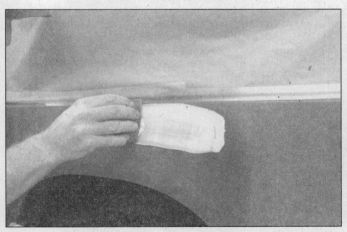

11   The primer will help reveal imperfections or scratches. Fill these with glazing compound. Follow the label instructions and sand it with 360 or 400-grit sandpaper until it's smooth. Repeat the glazing, sanding and respraying until the primer reveals a perfectly smooth surface

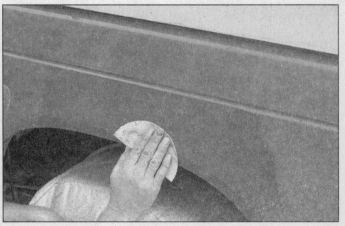

12   Finish sand the primer with very fine sandpaper (400 or 600-grit) to remove the primer overspray. Clean the area with water and allow it to dry. Use a tack rag to remove any dust, then apply the finish coat. Don't attempt to rub out or wax the repair area until the paint has dried completely (at least two weeks)

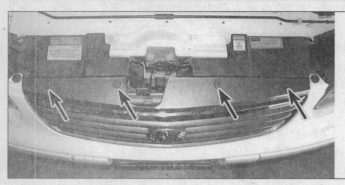

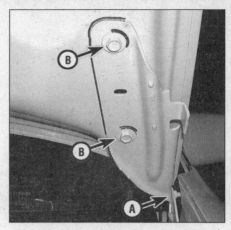

9.2 Remove the push-pins (arrows) and pull up the grille (Malibu model shown)

3   Due to the fact that all of the major body components (hood, fenders, etc.) are separate and replaceable units, any seriously damaged components should be replaced rather than repaired. Sometimes the components can be found in a wrecking yard that specializes in used vehicle components, often at considerable savings over the cost of new parts.

7   Hinges and locks - maintenance

Once every 3000 miles, or every three months, the hinges and latch assemblies on the doors, hood and trunk should be given a few drops of light oil or lock lubricant. The door latch strikers should also be lubricated with a thin coat of grease to reduce wear and ensure free movement. Lubricate the door and trunk locks with spray-on graphite lubricant.

8   Windshield and fixed glass - replacement

Replacement of the windshield and fixed glass requires the use of special fast-setting adhesive/caulk materials and some specialized tools and techniques. These operations should be left to a dealer service department or a shop specializing in glass work.

9   Radiator grille - removal and installation

Refer to illustration 9.2

1   The size and shape of the grille varies with the model, but the removal procedure is basically the same for all models that have a grille.

2   On Malibu models, open the hood and remove the plastic push-pins above the grille, then pull the grille straight up to release the tabs on the bottom **(see illustration)**.

3   On Cutlass and Grand Am models, there are two grille halves, each retained by a tab at one end and a screw at the other end. Remove the screw and pull each grille half out until the tab at the other end clears the fascia. **Note:** *On some models, there may also be clips on the back of the grille, accessed by removing the plastic splash*

shield above the front bumper fascia.

4   To install, place the grille in position and seat the tabs into the holes in the fascia.

5   The remainder of the installation is the reverse of removal.

10   Hood - removal, installation and adjustment

Removal and installation

Refer to illustration 10.2

Note: *The hood is somewhat awkward to remove and install; at least two people should perform this procedure.*

1   Open the hood, then place blankets or pads over the fenders and cowl area of the body. This will protect the body and paint as the hood is lifted off.

2   Make marks or scribe a line around the hood hinge to ensure proper alignment during installation **(see illustration)**.

3   Disconnect any cables or wires that will interfere with removal.

4   With an assistant supporting the weight of the hood, remove the hinge-to-body bolts and lift off the hood. **Note:** *Unless the hood or hinges are to be replaced, only remove the one bolt at the rear of each hinge, not the hinge-to-hood bolts. Hood realignment will be much easier because the original hinge-to-hood relationship is maintained.*

5   Installation is the reverse of removal. Align the hinge bolts with the marks made in Step 2.

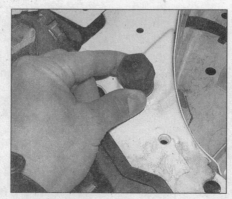

10.10 Screw the hood bumpers in or out to adjust the hood flush with the fenders

10.2 Mark the relationship of the hinges to the hood as shown, then, with the help of an assistant to hold the hood, remove the rear retaining bolts (A) from each hinge plate and lift off the hood/hinges - only remove the hinge-to-hood bolts (B) if the hood or hinges are to be replaced

Adjustment

Refer to illustrations 10.10 and 10.11

6   Fore-and-aft and side-to-side adjustment of the hood is done by moving the hinge plate slot after loosening the bolts or nuts.

7   Scribe a line around the entire hinge plate so you can determine the amount of movement **(see illustration 9.2)**.

8   Loosen the bolts or nuts and move the hood into correct alignment. Move it only a little at a time. Tighten the hinge bolts and carefully lower the hood to check the position.

9   If necessary after installation, the entire hood latch assembly can be adjusted up-and-down as well as from side-to-side on the radiator support so the hood closes securely and flush with the fenders.

10   Adjust the hood bumpers on the radiator support so the hood is flush with the fenders when closed **(see illustration)**.

11   The hood latch striker assembly can also be adjusted up-and-down and side-to-

10.11 Loosen the bolts (arrows) to adjust the position of the hood latch striker

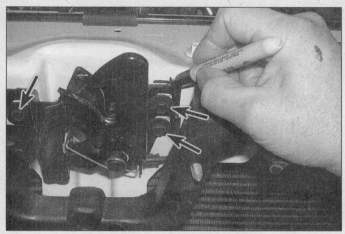

11.1 Mark around the edges of the latch assembly before removing the latch bolts (arrows)

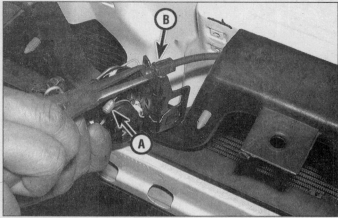

11.2 To disconnect the cable from the hood latch mechanism, pry the cable ferrule (A) out of the latch assembly and use pliers to disengage the cable housing end (B) from its slot in the latch

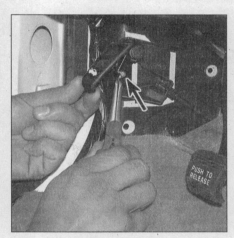

11.6 Use pliers to disconnect the hood cable end (arrow) from the release handle

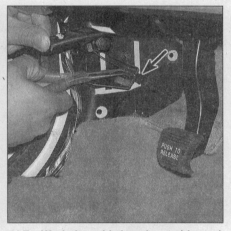

11.7a Work the cable housing end (arrow) out of the forward part of the handle frame

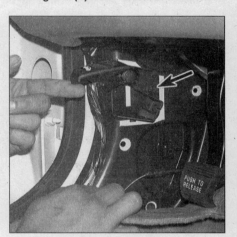

11.7b To detach the hood latch release cable handle, remove this retaining screw (arrow)

side after loosening the nuts **(see illustration)**. See Section 11 if the hood latch must be adjusted.

12 If the rear of the hood is too low, insert shims or washers of the correct thickness between the hood and the hinges.

13 The hood latch assembly, as well as the hinges, should be periodically lubricated with white lithium-base grease to prevent sticking and wear.

11 Hood latch and release cable - removal and installation

Hood latch

Refer to illustrations 11.1 and 11.2

1 Scribe a line around the latch to aid alignment when installing, then remove the retaining bolts to the radiator support **(see illustration)**. Remove the latch.

2 Disconnect the hood release cable by disengaging the cable from the latch assembly **(see illustration)**.

3 Installation is the reverse of removal. Adjust the latch so the hood engages securely when closed and the hood bumpers are slightly compressed (see Section 10).

Release cable

Refer to illustrations 11.6, 11.7a and 11.7b

4 Disconnect the release cable from the hood latch assembly as described in Step 2.

5 Unclip the release cable from the engine wiring harness. Attach a piece of wire to the cable.

6 Working in the passenger compartment, remove the kick panel to expose the hood latch release cable and handle **(see illustration)**.

7 Use pliers to pull the cable-housing end from the tab on the handle frame **(see illustration)**. **Note:** *If the handle assembly itself must be replaced, remove the one mounting bolt and remove the handle from the cowl* **(see illustration)**.

8 Trace the cable forward to the grommet where the cable goes through the firewall and pry the grommet out of the firewall. Pull the

handle and cable rearward into the passenger compartment.

9 Disconnect the guide wire from the old cable and fasten it to the new cable.

10 With the new cable attached to the wire, pull the wire back through the firewall until the new cable reaches the latch assembly. Make sure that the grommet is properly seated on both sides of the hole in the firewall. Push on the grommet with your fingers from the passenger compartment side to seat the grommet in the firewall correctly.

11 The remainder of installation is the reverse of removal.

12 Bumpers - removal and installation

Front bumper

Refer to illustrations 12.2a, 12.2b, 12.3, 12.4, 12.6 and 12.7

1 Raise the front of the vehicle and support it securely on jackstands. See Section 9

11

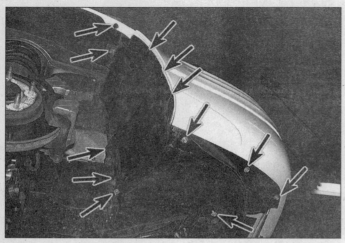

12.2a To detach the forward ends of the wheelwell liner, remove these fasteners (arrows) - some are pushpins and some are screws

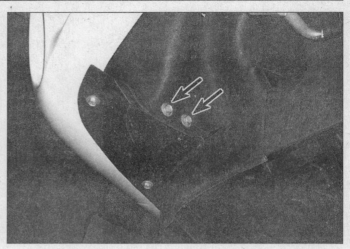

12.2b It will help to also remove these two bolts on each side (arrows), so the liner can be bent down for access to the fascia-to-fender fasteners

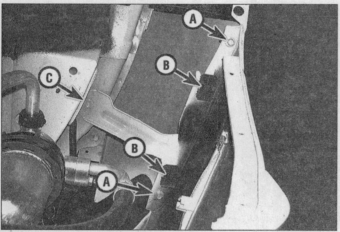

12.3 To detach the front bumper fascia from the fenders, remove the two bolts (A) and two plastic fasteners (B) from each fender (Malibu/Cutlass model shown; Grand Am and Alero models have four bolts) - C is the fender brace bolt

12.4 Remove the two bolts (arrows) from below the front edge of the fascia

12.6 Remove the two fasteners (arrow indicates the left side fastener) in the grille area and lift off the fascia

12.7 To detach the front impact bar, remove the nuts (arrows) - the two at the left here attach to a large U-bolt

and remove the grille.

2   Disconnect the fasteners retaining the splash apron below the radiator, which at each end forms part of the front of each wheelwell (see illustrations).

3   Detach the fascia from the fenders (see illustration).

4   Remove the two bolts from the bottom edge of the fascia (see illustration).

5   Before removing the bumper fascia, unplug the electrical connectors for the fog lights, if equipped (see Chapter 12).

6   Remove the two upper bolts in the grille area and remove the fascia (see illustration).

7   To remove the front bumper impact bar, simply remove the impact bar retaining nuts (see illustration).

8   Installation is the reverse of removal.

Rear bumper

Refer to illustrations 12.11a, 12.11b, 12.12, 12.13 and 12.15

9   Raise the rear of the vehicle and place it

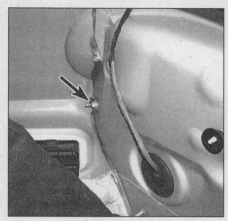

12.11a Inside the rear compartment, remove the one fascia nut (arrow indicates the right side nut) in the left and right taillight area

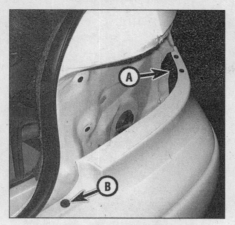

12.11b The nut shown in illustration 12.11a retains this bracket (A) between the body and the rear fascia - the bracket can stay with the fascia - B is the right-side fastener in the trunk lid opening, remove both left and right

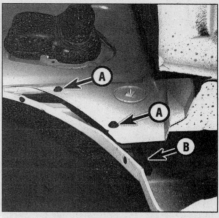

12.12 There are two rear bumper fascia retaining bolts (A) inside each rear wheelwell - the plastic bracket (B) stays with the fascia unless the fascia is to be replaced

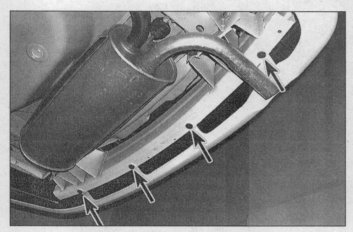

12.13 Remove the push-in retainers (arrows) from the underside of the rear bumper fascia, then remove the fascia

12.15 To detach the rear impact bar, remove the nuts (arrows)

securely on jackstands.

10   Refer to Chapter 12 and remove both taillight housings.

11   From inside the trunk compartment, remove the fascia nuts **(see illustrations)**.

12   In each rear wheelwell, remove the fascia bolts **(see illustration)**. **Note:** *Grand Am and Alero models will have three bolts in each wheelwell.*

13   Remove the push-in retainers along the bottom edge of the fascia **(see illustration)**.

14   Remove the rear bumper fascia.

15   To remove the rear bumper impact bar, remove the retaining nuts **(see illustration)**.

16   Installation is the reverse of removal.

13 Front fender - removal and installation

Refer to illustrations 13.4, 13.5a, 13.5b, 13.6, 13.7 and 13.8

1   Remove the hood and the hood hinge bracket from the rear of the fender (see Sec-

tion 9).

2   Loosen the front wheel lug nuts. Raise the vehicle, support it securely on jackstands and remove the front wheels.

3   Refer to Section 12 and remove the

front bumper fascia-to-fender bolts and the fender to body brace bolt **(see illustration 12.3)**.

4   Remove the main portion of the fenderwell liner **(see illustration)**.

13.4 Remove the fasteners (arrows) and the rear portion of the fenderwell liner

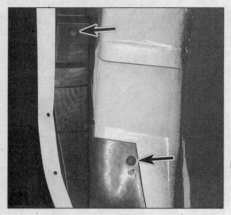

13.5a Remove the two pushpins (arrows) and this plastic insert . . .

13.5b . . . and you have access to the fender-to-body bolt (arrow) - Malibu model shown, Grand Am/Alero models have the same bolt, but lower on the body

13.6 Remove the two lower bolts (arrows) securing the fender to the rocker panel area (Malibu shown, Grand Am/Alero models have vertical bolts)

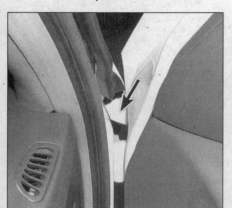

13.7 Open the door and remove the one fender bolt (arrow) at the top on Grand Am and Alero models only

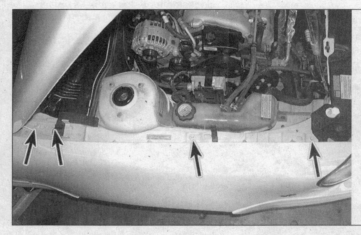

13.8 Remove the four bolts (arrows) along the upper edge of the fender

5   At the rear of the fenderwell, remove the plastic pushpins and the plastic insert, then access and remove the rear fender-to-body bolt **(see illustrations)**.

6   From below the vehicle, at the front rocker panel area, remove the two lower fender-to-body bolts **(see illustration)**.

7   On Grand Am and Alero models, open the front door and remove the one fender bolt

in the top of the door hinge area **(see illustration)**.

8   Remove the mounting bolts along the upper edge of the fender and remove the fender **(see illustration)**. **Caution:** *It will be helpful to first apply wide masking tape along the rear edge of the fender and the front edge of the door, to avoid scratching the paint during fender removal.*

9   Installation is the reverse of removal.

14   Cowl cover - removal and installation

Cowl cover

Refer to illustrations 14.2 and 14.3

1   Open the hood, then refer to Chapter 12 and remove the wiper arms, disconnecting the washer hoses (attached to each wiper

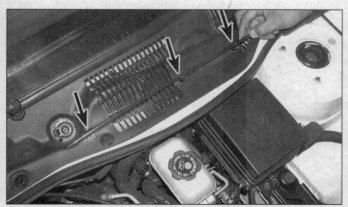

14.2 Release the washer tube from the clips (arrows) on the cowl cover

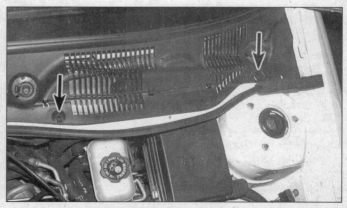

14.3 Remove the plastic push-pins (arrows indicate the two on the left side) and lift the cowl cover off

15.2 Pry the switch assembly out of the door panel and disconnect the electrical connectors

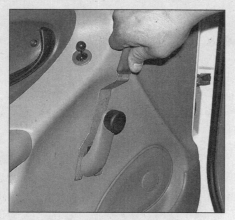

15.3 Manual window handles can be removed with an inexpensive tool, or work a clean rag between the handle and the door panel and pull on the rag to pop the retaining clip loose, then remove the handle

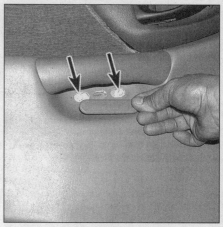

15.4a Pry out the plastic cover and remove the two door panel mounting screws (arrows, Malibu/Cutlass models)

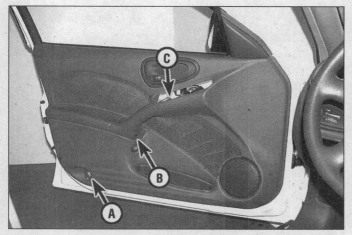

15.4b On Grand Am/Alero models, remove the reflector (A) and the screw behind it, then the screw (B) at the bottom of the armrest, and the screw in the window switch opening (C)

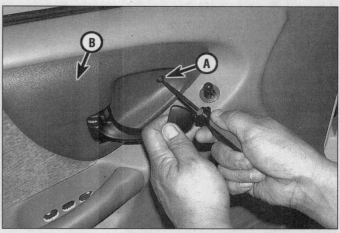

15.5 Remove the inside handle bezel screw (A), then pull the bezel (B) away from the door panel

arm) from the washer tube.

2   Snap the washer tube out of the clips on the cowl cover and swing it up and away from the cowl cover **(see illustration)**.

3   Remove the plastic push-pins securing the cowl cover to the body **(see illustration)**.

4   Remove the cowl cover.

5   Installation is the reverse of removal.

15 Door trim panels - removal and installation

Refer to illustrations 15.2, 15.3, 15.4a, 15.4b, 15.5, 15.6, 15.7a, 15.7b and 15.8

1   On models with power door locks and/or power windows, disconnect the cable from the negative terminal of the battery. **Caution:** *On models equipped with the Theft-lock audio system, be sure the lockout feature is turned off before performing any procedure which requires disconnecting the battery (see*

the front of this manual).

2   On models with power windows/door locks, pry the switch out of the door trim panel with a flat-blade screwdriver, unplug the electrical connector from the switch and remove the switch **(see illustration)**. **Note:** *On models with the trunk release button on the door panel, pry out the trim bezel around the release button.*

3   On models with manually-operated windows, remove the window regulator handle **(see illustration)**.

4   Behind the inside door pull, remove the plastic covers and remove the two panel-mounting screws **(see illustrations)**. **Note:** *The number and location of the screws varies slightly with model.*

5   Pull the inside door handle to the open position and remove the (Torx) handle bezel screw **(see illustration)**. Lift the bezel upward and remove it from the trim panel.

6   Remove the mirror trim panel **(see illustration)**.

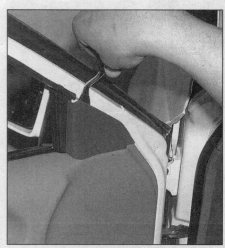

15.6 Remove the mirror cover by squeezing the bottom corners together, or pry at the top with a trim tool

11

7   Once all of the screws are removed, detach the trim panel from the door, working around the door with a plastic trim panel tool to disengage the clips. Disconnect any electrical connectors and remove the trim panel from the vehicle by lifting it up and away from the door **(see illustrations)**.

8 . If you're planning to repair or replace anything inside the door itself, you'll have to remove the water deflector between the trim panel and the door **(see illustration)**. Peel it off carefully, so that it can be reused.

9   After you're done servicing the door component(s), reattach the deflector to the door. Use extra sealant if necessary. Make sure that the deflector is securely attached to the door all the way around its perimeter.

10   Installation is otherwise the reverse of removal.

16   Door - removal, installation and adjustment

Removal and installation

Refer to illustrations 16.3, 16.5 and 16.7

Note: *The door is heavy and somewhat awkward to remove and install - at least two people should perform this procedure. This procedure applies to both front and rear doors.*

1   Raise the window completely and disconnect the negative cable from the battery if equipped with power windows. **Caution:** *On models equipped with the Theftlock audio system, be sure the lockout feature is turned off before performing any procedure which requires disconnecting the battery (see the front of this manual).*

2   Open the door all the way and support it on jacks or blocks covered with rags to prevent damaging the paint.

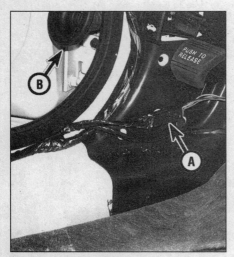

16.3 On Malibu and Cutlass models, pry up the plastic kick panel, pull back the carpeting, and disconnect the door wiring harness connector (A) and feed the wiring out through the grommet (B)

15.7a When all of the perimeter clips have been pried out using a trim tool (plastic wedge-shaped type being used here), lift up and away on the door panel

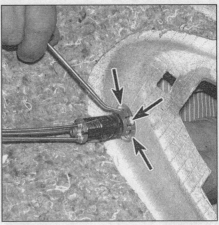

15.7b With the door panel pulled away from the door, disconnect any electrical connectors and pry the clips (arrows) inward to release the manual mirror control from the door panel

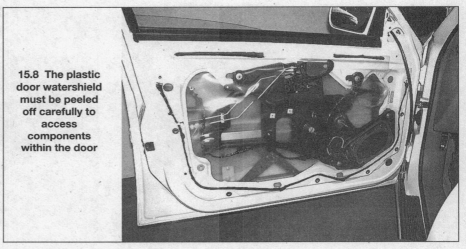

15.8 The plastic door watershield must be peeled off carefully to access components within the door

3   Remove the door trim panel and enough of the water deflector to access all electrical wiring harnesses (see Section 15). On Malibu and Cutlass models, pull back the rubber

16.5 Remove the bolt (arrow) in the door check strap

wire harness conduit and disconnect the electrical connector at the door, feeding it back out of the body and into the door **(see illustration)**.

4   On Grand Am and Alero models, unplug all electrical connectors, and detach all ground wires and harness retaining clips from the door. It's a good idea to label all connectors to aid the reassembly process. Working on the door side, detach the rubber conduit between the body and the door. Pull the wiring harness through the conduit hole and remove the wiring from the door, leaving the harness with the body.

5   Remove the bolt/pin in the door check strap **(see illustration)**.

6   Mark around the door hinges with a pen or a scribe to facilitate realignment during reassembly. **Note:** *It's a good idea at this point to apply a strip of wide masking tape along the edge of the fender or quarter-panel to protect the paint during door removal.*

7   Have an assistant hold the door, remove the hinge-to-door bolts **(see illustration)** and lift the door off.

8   Installation is the reverse of removal.

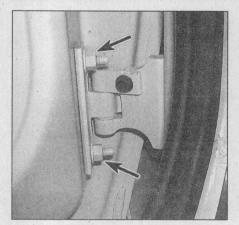

16.7 Door hinge nut locations (arrows indicate nuts on lower hinge, upper hinge similar)

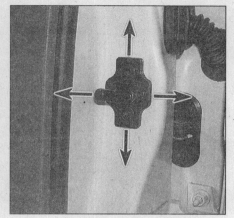

16.11 To adjust the door lock striker, loosen the Torx bolts (arrows) and move the striker slightly, then retighten and check the door alignment again

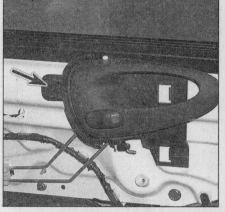

17.3 Remove the inside door handle mounting bolt (arrow), then slide the handle forward to release it from two clips in the door

Adjustment

Refer to illustration 16.11

9   Having proper door to body alignment is a critical part of a well-functioning door assembly. First check the door hinge pins and bushings for excessive play. If the door can be lifted (1/16-inch or more) without the car body lifting with it, the hinge pins and bushings should be replaced.

10   Door-to-body alignment adjustments are made by loosening the hinge-to-body bolts or hinge-to-door nuts and moving the door. Proper body alignment is achieved when the top of the doors are parallel with the roof section, the front door is flush with the fender, the rear door is flush with the rear quarter panel and the bottom of the doors are aligned with the lower rocker panel. If these goals can't be reached by adjusting the hinge-to-body or hinge-to-door bolts, body alignment shims may have to be purchased and inserted behind the hinges to achieve correct alignment.

11   To adjust the door-closed position, verify that the door latch is contacting the center of the striker. If it isn't, loosen the striker bolts **(see illustration)** and adjust the striker as necessary (up, down or sideways) to provide

optimal engagement with the latch mechanism. Tighten the striker bolts securely once the striker is adjusted.

17   Door latch, lock cylinder and handles - removal and installation

1   Remove the door trim panel (see Section 15) and peel away the water deflector in the vicinity of the component you're planning to remove. Disconnect the negative cable from the battery if equipped with power windows, mirrors or locks. **Caution:** *On models equipped with the Theftlock audio system, be sure the lockout feature is turned off before performing any procedure which requires disconnecting the battery (see the front of this manual).*

Inside handle

Refer to illustration 17.3

2   Disengage the actuating rods from the backside of the handle, making note of which rod attaches to which connector on the handle assembly.

3   Remove the bolts securing the handle assembly to the door **(see illustration)**.

4   Installation is the reverse of removal.

Outside handle

Refer to illustration 17.6

5   Refer to Section 15 to remove the door trim panel, then peel back the watershield in the area of the handle.

6   Disconnect the lock actuator rod from the door handle assembly, then remove the two mounting screws (inside the door), and remove the handle from the outside of the door **(see illustration)**.

7   Installation is the reverse of removal.

Lock cylinder

Refer to illustration 17.8

8   Pry off the lock cylinder retainer clip that secures the lock cylinder to the door **(see illustration)**.

9   Disengage the actuator rod from the lock cylinder lever and remove the lock cylinder **(see illustration 17.8)**.

10   Installation is the reverse of removal.

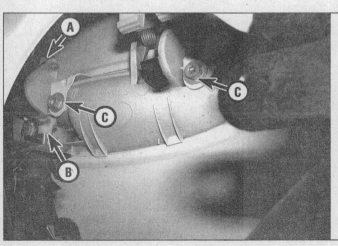

17.6 From inside the door, disconnect the handle rod (A) and the lock rod (B) at the outside door handle, then remove the handle mounting bolts (C)

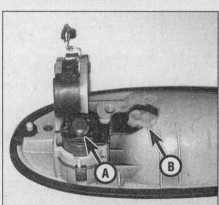

17.8 Pry off the lock cylinder retaining clip (A) and slip the cylinder out of the back of the handle assembly, then disconnect the rod (B) and remove the lever (shown removed from the door for clarity)

11

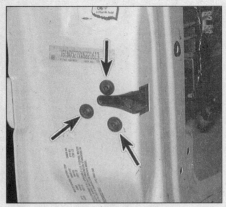

17.11 Remove the latch-retaining Torx screws (arrows)

18.2 Glass-to-channel retaining nuts (arrows) - Malibu/Cutlass models

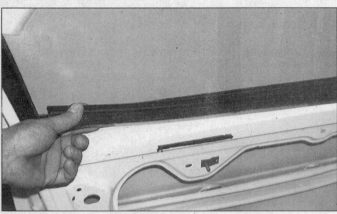

18.3 Pull up the interior glass sealing strip

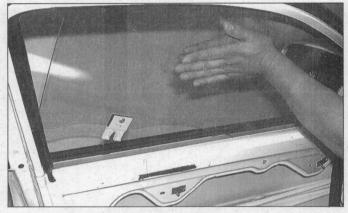

18.4 Carefully tilt the glass up and inward to remove it

Latch

Refer to illustration 17.11

11   Remove the screws securing the latch to the door **(see illustration)**.

12   Working through the large access hole, position the latch as necessary to disengage the actuator rods from the outside door handle, lock cylinder and the inside handle. Note how these rods are connected to the latch assembly. They must be installed exactly the same way during reassembly. Disconnect the electrical connectors and remove the latch

assembly. On Grand Am and Alero models, the rear glass run channel may need to be unbolted and moved to allow removal of the latch (see Section 19).

13   On Malibu and Cutlass models, the door latch is part of a "modular" door assembly that includes the speaker, latch, power window motor and regulator, all attached to a large plastic frame. After removing the screws from the latch **(see illustration 17.11)**, remove the whole modular assembly and the latch will come out with it, with the

rods still attached (see Section 19).

14   Installation is the reverse of removal.

18   Door window glass - removal and installation

Malibu and Cutlass models

Refer to illustrations 18.2, 18.3 and 18.4

1   Remove the door trim panel and the plastic water deflector (see Section 15).

2   Lower the window glass down enough to remove the window-to-channel nuts **(see illustration)**.

3   Remove the rubber sealing strip along the inside edge of the glass opening at the top of the door **(see illustration)**.

4   Remove the window from the door by pulling it up and out **(see illustration)**.

5   Installation is the reverse of removal.

Grand Am and Alero models

Refer to illustration 18.8

6   Remove the door trim panel and the plastic water deflector (see Section 15).

7   Lower the window glass enough to access the glass channel bolts.

8   Remove the regulator-to-window retaining bolts **(see illustration)**, then lower the glass (via the regulator, with temporary power

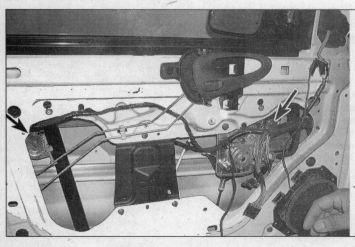

18.8 Remove the glass-to-channel bolts (arrows) - Grand Am/Alero models

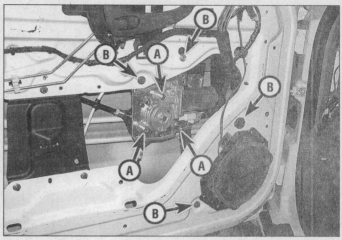

19.5a On Malibu/Cutlass models, the regulator/motor are part of a large inner door "modular" assembly - remove the nuts (A), bolts (B) and screw (C) to remove the assembly

19.5b On Grand Am and Alero models, remove the regulator bolts and rivets (B), then remove the motor/regulator assembly though the hole in the door. To replace the motor, remove the three bolts (A)

applied to the regulator motor, if a power window).

9   Remove the rubber sealing strip along the inside edge of the glass opening at the top of the door **(see illustration 18.3)**.

10   Remove the glass from the door.

11   Installation is the reverse of removal.

***

19   Door window glass regulator - removal and installation

Refer to illustrations 19.5a and 19.5b

1   On models with power windows, detach the negative battery cable. **Caution:** *On models equipped with the Theftlock audio system, be sure the lockout feature is turned off before performing any procedure which requires disconnecting the battery (see the front of this manual).*

2   Remove the door trim panel and the plastic water deflector (see Section 15).

3   Remove the window glass assembly (see Section 18).

4   On models with power windows, unplug the electrical connector from the window regulator motor.

5   Remove the regulator assembly **(see illustrations)**. On Malibu/Cutlass models, the door latch must be unbolted **(see illustration 17.11)**, as the latch comes out with the modular assembly.

6   Installation is the reverse of removal.

***

20   Mirrors - removal and installation

Outside mirrors

Refer to illustration 20.3

1   Remove the mirror trim panel (see Section 15).

2   On models with power mirrors, unplug the electrical connector from the power mir-

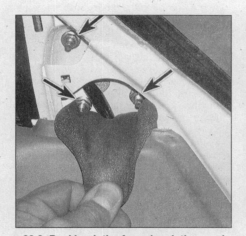

20.3 Peel back the foam insulation, and remove the mirror mounting nuts (arrows)

ror switch. **Note:** *On some models, the mirror trim panel is a housing for a small speaker. Remove the screw from the housing to remove the speaker, then disconnect the speaker electrical connector.*

3   Remove the three mirror retaining nuts **(see illustration)** and detach the mirror from the vehicle.

4   Installation is the reverse of removal.

Inside mirror

Refer to illustration 20.6

5   On models equipped with a rearview mirror reading lamp, disconnect the electrical connector at the windshield side of the mirror.

6   Loosen the setscrew at the base of the mirror stalk and slide the mirror up and off the support base on the windshield **(see illustration)**.

7   Installation is the reverse of removal.

8   If the support base for the mirror has come off the windshield, it can be reattached

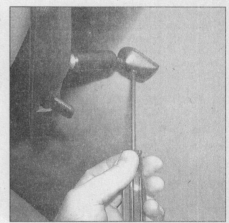

20.6 Disconnect the electrical connector (if equipped), then remove the setscrew

with a special mirror-adhesive kit available at auto parts stores. Clean the glass and support base thoroughly and follow the directions on the adhesive package, allowing the base to bond overnight before attaching the mirror.

***

21   Trunk lid - removal, installation and adjustment

Note: *The trunk lid is heavy and somewhat awkward to remove and install - at least two people should perform this procedure.*

Removal and installation

Refer to illustration 21.3

1   Open the trunk lid and cover the edges of the trunk compartment with pads or cloths to protect the painted surfaces when the lid is removed.

2   Disconnect any cables or wire harness

11

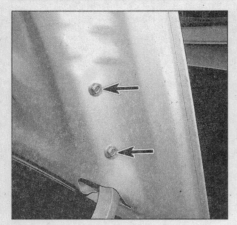

21.3 Before removing the four (two per side, right side shown) trunk lid hinge bolts, be sure to mark the relationship of the hinge bolts to the trunk lid to ensure correct alignment when the lid is installed

21.12 To readjust the tension on the trunk lid hinges, lever the ends of the torque rods up or down, one notch at a time, with a wrench as shown (raising the torque rods a notch makes it easier to raise the lid, and harder to lower it; moving the rods down a notch makes it harder to raise the lid, and easier to lower it)

21.15 Trunk latch striker - mounting nuts are on the back under the bumper fascia

connectors attached to the trunk lid that would interfere with removal, including the taillight connectors. Attach a long wire or string to the connectors and feed the harness down through the trunk lid until it comes out at the bottom-left corner of the trunk lid. Cut the guide string or wire with six inches hanging out.

3   Use a felt-tip marker or scribe to make alignment marks around the trunk lid hinge bolts **(see illustration)**.

4   While an assistant supports the lid, remove the hinge bolts from both sides and lift the trunk lid off the vehicle.

5   Installation is the reverse of removal. Tie the end of your guide string to the harness and feed it back through the trunk lid until it can be connected to the taillights and other components. **Note:** *When reinstalling the trunk lid, align the hinge with the marks made during removal.*

Adjustment

Refer to illustrations 21.12 and 21.15

6   Fore-and-aft and side-to-side adjustment of the trunk lid is done by moving the trunk lid in relation to the hinge plates after loosening the bolts or nuts.

7   Scribe a line around the entire hinge plate as described earlier in this section so you can judge the amount of movement.

8   Loosen the bolts or nuts and move the trunk lid into correct alignment. Move it only a little at a time. Tighten the hinge bolts or nuts and carefully lower the trunk lid to check the alignment.

9   If necessary after installation, the entire trunk lid latch assembly can be adjusted up and down as well as from side to side on the trunk lid so the lid closes securely and is flush with the rear quarter panels. To do this, scribe a line around the trunk lid latch mounting bolts to provide a reference point. Then loosen the bolts and reposition the latch assembly as necessary (see Section 22). Following adjustment, retighten the mounting bolts.

10   Adjust the bumpers on the trunk lid, so that the trunk lid is flush with the rear quarter panels when closed.

11   The trunk lid latch assembly , as well as the hinges, should be periodically lubricated with white lithium-based grease to prevent sticking and wear.

12   The effort required to raise or lower the trunk lid can also be adjusted by moving the ends of a pair of torque rods installed between the two trunk hinges **(see illustration)**. Each rod can be adjusted up or down in one of three positions. To increase the effort required to raise the trunk lid, or to decrease the effort to lower the lid, relocate the rods down a notch. To decrease the effort to raise the lid, or to increase the effort to lower the lid, raise the rods a notch. To move a rod up or down, use a box-end wrench or a short length of thick-wall tubing to lever the end of the rod.

13   To adjust the up/down closed position of the trunk lid, the latch striker can be adjusted.

14   Refer to Section 12 and remove the rear bumper fascia.

15   Loosen the two nuts at the striker plate, reposition the striker slightly, tighten the nuts and try the trunk lid fit again **(see illustration)**. When the fit is correct, fully tighten the striker mounting nuts, then reinstall the rear bumper fascia.

22   Trunk lid latch and lock cylinder - removal and installation

Latch

Refer to illustrations 22.1 and 22.2

1   The trunk latch cable is permanently attached at the latch end, and must be disconnected from the lock cylinder end **(see illustration)**.

2   Mark the relationship of the latch to the trunk lid and disconnect the electrical connector at the trunk latch **(see illustration)**.

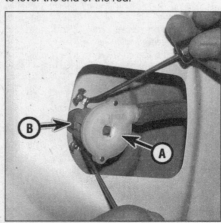

22.1 Use two hook tools to spread the two plastic tabs while pulling on the lock cylinder cable end (A) until it releases from the back of the lock cylinder (B)

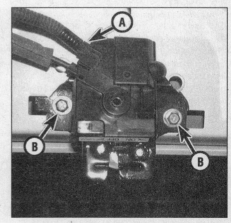

22.2 Mark the relationship of the trunk lid latch to the trunk lid, disconnect the electrical connector (A) and remove the bolts (B)

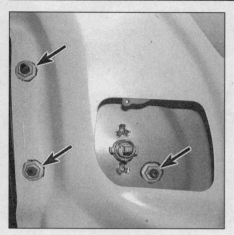

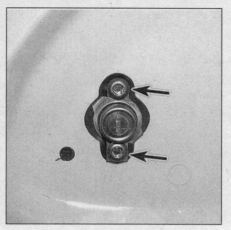

22.6 From the trunk side of the lid, remove the nuts (arrows) that retain the right-side exterior trunk lid trim (Malibu models shown, Grand Am models have a full-width trim panel with more nuts)

22.7 From the exterior of the trunk lid, drill out the rivets (arrows) retaining the lock cylinder

23.3 Pry the gear position indicator trim panel straight up to disengage the retaining clips (Malibu shown, arrow indicates general location of front console screws on Grand Am/Alero models)

3   Remove the latch retaining bolts and remove the latch.

4   Installation is the reverse of removal. Be sure to align the latch carefully with the marks you made prior to removal. Align the square hole in the lock-end of the cable to the square post on the back of the lock and snap the cable end in place.

Lock cylinder

Refer to illustrations 22.6 and 22.7

5   Disconnect the lock cylinder-to-latch cable end from the lock cylinder **(see illustration 22.1)**.

6   Remove the nuts from inside the trunk lid and remove the exterior trunk lid trim panel. On Malibu/Cutlass/Alero models, there are two taillight-like reflectors, one on each side of the trunk lid. Remove only the right-side one for access to the lock cylinder **(see illustration)**. On Grand Am models, there is one trim panel that goes across the whole

end of the trunk lid.

7   From the exterior of the trunk lid, drill out the two rivets that retain the lock cylinder to the trunk lid **(see illustration)**.

8   Installation is the reverse of removal. Don't forget the gasket between the lock cylinder and the trunk lid on Chevrolet models. Install the new lock cylinder with new rivets, if available, or use bolts and self-locking nuts.

23 Center console - removal and installation

Refer to illustrations 23.3, 23.4a, 23.4b and 23.4c

Warning: *The models covered by this manual are equipped with Supplemental Restraint Systems (SRS), more commonly known as airbags. Always disable the airbag system before working in the vicinity of any airbag system components to avoid the possibility of accidental deployment of the airbags, which*

could cause personal injury (see Chapter 12).

1   Disconnect the negative cable from the battery. **Caution:** *On models equipped with the Theftlock audio system, be sure the lock-out feature is turned off before performing any procedure which requires disconnecting the battery (see the front of this manual).*

2   Apply the parking brake lever and place the shift lever in the Neutral position. Refer to Section 27 and remove the front seats.

3   Pry out and remove the gear shift trim bezel **(see illustration)**. By rotating the bezel, it can be removed without taking off the shift knob.

4   Remove the console retaining screws, raise up the console **(see illustrations)**, unplug any electrical connectors and remove the console from the vehicle. On Grand Am and Alero models, the two rear console screws are at the bottom of the armrest/storage compartment. Remove the rubber mat at the bottom for access to the screws.

5   Installation is the reverse of removal.

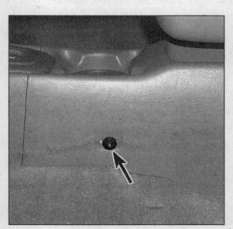

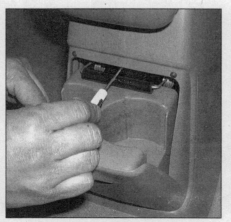

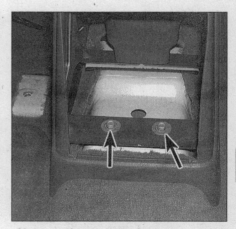

23.4a On either side of the console, remove the side screws (arrow) - seats must be removed for access to these two screws

23.4b On Malibu/Cutlass models, pull back the rear cupholder and detach it by depressing the spring clip with a screwdriver . . .

23.4c . . . then remove these two screws (arrows, Malibu shown)

11

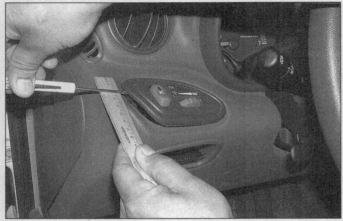

24.3 To detach the dimmer control switch, pry out with flat-bladed tool, then disconnect the electrical connector at the back (Grand Am model shown)

24.7 To detach the instrument cluster bezel from the instrument panel, remove the screws (arrows, Malibu shown, others similar, steering wheel removed for clarity)

24.9 Pry all around the cluster bezel (Malibu model shown) to release the clips from the instrument panel

24.10 Disconnect the electrical connectors (arrows, dimmer at left on this Malibu, hazard switch at right)

24   Dashboard trim panels - removal and installation

Warning: *The models covered by this manual are equipped with Supplemental Restraint Systems (SRS), more commonly known as airbags. Always disable the airbag system before working in the vicinity of any airbag system components to avoid the possibility of accidental deployment of the airbags, which could cause personal injury (see Chapter 12).*

1   Disconnect the negative battery cable.
Caution: *On models equipped with the Theft-lock audio system, be sure the lockout feature is turned off before performing any procedure which requires disconnecting the battery (see the front of this manual).*

2   Disable the airbag system (see Chapter 12).

Dash light dimmer switch bezel

Refer to illustration 24.3

3   At the left end of the instrument panel, use a flat-bladed tool to pry out the driver's-side switch bezel **(see illustration)**. **Note:** *On some models, this grille and the dimmer switch are part of the instrument cluster bezel.*

4   Pull the grille out far enough to disconnect the electrical connector from the instrument panel light dimmer control.

5   Installation is the reverse of the removal procedure.

Instrument cluster bezel

Refer to illustrations 24.7, 24.9 and 24.10

6   Refer to the Steps below and remove the center instrument panel trim bezel (Oldsmobile models).

7   Remove the screws securing the bezel at the top of the cluster **(see illustration)**. **Note:** *On Grand Am and Alero models, remove the steering column covers (see Section 25).*

8   Use a trim removal tool or a flat-blade screwdriver with the tip taped to pry around the complete edge of the instrument cluster bezel.

9   Grasp the bezel securely and pull out to detach the retaining clips from the instrument panel **(see illustration)**. **Note:** *On some models, the cluster bezel surrounds the cluster only, while on other models the bezel extends to the right across most of the instrument panel.*

10   Pull the panel out enough to disconnect the electrical connector from the instrument lighting dimmer switch and (on some models) the hazard warning flasher switch **(see illustration)**.

11   Installation is the reverse of removal.

Driver's knee bolster

Refer to illustration 24.12

12   The driver's knee bolster is a two-piece cover, part of which is actually a heating duct, under the instrument panel. Remove the plastic push-pins by popping out the center with a small screwdriver, then removing the pushpin **(see illustration)**. With the push-pins and screws removed, slide the two cover halves from any studs or clips and remove from under the dash.

13   Installation is the reverse of removal.

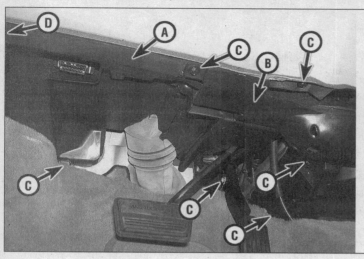

24.12 The panels under the driver's side include the knee bolster (A) and the column shaft cover (B) - fasteners are hex-head screws (C) and push-pins (D)

24.15 Pry around the center instrument panel bezel with a flat-bladed trim tool - pry off the ignition switch bezel (arrow) before removing the center bezel

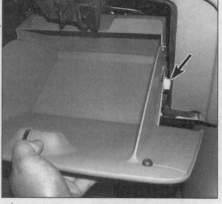

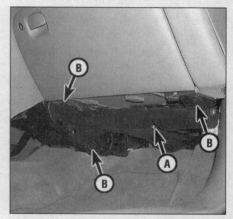

24.16 Disconnect the electrical connector at the rear of the cigar lighter, then squeeze the clips (arrows) to release it from the center bezel

24.19 Push in the two spring-loaded pins (arrow indicates the right side pin) to release the glove compartment from the instrument panel

24.21 To remove the passenger-side sound insulator (A), remove the screws (B)

Center instrument panel trim bezel

Refer to illustrations 24.15 and 24.16

14   On models where the ignition switch is mounted behind this bezel, use a small, taped screwdriver to pry off the bezel around the ignition switch. The switch will remain in the dash.

15   Pry around the bezel with a flat-bladed trim tool to release the clips **(see illustration)**.

16   Pull the panel away from the instrument panel enough to disconnect the electrical connectors behind it. On some models there will be a connector for just the cigar lighter; on other models, the hazard warning switch and enhanced traction switch are mounted to this bezel and must also be disconnected **(see illustration)**.

17   Installation is the reverse of removal.

Glove box

Refer to illustration 24.19

18   Open the glove box door.

19   Squeeze the two side tabs at the

lower/front of the of the glove compartment bin together and pull the door down until the bumpers have cleared the stops **(see illustration)**.

20   Installation is the reverse of removal.

Passenger-side lower panels

Refer to illustration 24.21

21   The sound insulation panel below the passenger's glove box is secured and removed in the same manner as the driver's knee bolster panels **(see illustration)**.

Defroster grilles and vent grilles

Refer to illustration 24.22

22   To remove the upper defroster grille, use a flat-bladed trim tool to pry around the circumference of the grille **(see illustration)**. Try not to scratch the plastic trim around the grille or the dash surface. To remove the smaller defroster grilles at the left and right ends of the instrument panel trim pad on some models, simply pry them out. **Note:** *The Grand Am defroster grille covers only the*

center area of the upper instrument panel, while on other models the grille extends the full width of the instrument panel.

23   To remove a left or right vent grille, remove the retaining screw and pry off the grille.

24   Installation is the reverse of removal.

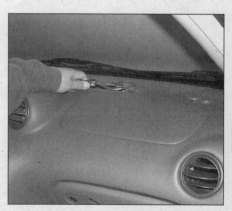

24.22 Pry around the edges with a flat trim tool to release the clips securing the defroster grille (Grand Am model shown)

11

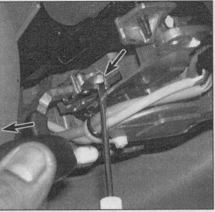

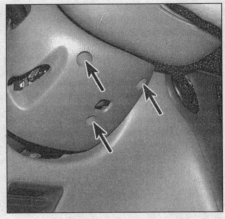

24.26 Remove the screws (arrows) and pull the ashtray housing forward until the lighter's electrical connector can be disconnected

25.1 Pry up on this spring clip (arrow) to release the pin securing the tilt lever (shown with lever removed and steering column covers pulled apart for clarity)

25.2 Remove the three Torx screws (arrows) from the lower cover and separate the upper and lower covers

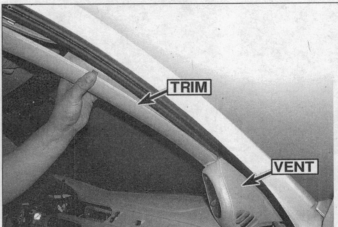

26.9a At the end of the dashboard, remove the fusebox cover and the screw (arrow) securing the lower end of the pillar trim

26.9b Use a trim tool to separate the pillar trim (the trim and vent are a unit)

Ashtray

Refer to illustration 24.26

25   On models with the ashtray below the center instrument panel trim bezel, slide the ashtray out and depress the top clip to remove it.

26   To remove the ashtray housing, remove the screws and disconnect the electrical connector **(see illustration)**.

27   Installation is the reverse of removal.

25   Steering column covers - removal and installation

Refer to illustrations 25.1 and 25.2

1   On models with tilt steering columns, remove the tilt lever first. It's possible to bend and remove the lower steering column cover with the lever in place, but there is a chance of damaging the cover. The factory recommends using a small screwdriver to release the spring and pin securing the lever to the column **(see illustration)**.

2   Remove the three screws from the lower steering column cover half **(see illustration)** and remove the lower steering column cover by unsnapping it from the upper cover.

3   Installation is the reverse of removal.

26   Instrument panel and cowl support - removal and installation

Warning: *The models covered by this manual are equipped with Supplemental Restraint Systems (SRS), more commonly known as airbags. Always disable the airbag system before working in the vicinity of any airbag system components to avoid the possibility of accidental deployment of the airbags, which could cause personal injury (see Chapter 12).*

Note: *This procedure is lengthy and difficult, even for an experienced mechanic. Due to the number of electrical connections, fasteners used, and the various safety systems involved, we don't recommend instrument panel removal for the home mechanic.*

1   Disconnect the negative battery cable.

Caution: *On models equipped with the Theft-lock audio system, be sure the lockout feature is turned off before performing any procedure which requires disconnecting the battery (see the front of this manual).*

2   Disable the airbag system (see Chapter 12).

Instrument panel trim pad

Refer to illustrations 26.9a, 26.9b, 26.10a, 26.10b, 26.10c and 26.10d

3   Remove the dashboard trim panels (see Section 24).

4   Remove the heater and air conditioning control assembly (see Chapter 3).

5   Refer to Chapter 12 and remove the stereo, ignition switch, fog lamp switch, hazard switch, trunk release switch (not all models), instrument cluster, and disconnect the electrical connectors from the fuse boxes at each end of the instrument panel.

6   Refer to Chapter 10 and remove the steering wheel and column.

7   Remove the center console (see Section 23).

8   On Grand Am and Alero models, remove the two round plastic discs at the top of the dash, on either side of the defroster grille. Dashboard trim panel screws are located

26.10a At the left end of the instrument panel, remove the screws (A), the cupholder (B, if applicable), and remove the bolt (C) to disconnect the electrical connector for the left fuse/relay box

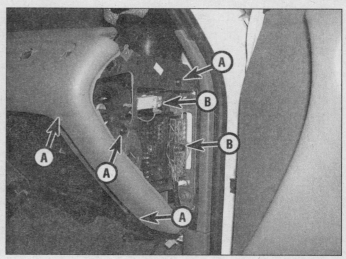

26.10b At the right side of the instrument panel, remove the screws (A) and the electrical connectors (B)

26.10c To detach the upper edge of the instrument panel, pry out the defroster grille or windshield trim, then remove the four upper bolts (socket is on one)

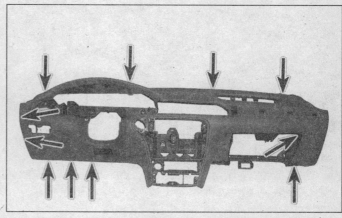

26.10d Overall view of instrument panel trim pad removed to show all screw locations (arrows, exact location may vary slightly with model)

under these discs.

9   Use a trim tool to remove the windshield post interior trim strips **(see illustrations)**.

10   Remove the instrument panel trim pad retaining screws **(see illustrations)**.

11   Carefully inspect the perimeter of the trim pad and remove any remaining retaining screws (some models have more screws along the upper front edge of the trim pad), then grasp the panel securely and detach it from the cowl support structure.

12   Installation is the reverse of removal.

Cowl support structure

Refer to illustrations 26.14, 26.16, 26.17, 26.18a, 26.18b, 26.19 and 26.20

13   Behind the instrument panel trim pad is a complicated, unitized structure that supports the body's cowl, the steering and other components. Refer to Chapter 3 and disconnect the coolant and refrigerant lines from the evaporator and heater core (be sure to read and follow the **Warnings** in Chapter 3).

14   Remove the underhood fuse/relay box

from its mounting and turn it over. From below, disconnect the electrical connectors **(see illustration)**.

15   Draw those connectors and their harnesses through the grommeted hole in the firewall.

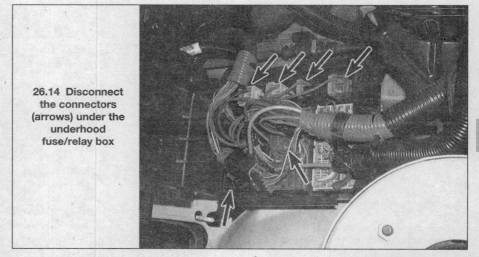

26.14 Disconnect the connectors (arrows) under the underhood fuse/relay box

11

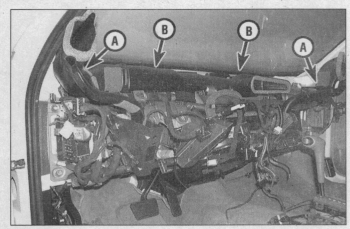

26.16 Remove the screws (A) and the push-pins (B) - remove the left end of the defroster ducting, but leave the rest in place for now

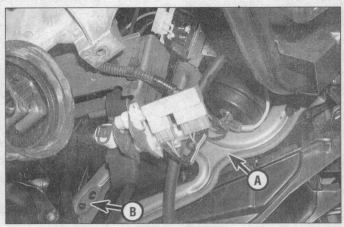

26.17 Remove the two sheetmetal braces (A indicates one) by removing the bolts (B)

26.18a At the left end of the support, remove these two bolts (arrows) . . .

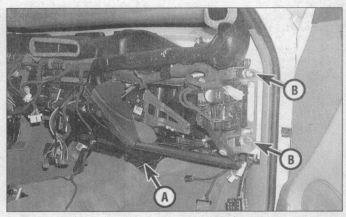

26.18b . . . then pull up the carpeting to disconnect the airbag harness (A) from the airbag control module below the passenger seat, the remove the two right-hand support structure bolts (B)

16   Along the top of the cowl support, remove the fasteners at the top of the cowl **(see illustration)**.

17   Under the center of the cowl support, remove the two stamped-sheetmetal braces **(see illustration)**.

18   Remove the support structure-to-cowl bolts at each side **(see illustrations)**.

19   On the engine side of the firewall, remove the three bolts securing the aluminum steering column support **(see illustration)**.

20   With an assistant to hold up the cowl support structure, remove the two upper cowl bolts **(see illustration)**. If you have disconnected all of the wiring connectors, and removed all of the fasteners, the cowl support can be pulled away from the cowl and lowered to the floor.

21   Installation is the reverse of the removal procedure. Do not reconnect electrical power

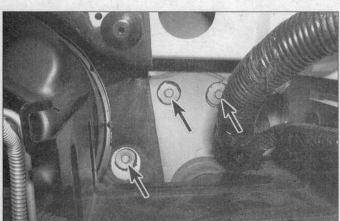

26.19 Remove the three bolts (arrows) securing the steering column support to the firewall (seen from the engine side)

26.20 Remove the two upper support-to-cowl bolts (arrow indicates the right bolt)

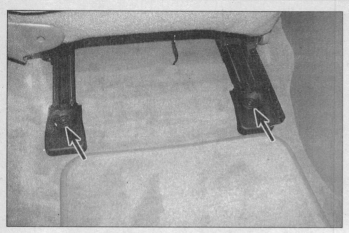

27.1 Remove the bolts (arrows) at the rear of each front seat track

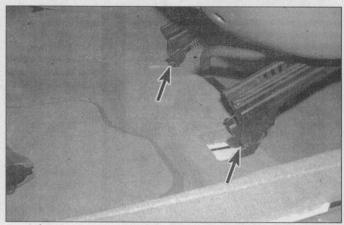

27.2 Tilt the front seat forward until the tabs (arrows) clear the slots in the floor

until you are certain that no harnesses have been pinched, all fasteners are tight and no connectors have been overlooked.

27 Seats - removal and installation

Warning: *The models covered by this manual are equipped with Supplemental Restraint Systems (SRS), more commonly known as airbags. Always disable the airbag system before working in the vicinity of any airbag system components to avoid the possibility of accidental deployment of the airbags, which could cause personal injury (see Chapter 12).*

Front seat

Refer to illustrations 27.1 and 27.2

1   At the rear of each front seat, remove the two bolts securing the seat tracks to the floor **(see illustration)**. The plastic covers around the ends of the seat tracks can stay in place. **Note:** *The main airbag module is beneath the carpeting under the right front seat. Refer to the Warning above and disable the airbag system before removing the passenger front seat.*

2   Tilt the seat up and toward the instrument panel, unplug any electrical connectors underneath and remove the seat **(see illustration)**. The front plastic track covers can remain in place while the seat track tabs are lifted out of the slots in the floor.

3   If you want to replace the seat adjuster assembly, or you need to remove the adjuster to remove or replace the carpet, unbolt the adjuster from the seat assembly.

4   Installation is the reverse of removal. Be sure to tighten all bolts securely.

Rear seat

Refer to illustrations 27.5 and 27.6

5   Lift the rear seat bottom cushion up sharply at the two front corners, until the two clips come out of the body, then slide the seat cushion forward and out of the vehicle **(see illustration)**.

6   At each lower corner, remove the nuts securing the two rear seat back tabs **(see illustration)**.

7   With the nuts removed, pull the seat back forward enough for the tabs to clear the studs, then pull the seat back sharply up and toward the instrument panel to disengage the

upper seat clips from the body.

8   Remove the seat back from the vehicle.

9   Installation is the reverse of removal.

28 Sunroof - adjustment

1   The electric drive system for the power sunroof is not adjustable. See Chapter 12 for troubleshooting the sunroof system. **Warning:** *Do not remove the sunroof motor unless the sunroof is completely closed. Otherwise a new motor will have to be installed and timed with a tool that comes with the new motor.*

2   The position of the glass panel can be adjusted in the following manner.

3   Open the interior sunshade all the way back and operate the sunroof until it is fully closed.

4   There are three glass-mounting screws on each side of the sunroof opening, inside the vehicle. Loosen the screws. Set the front edge of the glass to be just (1mm) below the surface of the roof.

5   Adjust the glass at the rear edge to be just *above* (1 mm) the roof at the rear.

6   Tighten the screws.

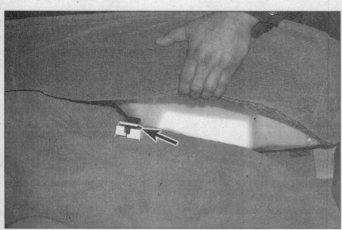

27.5 Disengage the clips at each front corner of the rear seat bottom cushion (arrow indicates one of the clips)

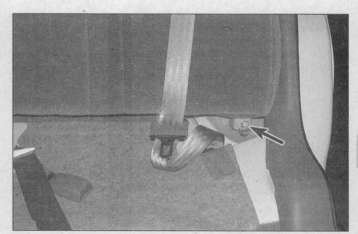

27.6 The bottom of the rear seat back is retained by tabs secured with nuts (arrow indicates one of the nuts)

11

Notes

Chapter 12
Chassis electrical system

Contents

1 General information

The electrical system is a 12-volt, negative ground type. A lead/acid-type battery that is charged by the alternator supplies power for the lights and all electrical accessories.

This Chapter covers the various electrical components not associated with the engine. Information on the battery, alternator, distributor and starter motor can be found in Chapter 5.

It should be noted that when portions of the electrical system are serviced, the cable should be disconnected from the negative battery terminal to prevent electrical shorts and/or fires.

Caution: *On models equipped with the Theft-lock audio system, be sure the lockout feature is turned off before performing any procedure which requires disconnecting the battery (see the front of this manual).*

Warning: *The models covered by this manual are equipped with Supplemental Restraint Systems (SRS), more commonly known as airbags. Always disable the airbag system before working in the vicinity of any airbag system components to avoid the possibility of accidental deployment of the airbags, which could cause personal injury (see Section 29).*

2 Electrical troubleshooting - general information

Refer to illustrations 2.5a, 2.5b, 2.6, 2.9 and 2.15

A typical electrical circuit consists of an electrical component, any switches, relays, motors, fuses, fusible links or circuit breakers related to that component and the wiring and connectors that link the component to both the battery and the chassis. To help you pinpoint an electrical circuit problem, wiring diagrams are included at the end of this Chapter.

Before tackling any troublesome electri-

12

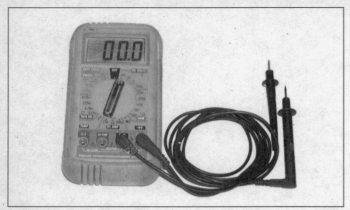

2.5a The most useful tool for electrical troubleshooting is a digital multimeter that can measure volts, amps and resistance

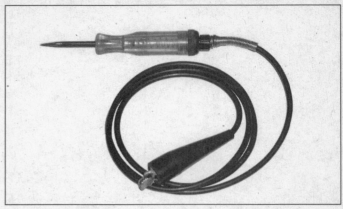

2.5b A test light is a very handy tool for checking voltage

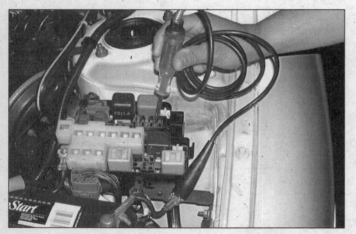

2.6 In use, the test light lead is clipped to a known good ground, then the pointed probe can test connectors, wires or electrical sockets - if the bulb lights, the circuit being tested has battery voltage

2.9 With a multimeter set to the ohms scale, resistance can be checked across two terminals - when checking for continuity, a low reading indicates continuity, a high reading indicates lack of continuity

cal circuit, first study the appropriate wiring diagrams to get a complete understanding of what makes up that individual circuit. Trouble spots, for instance, can often be narrowed down by noting if other components related to the circuit are operating properly. If several components or circuits fail at one time, chances are the problem is in a fuse or ground connection, because several circuits are often routed through the same fuse and ground connections.

Electrical problems usually stem from simple causes, such as loose or corroded connections, a blown fuse, a melted fusible link or a failed relay. Visually inspect the condition of all fuses, wires and connections in a problem circuit before troubleshooting the circuit.

If test equipment and instruments are going to be utilized, use the diagrams to plan ahead of time where you will make the necessary connections in order to accurately pinpoint the trouble spot.

The basic tools needed for electrical troubleshooting include a circuit tester or voltmeter (a 12-volt bulb with a set of test leads can also be used), a continuity tester, which includes a bulb, battery and set of test leads, and a jumper wire, preferably with a circuit breaker incorporated, which can be used to bypass electrical components **(see illustrations)**. Before attempting to locate a problem with test instruments, use the wiring diagram(s) to decide where to make the connections.

Voltage checks

Voltage checks should be performed if a circuit is not functioning properly. Connect one lead of a circuit tester to either the negative battery terminal or a known good ground. Connect the other lead to a connector in the circuit being tested, preferably nearest to the battery or fuse **(see illustration)**. If the bulb of the tester lights, voltage is present, which means that the part of the circuit between the connector and the battery is problem free. Continue checking the rest of the circuit in the same fashion. When you reach a point at which no voltage is present, the problem lies between that point and the last test point with voltage. Most of the time the problem can be traced to a loose connection. **Note:** *Keep in mind that some circuits receive volt-*

age only when the ignition key is in the Accessory or Run position.

Finding a short

One method of finding shorts in a circuit is to remove the fuse and connect a test light or voltmeter in place of the fuse terminals. There should be no voltage present in the circuit. Move the wiring harness from side-to-side while watching the test light. If the bulb goes on, there is a short to ground somewhere in that area, probably where the insulation has rubbed through. The same test can be performed on each component in the circuit, even a switch.

Ground check

Perform a ground test to check whether a component is properly grounded. Disconnect the battery and connect one lead of a continuity tester or multimeter (set to the ohms scale), to a known good ground. Connect the other lead to the wire or ground connection being tested. If the resistance is low (less than 5 ohms), the ground is good. If the bulb on a self-powered test light does not go on, the ground is not good.

2.15 To backprobe a connector, insert a small, sharp probe (such as a straight-pin) into the back of the connector alongside the desired wire until it contacts the metal terminal inside; connect your meter leads to the probes - this allows you to test a functioning circuit

Continuity check

A continuity check is done to determine if there are any breaks in a circuit - if it is passing electricity properly. With the circuit off (no power in the circuit), a self-powered continuity tester or multimeter can be used to check the circuit. Connect the test leads to both ends of the circuit (or to the "power" end and a good ground), and if the test light comes on the circuit is passing current properly **(see illustration)**. If the resistance is low (less than 5 ohms), there is continuity; if the reading is 10,000 ohms or higher, there is a break somewhere in the circuit. The same procedure can be used to test a switch, by connecting the continuity tester to the switch terminals. With the switch turned On, the test light should come on (or low resistance should be indicated on a meter).

Finding an open circuit

When diagnosing for possible open circuits, it is often difficult to locate them by sight because the connectors hide oxidation or terminal misalignment. Merely wiggling a connector on a sensor or in the wiring harness may correct the open circuit condition. Remember this when an open circuit is indicated when troubleshooting a circuit. Intermittent problems may also be caused by oxidized or loose connections.

Electrical troubleshooting is simple if you keep in mind that all electrical circuits are basically electricity running from the battery, through the wires, switches, relays, fuses and fusible links to each electrical component (light bulb, motor, etc.) and to ground, from which it is passed back to the battery. Any electrical problem is an interruption in the flow of electricity to and from the battery.

Connectors

Most electrical connections on these vehicles are made with multiwire plastic connectors. The mating halves of many connectors are secured with locking clips molded into the plastic connector shells. The mating halves of large connectors, such as some of those under the instrument panel, are held together by a bolt through the center of the connector.

To separate a connector with locking clips, use a small screwdriver to pry the clips apart carefully, then separate the connector halves. Pull only on the shell, never pull on the wiring harness as you may damage the individual wires and terminals inside the connectors. Look at the connector closely before trying to separate the halves. Often the locking clips are engaged in a way that is not immediately clear. Additionally, many connectors have more than one set of clips.

Each pair of connector terminals has a male half and a female half. When you look at the end view of a connector in a diagram, be sure to understand whether the view shows the harness side or the component side of the connector. Connector halves are mirror images of each other, and a terminal shown on the right side end-view of one half will be on the left side end view of the other half.

Backprobing a connector

It is often necessary to take circuit voltage measurements with a connector connected. Whenever possible, carefully insert a small straight pin (not your meter probe) into the rear of the connector shell to contact the terminal inside, then clip your meter lead to the pin. This kind of connection is called "backprobing" **(see illustration)**. When inserting a test probe into a male terminal, be careful not to distort the terminal opening. Doing so can lead to a poor connection and corrosion at that terminal later. Using the small straight pin instead of a meter probe results in less chance of deforming the terminal connector.

---

3   Fuses and fusible links - general information

Fuses

Refer to illustrations 3.1a, 3.1b and 3.3

The electrical circuits of the vehicle are protected by a combination of fuses, circuit breakers and fusible links. The interior fuse blocks are located at each end of the instrument panel and in the engine compartment **(see illustrations)**.

Each of the fuses is designed to protect a specific circuit, and the various circuits are identified on the fuse panel itself.

Miniaturized fuses are employed in the fuse blocks. These compact fuses, with blade terminal design, allow fingertip removal

3.1a   The engine compartment fuse box is located on the left side of the engine compartment and contains fuses and relays - the inside of the cover has a legend to identify the fuses and relays

3.1b   Under each dashboard endcap is a junction block or interior fuse center (left end fuse center shown), which contains fuses, circuit breakers and several mini-relays - the inside of the instrument panel end cap has the identification legends cast into the plastic (not visible in this photo)

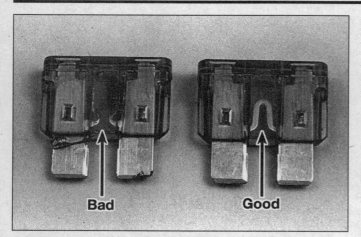

3.3 When a fuse blows, the element between the terminals melts - the fuse on the left is blown, the one on the right is good

3.7 The fusible link (arrow) is in the circuit from the alternator to the starter

and replacement. If an electrical component fails, always check the fuse first. The best way to check the fuses is with a test light. Check for power at the exposed terminal tips of each fuse. If power is present at one side of the fuse but not the other, the fuse is blown. A blown fuse can also be identified by visually inspecting it **(see illustration)**.

Be sure to replace blown fuses with the correct type. Fuses of different ratings are physically interchangeable, but only fuses of the proper rating should be used. Replacing a fuse with one of a higher or lower value than specified is not recommended. Each electrical circuit needs a specific amount of protection. The amperage value of each fuse is molded into the fuse body.

If the replacement fuse immediately fails, don't replace it again until the cause of the problem is isolated and corrected. In most cases, this will be a short circuit in the wiring caused by a broken or deteriorated wire.

Fusible links

Refer to illustrations 3.7 and 3.9

Some circuits are protected by fusible links. The links are used in circuits that are not ordinarily fused, such as the alternator circuit.

The fusible link for the alternator circuit is located on the front side of the engine, near the starter motor, and is easily identified **(see illustration)**. The link is a short length of green heavy wire spliced into the cable, and the ends are wrapped with tape.

To replace a fusible link, first disconnect the negative battery cable. **Caution:** *On models equipped with the Theftlock audio system, be sure the lockout feature is turned off before performing any procedure which requires disconnecting the battery (see the front of this manual).*

Although the fusible links appear to be a heavier gauge than the wires they're protecting, the appearance is due to the thick insulation. All fusible links are several wire gauges smaller than the wire they're designed to pro-

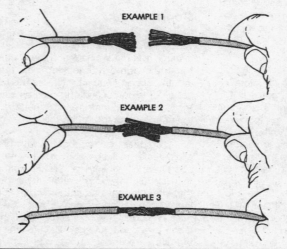

3.9 To repair a fusible link, cut out the damaged section, then join a new section by stripping the wire and twisting it together, as shown here - when securely joined, solder the connections and wrap them with electrical tape

tect. Fusible links can't be repaired, but a new link of the same size wire can be installed. The procedure is as follows:

a) *Cut the damaged fusible link out of the wire just behind the connector.*

b) *Strip the insulation back approximately 1-inch.*

c) *Spread the strands of the exposed wire apart, push them together and twist them in place* **(see illustration)**.

d) *Use rosin core solder at each end of the new link to obtain a good solder joint.*

e) *Use plenty of electrical tape around the soldered joint. No wires should be exposed.*

f) *Connect the negative battery cable. Test the circuit for proper operation.*

4 Circuit breakers - general information and check

Circuit breakers protect certain circuit, such as the power windows and power seats. There are two 25-amp circuit breakers, one located in each of the interior fuse/relay boxes at the ends of the dashboard.

Because the circuit breakers reset auto-

matically, an electrical overload in a circuit-breaker-protected system will cause the circuit to fail momentarily, then come back on. If the circuit does not come back on, check it immediately.

For a basic check, pull the circuit breaker up out of its socket on the fuse panel, but just far enough to probe with a voltmeter. The breaker should still contact the sockets.

With the voltmeter negative lead on a good chassis ground, touch each end prong of the circuit breaker with the positive meter probe. There should be battery voltage at each end. If there is battery voltage only at one end, the circuit breaker must be replaced.

5 Relays - general information and testing

General information

1   Several electrical accessories in the vehicle, such as the fuel injection system, horns, starter, and fog lamps use relays to transmit the electrical signal to the compo-

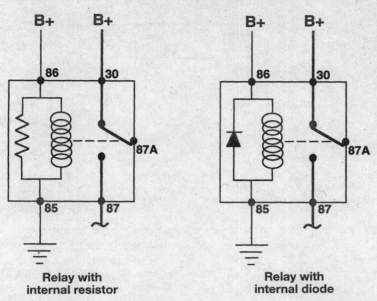

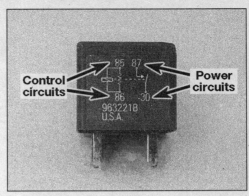

5.2b Most relays are marked on the outside to easily identify the control circuit and power circuits - this one is of the four-terminal type

Relay with
internal resistor

Relay with
internal diode

24053-12-5.2a HAYNES

5.2a Typical ISO relay designs, terminal numbering and circuit connections

nent. Relays use a low-current circuit (the control circuit) to open and close a high-current circuit (the power circuit). If the relay is defective, that component will not operate properly. Most relays are mounted in the engine compartment and interior fuse/relay boxes. If a faulty relay is suspected, it can be removed and tested using the procedure below or by a dealer service department or a repair shop. Defective relays must be replaced as a unit.

Testing

Refer to illustrations 5.2a and 5.2b
2   Most of the relays used in these vehicles are often called "ISO" relays, which refers to the International Standards Organization. The terminals of ISO relays are numbered to indicate their usual circuit connections and functions. There are two basic layouts of terminals on the relays used in the covered vehicles **(see illustrations)**.
3   Refer to the wiring diagram for the circuit to determine the proper connections for the relay you're testing. If you can't determine the correct connection from the wiring diagrams, however, you may be able to determine the test connections from the information that follows.
4   Two of the terminals are the relay control circuit and connect to the relay coil. The other relay terminals are the power circuit. When the relay is energized, the coil creates a magnetic field that closes the larger contacts of the power circuit to provide power to the circuit loads.
5   Terminals 85 and 86 are normally the control circuit. If the relay contains a diode, terminal 86 must be connected to battery positive (B+) voltage and terminal 85 to

ground. If the relay contains a resistor, terminals 85 and 86 can be connected in either direction with respect to B+ and ground.
6   Terminal 30 is normally connected to the battery voltage (B+) source for the circuit loads. Terminal 87 is connected to the ground side of the circuit, either directly or through a load. If the relay has several alternate terminals for load or ground connections, they usually are numbered 87A, 87B, 87C, and so on.
7   Use an ohmmeter to check continuity through the relay control coil.

   a) *Connect the meter according to the polarity shown in **illustrations 5.2a or 5.2b** for one check; then reverse the ohmmeter leads and check continuity in the other direction.*
   b) *If the relay contains a resistor, resistance should be the specified value with the ohmmeter in either direction.*
   c) *If the relay contains a diode, resistance should be the specified coil resistance value with the ohmmeter in the forward polarity direction. With the meter leads reversed, resistance should be lower.*
   d) *If the ohmmeter shows infinite resistance in both directions, replace the relay.*

8   Remove the relay from the vehicle and use the ohmmeter to check for continuity between the relay power circuit terminals. There should be no continuity between terminal 30 and 87 with the relay de-energized. On the smaller micro-relays, make sure you have the polarity correct before testing them.
9   Connect a fused jumper wire to terminal 86 and the positive battery terminal. Connect another jumper wire between terminal 85 and ground. When the connections are made, the

relay should click.
10   With the jumper wires connected, check for continuity between the power circuit terminals. Now, there should be continuity between terminals 30 and 87.
11   If the relay fails any of the above tests, replace it.

6   Turn signal/hazard flashers - check and replacement

Warning: *The models covered by this manual are equipped with Supplemental Restraint Systems (SRS), more commonly known as airbags. Always disable the airbag system before working in the vicinity of any airbag system components to avoid the possibility of accidental deployment of the airbags, which could cause personal injury (see Section 29).*
1   The turn signal and hazard flasher on these models are not separate components, but the function is handled by a timer within the hazard warning switch.
2   When the flasher unit is functioning properly, an audible click can be heard during its operation. If the turn signal indicator on one side of the vehicle flashes much more rapidly than normal, a faulty turn signal bulb is indicated.
3   If both turn signals fail to blink, the problem may be due to a blown fuse, a faulty hazard switch or a loose or open connection. If a quick check of the fuse box indicates that the turn signal fuse has blown, check the wiring for a short before installing a new fuse.
4   Refer to Section 9 and disconnect the hazard switch electrical connector. Probe the B terminal with a test light connected to a body ground. If the test light doesn't light, check for a ground fault in the circuit (refer to the wiring diagrams at the end of this Chapter for terminal designations).
5   Check with a test light between the B and A terminals of the connector. There should be power.
6   With the switch On, there should be power on the output side with the connector in place. Backprobe the connector at G and F

12

with test light connected to ground. With the ignition key On and the turn signal switch turned left or right, the test light should flash. If it doesn't, replace the hazard switch.

7   Steering column switches - check and replacement

Warning: *The models covered by this manual are equipped with Supplemental Restraint Systems (SRS), more commonly known as airbags. Always disable the airbag system before working in the vicinity of any airbag system components to avoid the possibility of accidental deployment of the airbags, which could cause personal injury (see Section 29).*

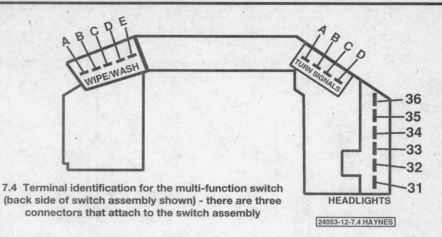

7.4 Terminal identification for the multi-function switch (back side of switch assembly shown) - there are three connectors that attach to the switch assembly

24053-12-7.4 HAYNES

Multi-function switch

Check

Refer to illustrations 7.4

1   Disconnect the negative battery cable. **Caution:** *On models equipped with the Theft-lock audio system, be sure the lockout feature is turned off before performing any procedure which requires disconnecting the battery (see the front of this manual).*

2   Refer to the replacement procedure below to remove the multi-function switch for testing. **Note:** *The turn signal switch, headlight dimmer switch and wiper/washer switch are all part of the same multi-function switch assembly, and can't be replaced individually.*

3   Problems with the left side of the multi-function switch (headlight/turn signal/dimmer) can sometimes be determined by accessing any diagnostic trouble codes (DTCs) from the Body Control Module (see Chapter 6).

4   If you don't have access to a scan tool, use an ohmmeter and test light to check the terminals **(see illustration)**. To check the headlight portion of the assembly, backprobe terminal 34 with the connector in place. A grounded test light should show power there. If not, check the circuit (refer to the wiring diagrams at the end of this Chapter). If power was at this terminal, yet the headlights do not work, replace the entire multi-function switch assembly. Backprobe the same connector at

terminal 33. If there is voltage, replace the switch assembly.

5   If the problem is that normal headlight operation is non-operational, yet the daytime running lights operate, then backprobe terminal 35. If there is no voltage there, replace the switch assembly.

6   To check the washer/wiper side of the switch assembly, backprobe the connector (in place) at terminal B while pushing the washer switch (the ignition switch should be in the Run position). If the test light doesn't illuminate, replace the multi-function switch assembly. If there is no voltage at the C terminal, set the switch to the High position and check for voltage at terminal A. If there is no voltage, check the wiring diagrams and search for a short or open in the circuit. If the circuit is ok, replace the multi-function switch assembly. If the problem is that the wipers won't park properly, turn the wiper switch to Off (ignition key in Run) and disconnect the wiper switch electrical connector. If the wipers now park properly, replace the multi-function switch assembly.

7   To check the turn signal portion of the switch assembly, backprobe the D terminal with a test light connected to a body ground. With the left turn signal and hazard switch On, the test light should flash. If it doesn't, replace the multi-function switch assembly.

Replacement

Refer to illustrations 7.9 and 7.11

8   Disconnect the negative battery cable. **Caution:** *On models equipped with the Theft-lock audio system, be sure the lockout feature is turned off before performing any procedure which requires disconnecting the battery (see the front of this manual).*

9   Remove the steering column covers (see Chapter 11) **(see illustration).**

10   Remove the multi-function switch retaining screw. **Note:** *The screws are externally male hex, and internally Torx type.*

11   Remove the multi-function switch. Slide the switch straight up off the column and unplug the connectors **(see illustration).**

12   Insert the connectors into the new multi-function switch, pushing in until they are securely locked in place.

13   The remainder of installation is the reverse of removal.

Cruise control switches

Check

Refer to illustration 7.15

14   The cruise control switch pods are located on the steering wheel. Remove the airbag module for access to the electrical connector for the cruise control switches (see Chapter 10, Section 14).

15   Unplug the electrical connector for the cruise control switches and, using an ohmmeter, probe terminals A and D of the connector **(see illustration)**. Continuity should alternate as the switch is actuated On and Off.

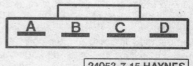

24053-7.15 HAYNES

7.15 Terminal identification for the cruise control switch electrical connector (switch side of harness)

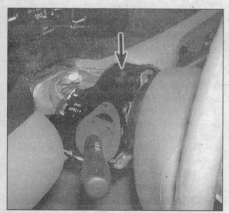

7.9 The multi-function switch is held in place by one screw (arrow)

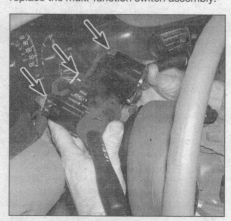

7.11 Unplug the electrical connectors (arrows) from the multi-function switch

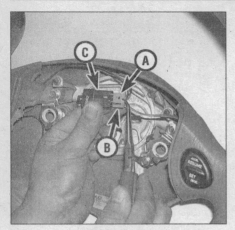

7.21 Depress the tab (A) and pull out the connector insert (B), then pull the individual cruise control switch wires out of the connector (C)

8.4 Remove the ignition switch mounting screws (arrows)

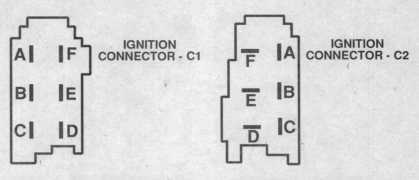

SWITCH POSITION	CONTINUITY BETWEEN
Lock	None
Acc	C1-A to C2-C, C2-A to C1-E
Run	C1-A to C2-C, C1-A to C2-B C2-A to C1-C, C2-A to C1-E
Start	C1-A to C2-B, C2-A to C1-B C2-A to C1-E

24053-12-8.6 HAYNES

8.6 Ignition switch terminal identification and continuity chart

16   With the cruise On/Off switch in the On position (continuity existing between A and D), probe terminals B and D with the ohmmeter. Depress the Set/Decel button - the meter should register continuity when the button is depressed and no continuity when the button is released.

17   With the cruise On/Off switch in the On position (continuity existing between A and D), probe terminals C and D with the ohmmeter. Depress the Accel/Resume button - the meter should register continuity when the button is depressed and no continuity when the button is released.

18   If one of the switches doesn't work as described, replace both of them as a set (at the time of writing the switches were not available separately).

Replacement

Refer to illustration 7.21

19   Disconnect the negative battery cable. **Caution:** *On models equipped with the Theft-lock audio system, be sure the lockout feature is turned off before performing any procedure which requires disconnecting the battery (see the front of this manual).*

20   Refer to Chapter 10, Section 14 and remove the driver's airbag module from the steering wheel.

21   Take apart the connector by removing the insert, then separate the individual wires **(see illustration). Note:** *The connector is marked with letters for the wires, but note the colors of the wires in each socket and make a sketch before removing them.*

22   Using a flat tool or a screwdriver tip covered with tape to protect the steering wheel covering, pry the switch up out of the steering wheel and feed the wires out.

23   Installation is the reverse of removal.

Audio controls (Grand Am only)

24   These models have driver controls for the audio system located on the steering wheel where other models have the cruise control switches. The cruise control switches on these models are located at the center/bottom of the steering wheel.

25   The audio controls are removed from the steering wheel in basically the same procedure as described above for the cruise control switches.

8   Ignition switch and key lock cylinder - check and replacement

Warning: *The models covered by this manual are equipped with Supplemental Restraint Systems (SRS), more commonly known as airbags. Always disable the airbag system before working in the vicinity of any airbag system components to avoid the possibility of*

accidental deployment of the airbags, which could cause personal injury (see Section 29).

Ignition switch

Check

Refer to illustrations 8.4 and 8.6

1   Disconnect the negative battery cable. **Caution:** *On models equipped with the Theft-lock audio system, be sure the lockout feature is turned off before performing any procedure which requires disconnecting the battery (see the front of this manual).*

2   Before removing the ignition switch for testing, check the main ignition fuse in the underhood fuse/relay box. To access the ignition switch, start by referring to Chapter 11 and remove the center instrument panel trim bezel and the instrument cluster bezel.

3   Refer to Section 11 and remove the instrument cluster.

4   Remove the two screws securing the ignition switch **(see illustration)**.

5   Push the switch about an inch straight back to clear the mounting bracket, then lift it up and out from behind the instrument panel.

6   Check the switch for continuity between the indicated terminals with the key in each position **(see illustration)**.

7   If the continuity is not as specified, replace the switch.

8   Check the lock cylinder in each position to make sure it isn't worn or loose and that the key position corresponds to the markings on the housing.

Replacement

Refer to illustrations 8.10 and 8.11

9   Follow Steps 1 through 5 to access the ignition switch.

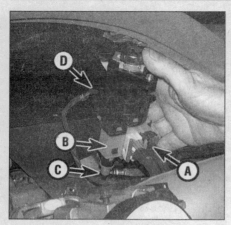

8.10 Disconnect the electrical connectors (A), then push in the tab (B) to release the interlock cable (C) - the two-wire connector (D) can only be disconnected when the lock cylinder is removed

10   Disconnect the ignition switch electrical connectors and the ignition interlock cable **(see illustration)**.
11   Using the key, turn the lock cylinder to Run and press in the pin to release the cylinder, then withdraw the cylinder with the key **(see illustration). Note:** *The cylinder must be removed even if only the switch portion is to be replaced, in order to release the two-wire connector.*
12   The remainder of the installation is the reverse of the removal procedure.

---

9   Instrument panel switches - check and replacement

Warning: *The models covered by this manual are equipped with Supplemental Restraint Systems (SRS), more commonly known as airbags. Always disable the airbag system before working in the vicinity of any airbag system components to avoid the possibility of*

8.11 Turn the key to Run, then depress the pin (arrow) and withdraw the lock cylinder

accidental deployment of the airbags, which could cause personal injury (see Section 29).

Instrument panel dimmer switch

Refer to illustrations 9.2 and 9.4

Check

1   Refer to Steps below to remove the dimmer switch for testing.
2   Test the switch terminals with a test light and voltmeter **(see illustration)**. If the voltage output doesn't vary as the switch is turned, replace the switch.

Replacement

3   Refer to Chapter 11 for removal of the instrument cluster trim bezel (which includes the headlight switch on most models). On Alero models, use a flat-bladed trim tool to pry out the small trim plate the switch is mounted to.
4   Squeeze the clips on the backside of the trim bezel to release the dimmer switch **(see illustration)**.

`24053-12-9.2-HAYNES`

9.2 Dashboard dimmer switch terminal identification - With the connector in place, backprobe at C to test for power when the headlights are on, and then test at B, where voltage should vary as the dimmer switch is turned

5   Installation is the reverse of the removal procedure.

Trunk release switch

Refer to illustration 9.6

6   Use pliers or a flat-bladed trim tool to pry the trunk release switch from the instrument panel **(see illustration)**.
7   Disconnect the electrical connector on the back of the switch.
8   Using an ohmmeter, check that there is continuity between the two switch terminals only when the switch is depressed.
9   To replace the switch, it must be pushed into the panel with a tab on the side aligned with a notch in the panel.

Hazard warning switch

Refer to illustrations 9.11 and 9.12

10   Refer to Chapter 11 for removal of the center instrument panel bezel on models where the hazard switch is mounted to the to the right side of the bezel. On Pontiac models, the hazard switch is mounted at the top of the center instrument panel trim bezel (refer to Chapter 11 to remove the bezel).
11   With the bezel pulled away from the instrument panel, disconnect the electrical connector, squeeze the two tabs and push

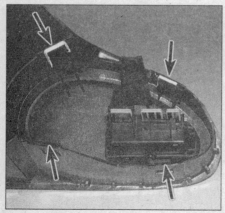

9.4 Squeeze the four clips (arrows, Malibu model shown) to release the dimmer switch from the trim bezel - push the switch out to the front

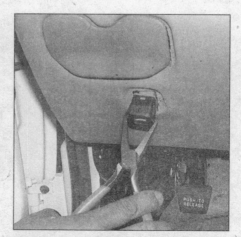

9.6 Carefully pull the trunk release switch straight out of the dash

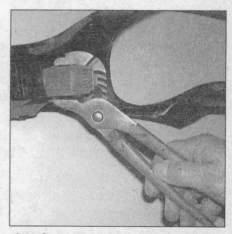

9.11 Squeeze the two tabs (arrows) with pliers and push the hazard switch out through the front of the bezel

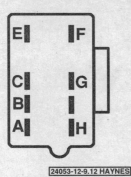

24053-12-9.12 HAYNES

9.12 Terminal identification for the hazard warning switch

11.3a Remove the mounting screws (arrows) and pull the instrument cluster away from the dash

the switch out the front of the bezel **(see illustration)**.

12   Using a test light, backprobe the hazard switch with the connector in place **(see illustration)**. With the switch On, check for voltage at B and C (C only with the key at Run or Start). Next, check that the test light blinks at F, G and K. If the switch fails the tests, replace it.

13   Installation is the reverse of the removal procedure.

10   Instrument panel gauges - check

1   All tests below require the ignition switch to be turned to Off position before testing.

2   If the gauge pointer does not move from the empty or cold positions when the key is turned On, check the fuses, referring to the wiring diagrams at the end of this chapter for the fuses controlling the circuit in question. If the fuse is OK, locate the particular sending unit for the circuit you're working on (see Chapter 2 for the oil pressure sending unit location, Chapter 3 for the temperature gauge sending unit location, and Chapter 4 for fuel sending unit location). Each of these Chapters has a basic test for the sending unit.

3   If the fuses are OK, check that the cooling system is in good condition (see Chapter 3), the fuel tank does have fuel, and that there is proper oil pressure with the engine running. If the gauges do not accurately reflect actual conditions and the fuses and sending units are OK, have the instrument panel diagnosed with a factory scan tool at a dealership.

11   Instrument cluster - removal and installation

Refer to illustrations 11.3a and 11.3b
Warning: *The models covered by this manual are equipped with Supplemental Restraint Systems (SRS), more commonly known as airbags. Always disable the airbag system before working in the vicinity of any airbag system components to avoid the possibility of accidental deployment of the airbags, which could cause personal injury (see Section 29).*

1   Disable the airbag system (see Section 29).

2   Remove the instrument cluster bezel (see Chapter 11).

3   Remove the retaining screws and pull the cluster forward enough to disconnect the one electrical connector **(see illustrations)**.

4   Installation is the reverse of the removal procedure.

12   Radio and speakers - removal and installation

Warning: *The models covered by this manual are equipped with Supplemental Restraint Systems (SRS), more commonly known as airbags. Always disable the airbag system before working in the vicinity of any airbag system components to avoid the possibility of accidental deployment of the airbags, which could cause personal injury (see Section 29).*

Radio/CD player

Refer to illustrations 12.3 and 12.4

1   Disable the airbag system (see Section 29).

2   Remove the center bezel panel from the dash (see Chapter 11).

3   Remove the screws and pull the radio/CD player assembly away from the dash **(see illustration)**.

4   Disconnect the antenna lead and the electrical connectors, then remove the audio unit **(see illustration)**.

5   Installation is the reverse of removal.

Speakers

Front

Refer to illustration 12.7

6   Refer to Chapter 11 and remove the door trim panel.

11.3b Disconnect the electrical connector (arrow) at the right end of the cluster

12.3 Remove the audio unit mounting screws (arrows)

12.4 Disconnect the antenna lead (A) and the electrical connectors (B)

12

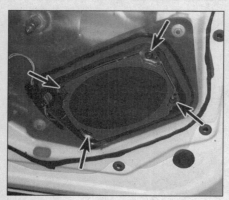

12.7 With the door trim panel removed, remove the front speaker mounting screws (arrows)

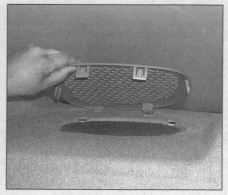

12.9 Pry up the rear speaker grilles until all clips are released

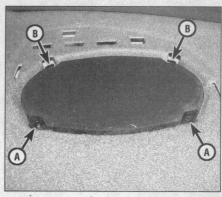

12.11 Rear speaker mounting screws (A) and clips (B)

7   Remove the speaker retaining screws. Unplug the electrical connector and remove the speaker **(see illustration)**.

8   Installation is the reverse of removal.

Rear

Refer to illustrations 12.9 and 12.11

9   Use a trim tool to pry up the speaker grilles from the rear package shelf **(see illustration)**.

10   Open the trunk and disconnect the electrical connectors from the speakers.

11   Remove the screws securing the speakers, pull them forward to disengage the two clips at the rear, then lift them out of the package shelf **(see illustration)**.

12   Installation is the reverse of removal.

---

13   Antenna - check and replacement

Malibu/Cutlass/Grand Am models

Refer to illustrations 13.1 and 13.3

1   On these models with a fixed external antenna, remove the antenna mast **(see illustration)**. Apply masking tape around the antenna mount to avoid scratching the paint.

2   Working in the trunk, pry out the plastic clips securing the passenger side trunk finishing panels to allow access to the antenna connections.

3   Detach the antenna lead and remove the antenna from the vehicle **(see illustration)**.

4   Installation is the reverse of removal.

Alero models

5   These models have a wire-grid antenna incorporated into the defogger grid in the rear window.

6   The antenna grid can be tested for continuity in the same manner as outlined in Section 21, and if there is a break in the grid, it can be repaired in the same manner as the rear defogger grid (see Section 21).

7   Models with a grid antenna also have a rear window antenna module, mounted

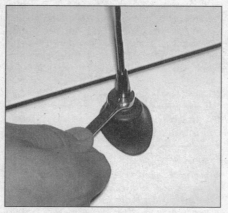

13.1 Use a small wrench to remove a fixed antenna mast

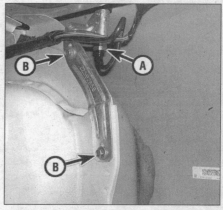

13.3 From inside the trunk, remove the radio cable (A), then remove the antenna base mounting bolts (B)

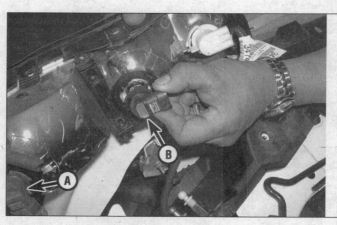

14.2 Disconnect the electrical connector, then twist and pull out the bulb holder and replace the bulb - A is the low-beam bulb, B is the high-beam bulb

behind the top of the rear seat back, below the package shelf. On these models, the antenna cable from the radio attaches to the antenna module.

---

14   Headlight bulb - replacement

Refer to illustration 14.2

Warning: *These models are equipped with halogen gas-filled bulbs, which are under pressure and may shatter if the surface is scratched or the bulb is dropped. Wear eye*

protection and handle the bulbs carefully, grasping only the base whenever possible. Do not touch the surface of the bulb with your fingers because the oil from your skin could cause it to overheat and fail prematurely. If you do touch the bulb surface, clean it with rubbing alcohol.

1   The headlight housing must be removed to access the bulbs on the back of the housing (see Section 16).

2   Twist and withdraw the bulb assembly from the back of the headlight housing **(see illustration)**.

15.1 Use a Torx socket and extension to adjust the headlights - the adjuster closest to the fender (A) controls the horizontal movement and the one closest to the radiator (B), vertical movement (Grand Am model shown, others similar)

3   Without touching the glass with your bare fingers, insert the new bulb into the socket, then insert the assembly into the headlight housing and twist to secure it.
4   Reinstall the headlight housing (see Section 16), then test headlight operation.

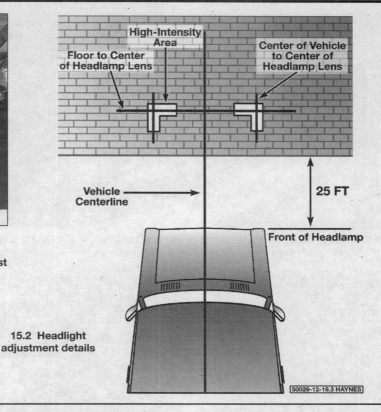

15.2 Headlight adjustment details

15   Headlights and fog lights - adjustment

Headlights

Refer to illustrations 15.1 and 15.2
Note: *It is important that the headlights are aimed correctly. If adjusted incorrectly they could blind the driver of an oncoming vehicle and cause a serious accident or seriously reduce your ability to see the road. The headlights should be checked for proper aim every 12 months and any time a new headlight is installed or front end body work is performed. It should be emphasized that the following procedure is only an interim step that will provide temporary adjustment until the headlights can be adjusted by a properly equipped shop.*
1   These models are equipped with composite headlights with two adjustment screws, one controlling left-and-right movement and one for up-and-down movement **(see illustration)**.
2   There are several methods of adjusting the headlights. The simplest method requires a blank wall 25 feet in front of the vehicle and a level floor **(see illustration)**.
3   Position masking tape vertically on the wall in reference to the vehicle centerline and the centerlines of both headlights.
4   Position a horizontal tape line in reference to the centerline of all the headlights.
Note: *It may be easier to position the tape on the wall with the vehicle parked only a few inches away.*
5   Adjustment should be made with the vehicle sitting level, the gas tank half-full and no unusually heavy load in the vehicle.
6   Starting with the low beam adjustment, position the high intensity zone so it is two inches below the horizontal line and two inches to the side of the vertical headlight line away from oncoming traffic. Twist the adjustment screws until the desired level has been achieved.
7   With the high beams on, the high intensity zone should be vertically centered with the exact center just below the horizontal line. **Note:** *It may not be possible to position the headlight aim exactly for both high and low beams. If a compromise must be made, keep in mind that the low beams are the most used and have the greatest effect on driver safety.*
8   Have the headlights adjusted by a dealer service department at the earliest opportunity.

Fog lights

9   Some models have optional fog lights that can be aimed just like headlights.
10   Position tape on a wall 25 feet in front of the vehicle **(see illustration 15.2)**. Tape a horizontal line on the wall that represents the height of the fog lamps, and another tape line four inches below that line.
11   Using the adjusting screws on the fog lamps, adjust the pattern on the wall so that the top of the fog lamp beam meets the lower line on the wall, and that the beam is centered horizontally in front of the fog lamps.

16   Headlight housing - replacement

Refer to illustrations 16.1a, 16.1b and 16.2
1   Open the hood and remove the plastic pushpins securing the shield over the head-

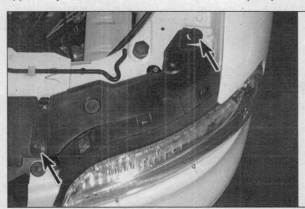

16.1a Remove the pins (arrows) and the plastic sight shield over the headlight housing (Malibu/Cutlass models)

12

16.1b On Grand Am/Alero models, the sight shield is full-width, remove the plastic pushpins (arrows indicate pins on one side)

16.2 Pull up on these two clips (arrows), then pull the headlight housing forward (Grand Am model shown)

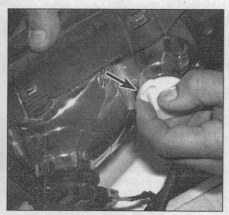

17.2 The front turn signal bulb is accessed from the back of the headlight housing - squeeze the clip (arrow) to withdraw the bulb holder

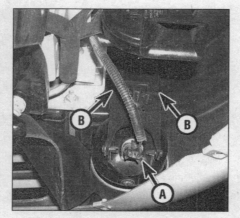

17.6 Disconnect the fog light electrical connector (A), the remove the bracket-to-bumper beam bolts (B)

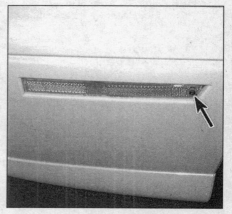

17.9 Remove the screw (arrow) then tilt the lens out of the fender to access the bulb holder

light housing (see illustrations).

2   Pull straight up on the two clips to release the headlight housing from the ball-studs on the body (see illustration).

3   Disconnect the electrical connectors.

4   Installation is the reverse of removal.

17   Bulb replacement

Front turn signal lights

Refer to illustration 17.2

1   The park, turn and side marker light is part of the headlight housing. See Section 16 for removal of the headlight housing to access this bulb, which is the innermost bulb on each headlight housing.

2   Depress the clip next to the bulb holder and twist the bulb holder out of the headlight housing (see illustration).

3   Remove the bulb from the holder, then use a glove or rag to hold the new bulb when inserting it into the holder.

4   Reinstall the headlight housing as in Section 16.

Fog lights

Refer to illustration 17.6

Warning: *The optional fog lights are equipped with halogen gas-filled bulbs, which are under pressure and may shatter if the surface is scratched or the bulb is dropped. Wear eye protection and handle the bulbs carefully, grasping only the base whenever possible. Do not touch the surface of the bulb with your fingers because the oil from your skin could cause it to overheat and fail prematurely. If you do touch the bulb surface, clean it with rubbing alcohol.*

5   From under the front bumper fascia, remove the pushpins and take down the splash shield to access the back of the fog lamps (see Chapter 11).

6   Remove the two bolts retaining the fog lamp to the back of the front bumper reinforcement beam and lower the fog lamp (see illustration).

7   Twist the bulb holder out, pull out the old bulb, then install the new bulb, using gloves or a rag while handling the new bulb.

8   Installation of the fog lamp is the reverse of the removal procedure.

Front side marker lights

Refer to illustrations 17.9 and 17.10

9   Remove the screw in the side marker light and tilt the lens/housing out of the front fender (see illustration).

10   Twist the bulb holder out, then replace the bulb (see illustration).

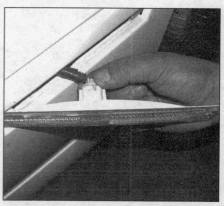

17.10 Twist the bulb holder out of the side marker housing to replace the bulb

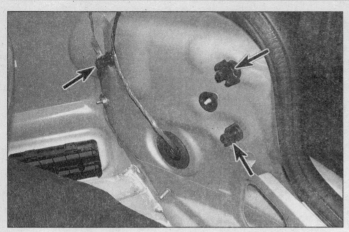

17.12 Pull back the trunk liner to access the taillight housing wingnuts (arrows)

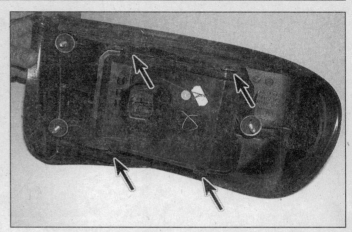

17.13 Disconnect the electrical connector, then release the four clips (arrows) to remove the bulb holder

11   Installation is the reverse of the removal procedure.

Tail/brake/turn and back-up lights

Refer to illustrations 17.12, 17.13 and 17.14

12   Open the trunk and pull back the trim cover (remove the plastic nuts first) for access to the fasteners on the back of the taillight housing **(see illustration)**. Unscrew the wing nuts.

13   Pull the taillight housing out and release the four clips to disengage the bulb holder from the taillight housing **(see illustration)**. **Note:** *All four of the rear bulbs are mounted on this one bulbholder plate.*

14   Carefully wiggle the bulbs straight out to remove them from the holder **(see illustration)**.

15   The remainder of the installation is the reverse of the removal procedure.

License plate light

Refer to illustration 17.17

16   The license plate light assembly is mounted above the license plate.

17   On Malibu and Cutlass models, twist the bulb holder out and pull it down for access to the bulb **(see illustration)**.

18   On Grand Am and Alero models, remove the two license lamp screws from outside, above the license plate, then from inside the trunk pull up the license lamp housing and replace the bulb.

19   Pull the bulb straight out to replace it.

High-mounted brake light

Grand Am and Alero models

Refer to illustrations 17.21

20   On these models, the high-mounted brake light is part of a trim panel across the back of the trunk lid.

21   On the Grand Am remove the screws from the backside of the trunk lid securing the bulbholder retaining strap, then twist and remove the bulbholders to replace the bulbs **(see illustration)**.

22   Installation is the reverse of the removal procedure.

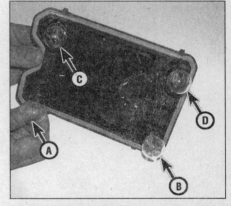

17.14 All four rear bulbs are on one holder - A and B are the taillight/brake light bulbs, C is the turn signal and D is the back-up bulb

Malibu and Cutlass models

Refer to illustration 17.24

23   On these models, open the trunk and access the high-mounted brake light housing

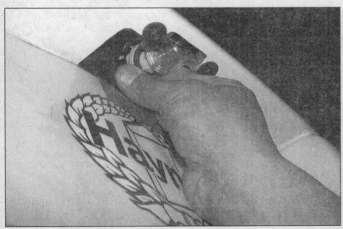

17.17 The license light bulb can be accessed by twisting the bulb holder out of the rear of the license lamp (Malibu and Cutlass models)

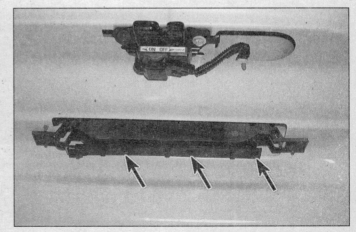

17.21 On Grand Am/Alero models, remove the screws (arrows) and pull back the bulb holder for access to the high-mounted brake light bulbs

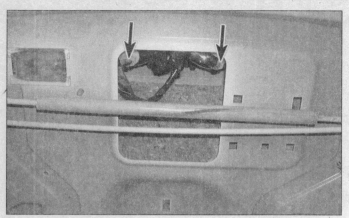

17.24 The bulb holders (arrows) for the high-mounted brake light on Malibu and Cutlass models are under the rear package shelf, and accessed through the trunk

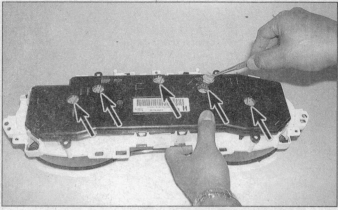

17.25 To remove an instrument cluster light bulb (arrows), depress it and turn it counterclockwise to release it

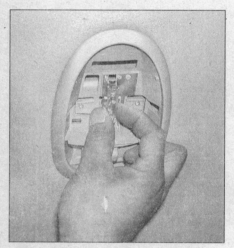

17.27 With the dome light lens removed, pull out the dome light bulb

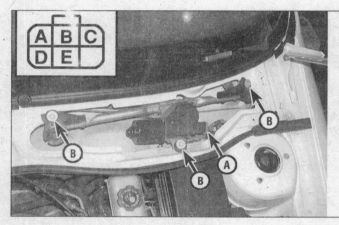

18.2 Backprobe the connector (arrow A) at the wiper motor to check for power at terminal B (low) or C (high), then check for a good continuity to ground at terminal A - (the B arrow indicates wiper assembly mounting bolts)

from below the rear package shelf.

24   Twist out the two bulb holders from beneath the high-mounted brake light housing and replace the bulbs **(see illustration).**

Instrument cluster illumination

Refer to illustration 17.25

25   To gain access to the instrument cluster illumination bulbs, the instrument cluster will have to be removed (see Section 11). The bulbs can then be removed and replaced from the rear of the cluster **(see illustration).**

Interior lights

Refer to illustration 17.27

26   Remove the lenses for the rearview mirror or vanity mirror lights by prying the cover off with a small screwdriver.

27   To replace the dome light, use a small, taped screwdriver to release the cover from the housing, then remove the bulb from the holder **(see illustration).**

28   Installation is the reverse of the removal procedure.

18   Wiper motor - check and replacement

Check

Refer to illustration 18.2

Note: *Refer to the wiring diagrams for wire colors and locations in the following checks. When checking for voltage, probe a grounded 12-volt test light to each terminal at a connector until it lights; this verifies voltage (power) at the terminal. If the following checks fail to locate the problem, have the system diagnosed by a dealer service department or other properly equipped repair facility.*

1   If the wipers work slowly, make sure the battery is in good condition and has a strong charge (see Chapter 1). If the battery is in good condition, remove the wiper motor (see below) and operate the wiper arms by hand. Check for binding linkage and pivots. Lubricate or repair the linkage or pivots as necessary. Reinstall the wiper motor. If the wipers still operate slowly, check for loose or corroded connections, especially the ground connection. If all connections look OK, replace the motor.

2   If the wipers fail to operate when activated, check the fuse. If the fuse is OK, con-

nect a jumper wire between the wiper motor and ground, then retest. If the motor works now, repair the ground connection. If the motor still doesn't work, turn the wiper switch to the HI position and check for voltage at the motor **(see illustration)**. **Note:** *The cowl cover will have to be removed (see Chapter 11).* If there's voltage at the connector, remove the motor and check it off the vehicle with fused jumper wires from the battery. If the motor now works, check for binding linkage (see Step 1 above). If the motor still doesn't work, replace it. If there's no voltage to the motor, check for voltage at the wiper control relays in the power distribution center (see Section 5 for relay testing). If there's voltage at the wiper control relays and no voltage at the wiper motor, check the multi-function switch for continuity (see Section 7).

3   If the wipers stop at the position they're in when the switch is turned off (fail to park), check for a good ground on the connector side at the motor. With an ohmmeter connected between any of the black wire terminals and a known ground, resistance should be 0 to 2 ohms.

4   If the wipers won't shut off unless the ignition is OFF, disconnect the wiring from the wiper control switch. If the wipers stop, replace the switch. If the wipers keep run-

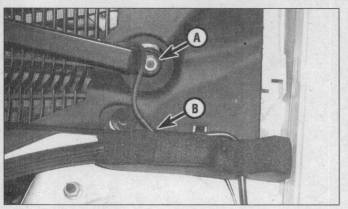

18.6 Use a small screwdriver to pry off the wiper arm nut cover, then remove the nut (A), disconnect the washer hose (B) and pull the arm straight off its splined shaft

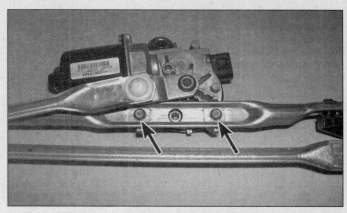

18.10 With the wiper assembly out of the vehicle, remove the motor mounting bolts (arrows)

ning, there's a defective limit switch in the motor; replace the motor.

5   If the wipers won't retract below the hood line, check for mechanical obstructions in the wiper linkage or on the vehicle's body that would prevent the wipers from parking. If there are no obstructions, check the wiring between the switch and motor for continuity (see Section 7). If the wiring and switch are OK, replace the wiper motor.

Replacement

Refer to illustrations 18.6 and 18.10

6   Remove the windshield wiper arms **(see illustration)**.
7   Remove the cowl cover (see Chapter 11).
8   Disconnect the electrical connector from the wiper motor **(see illustration 18.2)**.
9   Detach the wiper motor/linkage assembly from the cowl **(see illustration 18.2)**.
10   Remove the wiper motor retaining bolts and remove the motor **(see illustration)**.
11   Installation is the reverse of removal.

19   Horn - check and replacement

Refer to illustration 19.1
Note: *Check the fuse before beginning electrical diagnosis.*
1   Unplug the electrical connector from the horns **(see illustration)**. **Note:** *The horn is easiest to access by removing the right-side headlight housing (see Section 16).*
2   To test the horns, refer to the wiring diagrams and connect battery voltage to the dark green-wire terminal (at the horn side of the connector), and temporarily ground the black-wire terminal with a pair of jumper wires. If the horns don't sound, replace them. If they do sound, the problem lies in the switch, relay or the wiring between the components.
3   Refer to Section 5 for testing of the relay, which is located in the engine compartment fuse/relay box.
4   To replace the horn, unplug the electrical connector and remove the bracket bolt

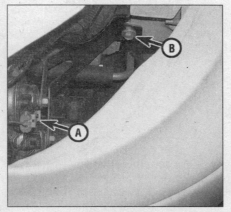

19.1 Unplug the electrical connector (A) and remove the bolt (B), then detach the horns

(see illustration 19.1).
5   Installation is the reverse of removal.

20   Daytime Running Lights (DRL) - general information

The Daytime Running Lights (DRL) system used on all models turns the headlights on whenever the engine is started. The only exception is when the engine is turned on when the parking brake is engaged. Once the parking brake is released, the lights will remain on as long as the ignition switch is on, even if the parking brake is later applied.

The DRL system supplies reduced power to the headlights so they won't be too bright for daytime use while prolonging headlight life.

21   Rear window defogger - check and repair

1   The rear window defogger consists of a number of horizontal heating elements baked onto the inside surface of the glass. Power is supplied through a mini-relay in the engine

21.5 When measuring the voltage at the rear window defogger grid, wrap a piece of aluminum foil around the negative probe of the voltmeter and press the foil against the wire with your finger

compartment fuse/relay box. The heater is controlled by the instrument panel switch, which is part of the heating/air-conditioning controls. Test the switch for continuity (see Chapter 3). **Note:** *On most models, the rear window grid is used as a defogger only, while on Alero models, the defogger grid also serves as the antenna for the radio.*
2   Small breaks in the element can be repaired without removing the rear window.

Check

Refer to illustrations 21.5, 21.6 and 21.8
3   Turn the ignition switch and defogger switches to the ON position.
4   Using a voltmeter, place the positive probe against the defogger grid positive terminal and the negative probe against the ground terminal. If battery voltage is not indicated, check the fuse, defogger switch, defogger relay and related wiring. If voltage is indicated, but all or part of the defogger doesn't heat, proceed with the following tests.
5   When measuring voltage during the next two tests, wrap a piece of aluminum foil around the tip of the voltmeter positive probe

12

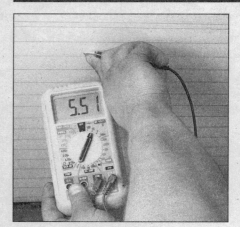

21.6 To determine if a heating element has broken, check the voltage at the center of each element - if the voltage is 6-volts, the element is unbroken

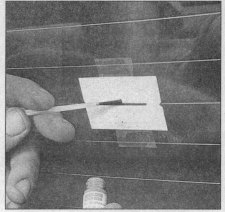

21.8 To find the break, place the voltmeter negative lead against the defogger ground terminal, place the voltmeter positive lead with the foil strip against the heat wire at the positive end and slide it toward the negative terminal end - the point at which the meter deflects is where the wire is broken

21.14 To use a defogger repair kit, apply masking tape to the inside of the window at the damaged area, then brush on the special conductive coating

and press the foil against the heating element with your finger **(see illustration)**. Place the negative probe on the defogger grid ground terminal.

6   Check the voltage at the center of each heating element **(see illustration)**. If the voltage is 5 to 6 volts, the element is okay (there is no break). If the voltage is 0 volts, the element is broken between the center of the element and the positive end. If the voltage is 10 to 12 volts the element is broken between the center of the element and the ground side. Check each heating element.

7   If none of the elements are broken, connect the negative probe to a good chassis ground. The voltage reading should stay the same, if it doesn't the ground connection is bad.

8   To find the break, place the voltmeter negative probe against the defogger ground terminal. Place the voltmeter positive probe with the foil strip against the heating element at the positive side and slide it toward the negative side. The point at which the voltmeter deflects from several volts to zero is the point where the heating element is broken **(see illustration)**.

Repair

Refer to illustration 21.14

9   Repair the break in the element using a repair kit specifically for this purpose, such as Dupont paste No. 4817 (or equivalent). The kit includes conductive plastic epoxy.

10   Before repairing a break, turn off the system and allow it to cool for a few minutes.

11   Lightly buff the element area with fine steel wool; then clean it thoroughly with rubbing alcohol.

12   Use masking tape to mask off the area being repaired.

13   Thoroughly mix the epoxy, following the kit instructions.

14   Apply the epoxy material to the slit in the masking tape, overlapping the undamaged area about 3/4-inch on either end **(see illustration)**.

15   Allow the repair to cure for 24 hours before removing the tape and using the system.

22   Cruise control system - description and check

Refer to illustrations 22.5a, 22.5b and 22.8

1   The cruise control system maintains vehicle speed with a servo motor (located on the right fenderwell) that is connected to the throttle linkage by a cable. The system consists of the servo motor, brake switch, control switches and a relay. Some features of the system require special testers and diagnostic procedures that are beyond the scope of the home mechanic. Listed below are some general procedures that may be used to locate common problems.

2   Check the cruise control fuse in the interior fuse/relay box at the right-hand end of the instrument panel (see Section 3).

3   The cruise control brake release switch (to the right of the brake light switch) deactivates the cruise control system when the brake pedal is depressed.

4   To check the brake release switch, remove the driver's side under-dash panel and heater duct, then check the continuity of the switch with an ohmmeter. With the brake pedal at rest, continuity should exist across the switch terminals. With the brake pedal depressed, there should be no continuity. If the switch doesn't work like this, replace it (see Chapter 9, Section 17; it's replaced and adjusted just like the brake light switch).

5   Check the control cable between the cruise control servo/amplifier and the throttle linkage and replace as necessary **(see illustration)**. To adjust the cable at the throttle end, make sure the throttle linkage is at the full-closed position. On 2.4L engines, turn the thumbscrew to remove any slack in the cruise cable **(see illustration)**. On V6 engines, release the lock-clip at the cable housing end (at the throttle linkage), and move the cable by hand to remove any slack, then push the lock-clip back in.

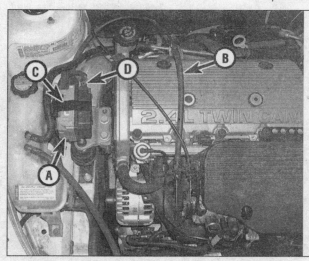

22.5a The cruise control servo (A) is located near the underhood fuse/relay box - make sure the cruise control cable (B) mounted on the throttle body is not damaged and that it operates smoothly when the throttle is opened C indicates the servo mounting strap, and D is the servo electrical connector

22.5b Adjust the cruise cable on 2.4L engines by turning the thumbscrew (arrow) to remove slack

6　The cruise control system uses information from the PCM, including the Vehicle Speed Sensor, which is located in the transmission. To test the speed sensor, see Chapter 6.

7　The testing of the steering-wheel-mounted cruise control switches is covered in Section 7.

8　Some tests of the servo can be made by the home mechanic. Turn the ignition key to On (engine not running). Disconnect the electrical connector at the servo and use a grounded test light to check for battery power at terminal F on the harness side **(see illustration)**. If power isn't there, refer to the wiring diagrams at the end of this Chapter and check the circuit.

9　With the cruise control switch in the On position, there should be battery voltage present in terminal A, and at terminal B in the Set/Coast position.

10　Test-drive the vehicle to determine if the cruise control is now working. If it isn't, take it to a dealer service department or an automotive electrical specialist for further diagnosis.

23 Power window system - description and check

Refer to illustrations 23.12 and 23.16

1　The power window system operates the electric motors mounted in the doors which lower and raise the windows. The system consists of the control switches, the motors (regulators), glass mechanisms and associated wiring.

2　Power windows are wired so they can be lowered and raised from the master control switch by the driver or by remote switches located at the individual windows. Each window has a separate motor that is reversible. The position of the control switch determines the polarity and therefore the direction of operation. Some systems are equipped with relays that control current flow

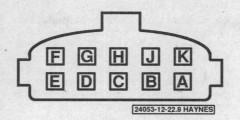

22.8 Pin identification for the cruise control servo connector

to the motors.

3　Some vehicles are equipped with a separate circuit breaker for each motor in addition to the fuse or circuit breaker protecting the whole circuit. This prevents one stuck window from disabling the whole system.

4　The power window system will only operate when the ignition switch is ON. In addition, many models have a window lockout switch at the master control switch which, when activated, disables the switches at the rear windows and, sometimes, the switch at the passenger's window also. Always check these items before troubleshooting a window problem.

5　These procedures are general in nature, so if you can't find the problem using them, take the vehicle to a dealer service department.

6　If the power windows don't work at all, check the fuse or circuit breaker.

7　If only the rear windows are inoperative, or if the windows only operate from the master control switch, check the rear window lockout switch for continuity in the unlocked position. Replace it if it doesn't have continuity.

8　Check the wiring between the switches and fuse panel for continuity. Repair the wiring, if necessary.

9　If only one window is inoperative from the master control switch, try the other control switch at the window. **Note:** *This doesn't*

apply to the drivers door window.

10　If the same window works from one switch, but not the other, check the switch for continuity.

11　If the switch tests OK, check for a short or open in the wiring between the affected switch and the window motor.

12　If one window is inoperative from both switches, remove the trim panel from the affected door and check for voltage at the switch and at the motor while the switch is operated. First check for voltage at the connectors **(see illustration)**. With the ignition key On and connectors in place on the switch, backprobe at the designated wire with a grounded test light. Pushing the driver's window switch Down, there should be voltage at terminal B of connector C2. Pushing the same switch Up, there should be voltage at C2-F. If these voltage are OK, disconnect the electrical connector at the driver's motor, and check for voltage there when the switch is operated.

13　If voltage is reaching the motor and the switch is OK, disconnect the glass from the regulator (see Chapter 11). Move the window up and down by hand while checking for binding and damage. Also check for binding and damage to the regulator. If the regulator is not damaged and the window moves up and down smoothly, replace the motor. If there's binding or damage, lubricate, repair or replace parts, as necessary.

14　If voltage isn't reaching the motor, check the wiring in the circuit for continuity between the switches and motors. You'll need to consult the wiring diagram for the vehicle. Some power window circuits are equipped with relays. If equipped, check that the relays are grounded properly and receiving voltage from the switches. Also check that each relay sends voltage to the motor when the switch is turned on. If it doesn't, replace the relay.

15　Test the windows after you are done to confirm proper repairs. If the main power window switch is to be replaced, pry the switch unit out of the door panel then disconnect the two connectors.

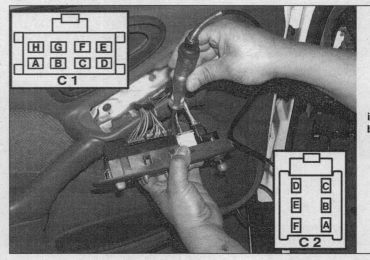

23.12 Pin identification for both connectors at the master power window switch

12

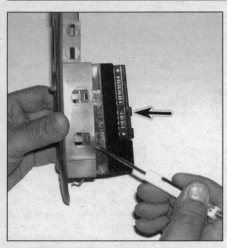

23.16 Pry the switch assembly from the driver's door panel, disconnect the electrical connectors, then use a small screwdriver to release the clips holding the window switch (arrow) to the assembly

16   The power window switch is part of a larger switch assembly on the driver's door, that also includes the door lock and power mirror switches. To replace the window switch, remove the assembly from the door panel (see Step 15), then release the tabs to remove the switch **(see illustration)**.

24   Power door lock system and keyless entry - description and check

Refer to illustration 24.10

1   The power door lock system operates the door lock actuators mounted in each door. The system consists of the switches, actuators and associated wiring. Diagnosis can usually be limited to simple checks of the wiring connections and actuators for minor faults that can be easily repaired.

2   Power door lock systems are operated by bi-directional solenoids located in the doors. The lock switches have two operating positions: Lock and Unlock. These switches activate a relay, which in turn connects voltage to the door lock solenoids. Depending on which way the relay is activated, it reverses polarity, allowing the two sides of the circuit to be used alternately as the feed (positive) and ground side.

3   Some vehicles may have keyless entry, electronic control modules and anti-theft systems incorporated into the power locks. If you are unable to locate the trouble using the following general steps, consult your dealer service department. **Note:** *Some vehicles also have control switches connected to the key locks in the doors, which unlock all the doors when one is unlocked.*

4   Always check the circuit protection first. Some vehicles use a combination of circuit

breakers and fuses. Refer to the wiring diagrams at the end of this Chapter.

5   Operate the door lock switches in both directions (Lock and Unlock) with the engine off. Listen for the faint click of the relay operating.

6   If there's no click, check for voltage at the switches. If no voltage is present, check the wiring between the fuse panel and the switches for shorts and opens.

7   If voltage is present but no click is heard, test the switch for continuity. Replace it if there's not continuity in both switch positions. To remove the switch on Malibu and Cutlass models, remove the screw at the door handle bezel and remove the bezel. The switch is clipped to this bezel. On Alero models, use a flat-bladed trim tool to pry out the power door lock bezel from the upper door panel, and on Grand Am models, the door lock switch is a part of the power window switch assembly (see Section 23).

8   If the switch has continuity but the relay doesn't click, check the wiring between the switch and relay for continuity. Repair the wiring if there's no continuity.

9   If the relay is receiving voltage from the switch but is not sending voltage to the solenoids, check for a bad ground at the relay case. If the relay case is grounding properly, replace the relay.

10   If all but one lock solenoids operate, remove the trim panel from the affected door (see Chapter 11) and check for voltage at the solenoid while the lock switch is operated **(see illustration)**. One of the wires should have voltage in the Lock position; the other should have voltage in the Unlock position.

11   If the inoperative solenoid is receiving voltage, replace the solenoid.

12   If the inoperative solenoid isn't receiving voltage, check for an open or short in the wire between the lock solenoid and the relay. **Note:** *It's common for wires to break in the portion of the harness between the body and door (opening and closing the door fatigues and eventually breaks the wires).*

13   On the models covered by this manual, power door lock system communication goes through the Body Control Module. If the above tests do not pinpoint a problem, take the vehicle to a dealer or qualified shop with the proper scan tool to retrieve trouble codes from the BCM.

Keyless entry system

14   The keyless entry system consists of a remote control transmitter that sends a coded infrared signal to a receiver, which then operates the door lock system.

15   Replace the battery when the transmitter doesn't operate the locks at a distance of 10 feet. Normal range should be about 30 feet.

16   Use a small screwdriver to carefully separate the case halves.

17   Replace the three-volt, CR2032 lithium battery.

18   Snap the case halves together.

24.10 Check for power at the door lock actuator connector (arrow) with the switch depressed - check the door lock actuator itself by disconnecting the connector and using jumper wires to temporarily apply battery voltage and ground directly

25   Electric side view mirrors - description and check

Refer to illustration 25.9

1   Most electric side view mirrors use two motors to move the glass; one for up and down adjustments and one for left-right adjustments.

2   The control switch has a selector portion that sends voltage to the left or right side mirror. With the ignition ON but the engine OFF, roll down the windows and operate the mirror control switch through all functions (left-right and up-down) for both the left and right side mirrors.

3   Listen carefully for the sound of the electric motors running in the mirrors.

4   If the motors can be heard but the mirror glass doesn't move, there's probably a problem with the drive mechanism inside the mirror. Remove and disassemble the mirror to locate the problem.

5   If the mirrors don't operate and no sound comes from the mirrors, check the fuse (see Chapter 1).

6   If the fuse is OK, remove the mirror control switch from its mounting without disconnecting the wires attached to it (see Chapter 11 for door panel removal or Section 23 for models where the switches are grouped). Turn the ignition ON and check for voltage at the switch. There should be voltage at the orange wire terminal. If there's no voltage at the switch, check for an open or short in the wiring between the fuse panel and the switch.

7   Re-connect the switch. Locate the wire going from the switch to ground. Leaving the switch connected, connect a jumper wire between this wire and ground. If the mirror works normally only with this jumper in place, repair the faulty ground connection.

8   If the mirror still doesn't work, remove the mirror and check the wires at the mirror

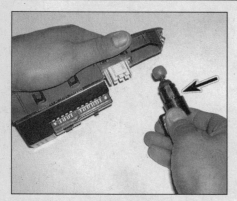

25.9 Remove the power mirror switch (arrow) from the driver's door switch panel by releasing the clips

for voltage. Check with ignition ON and the mirror selector switch on the appropriate side. Operate the mirror switch in all its positions. There should be voltage at one of the switch-to-mirror wires in each switch position (except the neutral "off" position).

9   If voltage isn't present in each switch position, check the wiring between the mirror and control switch for opens and shorts. If there are no shorts, replace the switch. The mirror switch is part of the power window switch assembly **(see illustration)**.

10   If there's voltage, remove the mirror and test it off the vehicle with jumper wires. Replace the mirror if it fails this test.

26   Electric sunroof switch - check and replacement

Check

Refer to illustration 26.6

1   The electric sunroof is powered by a single motor located in the roof behind the headliner, near the overhead console. When sunlight isn't desired, an interior sliding panel can be closed.

2   The control switch (tilt and slide) sends a ground signal to the sunroof motor when the switch is pressed. With the ignition On but the engine Off, operate the sunroof control switch through the tilt and slide functions.

3   Listen carefully for the sound of the sunroof motor running in the roof.

4   If the motors can be heard but the sunroof glass doesn't move, there's probably a problem with the drive mechanism or drive cables.

5   If the sunroof does not operate and no sound comes from the motor, check the fuse (fuse H in the right-hand interior fuse panel).

6   If the fuse is OK, check the switch. Use a trim tool to pry the switch out of the headliner and disconnect the connector. Using an ohmmeter, check for continuity between terminals A and B in the Open position, and between C and B in the Close position **(see illustration)**. If continuity isn't as described, replace the switch.

26.6 Sunroof switch terminals (shown from connector side)

7   If the switch has continuity re-connect the switch. Locate the wire going from the switch to ground. Leaving the switch connected, connect a jumper wire between this wire and ground. If the motor works normally with this wire in place, repair the faulty ground connection.

Replacement

8   If the switch is the problem, disconnect the electrical connector and connect it to the new switch, then press it into position in the headliner opening.

9   The sunroof must be fully closed if the motor or module is to be removed. **Warning:** *Do not remove the sunroof motor unless the sunroof is completely closed. Otherwise a new motor will have to be installed and timed with a tool that comes with the new motor.*

10   The motor and the sunroof express module are both accessible once the headliner is removed. However, the removal of the headliner is a tedious procedure, requires two people, and may be beyond the abilities of some weekend mechanics. If the switch isn't the problem in the sunroof circuit, take the vehicle to a qualified repair shop for further diagnosis and repair.

27   Power seats - description and check

Refer to illustration 27.9

1   Power seats allow you to adjust the position of the seat with little effort. The optional power seats on these models adjust forward and backward, up and down and tilt forward and backward.

2   The power seat system consists of a motor, a switch on the seat, and the circuit breaker (in the interior fuse/relay panel at the left end of the instrument panel).

3   Look under the seat for any objects which may be preventing the seat from moving.

4   If the seat won't work at all, check the circuit breaker (see Section 4).

5   With the engine off to reduce the noise level, operate the seat controls in all directions and listen for sound coming from the seat motors.

6   If the motors run or click but the seat doesn't move, the seat drive mechanism is damaged and the motor assembly must be replaced.

7   If the motor doesn't work or make noise, check for voltage at the motor while an assistant operates the switch.

8   If the motor is getting voltage but doesn't run, test it off the vehicle with jumper wires. If it still doesn't work, replace it (see Chapter 11 for seat removal).

9   If the motor isn't getting voltage, check for voltage at the switch, at the A terminal **(see illustration)**. If there's no voltage at that terminal, check the wiring between the fuse panel and the switch. If there's voltage at the switch, check the switch for continuity in all its operating positions. Replace the switch if it fails any test.

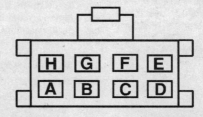

SWITCH POSITION	CONTINUITY BETWEEN
Forward	A and H
Back	A and G
Front up	A and E
Front down	A and F
Rear up	A and B
Rear down	A and C

27.9 Power seat switch terminal identification and continuity tests

12

29.1 The airbag Control Module (arrow) is located on the floor under the front passenger seat (carpeting pulled back)

29.8 Disconnect the driver's airbag connector (arrow) here above the left fuse/relay panel

10   If the switch is OK, check for a short or open in the wiring between the switch and motor.

11   Test the completed repairs.

28   Data Link Communication system - description

1   The vehicles covered by this manual have a complex electrical system, encompassing many power accessories, and a number of separate electronic modules.

2   The Powertrain Control Module (PCM) is mainly responsible for engine and transaxle control, but also communicates with other modules around the vehicle through a Data Link Communication system, which sends serial port data very quickly between the various modules. Many of the computer functions involved in the operation of body systems are routed through the Body Control Module (BCM), which communicates with the PCM.

3   Among the modules in the Data Link system besides the BCM and PCM are the Sensing Diagnostic Module (airbag system), the Electronic Brake Control Module, and the instrument panel cluster. The BCM further communicates with various body subsystems.

4   All of the modules in the vehicle have associated trouble codes. When other troubleshooting procedures fail to pinpoint the problem, check the wiring diagrams at the end of this Chapter to see if the BCM or PCM are involved in the circuit. If so, bring your vehicle to a dealer or other repair facility with the diagnostic tools required to extract the trouble codes.

29   Airbag system - general information

Refer to illustrations 29.1, 29.8 and 29.9

1   These models are equipped with a Supplemental Restraint System (SRS), more

commonly known as airbags, designed to protect the driver and front seat passenger from serious injury in the event of a head-on or frontal collision. All models have a sensing/diagnostic control unit, located under the passenger seat, below the carpeting **(see illustration)**. **Warning:** *If your vehicle is ever involved in a flood, or the interior carpeting is soaked for any reason, disconnect the battery and do not start the vehicle until the airbag system can be checked by your dealer. If the SRS system is subjected to flooding, the airbags could go off upon starting the vehicle, even without an accident taking place.*

Airbag modules

2   The airbag modules consist of a housing incorporating the cushion (airbag) and inflator unit. The inflator assembly is mounted on the back of the housing over a hole through which gas is expelled, inflating the bag almost instantaneously when an electrical signal is sent from the system. The specially-wound wire on the driver's side that carries this signal to the module is called a clockspring. The clockspring is a flat, ribbon-like electrically conductive tape that is wound many times so that it can transmit an electrical signal regardless of steering wheel position.

Sensing/diagnostic control unit and sensors

3   The sensing/diagnostic control unit contains an on-board microprocessor which monitors the operation of the system, and also contains a crash sensor. It checks this system every time the vehicle is started, causing the "AIRBAG" light to flash seven times then go off, if the system is operating properly. If there is a fault in the system, the light will go on and continue, either illuminated steadily or blinking, and the unit will store fault codes indicating the nature of the fault.

Operation

4   For the airbag(s) to deploy, an accelerometer in the diagnostic control unit is activated. The control unit then compares this

force to a value stored in its memory. If the control unit determines the force is in excess of the value, the circuits to the airbag inflators are closed and the airbags inflate. If the battery is destroyed by the impact, or is too low to power the inflator, a back-up power unit inside the control unit provides power.

Self-diagnosis system

5   A self-diagnosis circuit in the control unit displays a light on the instrument panel when the ignition switch is turned to the On position. If the system is operating normally, the light should go out after about seven blinks. If the light doesn't come on, or doesn't go out after a short time, or if it comes on while you're driving the vehicle, or if it blinks at any time, there's a malfunction in the SRS system. Have it inspected and repaired as soon as possible. Do not attempt to troubleshoot or service the SRS system yourself. Even a small mistake could cause the SRS system to malfunction when you need it.

Servicing components near the SRS system

6   Nevertheless, there are times when you need to remove the steering wheel, radio or service other components on or near the dashboard. At these times, you'll be working around components and wire harnesses for the SRS system. The SRS wiring harnesses are easy to identify: They're all covered with a bright yellow conduit. Do not unplug the connectors for these wires. And do not use electrical test equipment on airbag system wires; it could cause the airbag(s) to deploy. *ALWAYS DISABLE THE SRS SYSTEM BEFORE WORKING NEAR THE SRS SYSTEM COMPONENTS OR RELATED WIRING.*

Disabling the SRS system

Warning: *Any time you are working in the vicinity of airbag wiring or components, DISABLE THE SRS SYSTEM.*

7   In the driver's side fuse/relay panel at the left end of the instrument panel, remove the airbag fuse.

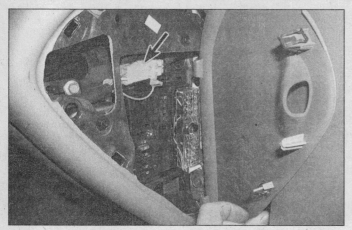

29.9 Disconnect the passenger airbag connector (arrow) here above the right fuse/relay panel

29.16 Remove the screws (arrows), then push the passenger airbag module up out of the instrument panel (shown with dashboard trim panel removed for clarity)

Driver's side airbag

8   Just above the driver's side fuse/relay panel at the left end of the instrument panel, disconnect the yellow airbag connector. You'll have to remove a CPA (Connector Position Assurance) clip first, then disconnect the connector **(see illustration)**.

Passenger's side airbag

9   At the right end of the instrument panel, disconnect the two-pin electrical connector (yellow harness) just above the passenger-side fuse/relay box **(see illustration)**. **Note:** *On 1997 Malibu and Cutlass models, remove the glovebox (see Chapter 11) to access the passenger airbag connector.*

Enabling the system

10   After you've disabled the airbag and performed the necessary service, reconnect the passenger and driver's airbag connectors (ignition key Off and key out).

11   Install the airbag fuse.

12   When starting the vehicle for the first time after performing service that required SRS disabling, the manufacturer suggests inserting the key and starting the vehicle with your body away from either airbag.

Removal and installation

Driver's side airbag

13   Refer to Chapter 10 for removal and installation of the driver's side airbag.

Passenger side airbag

Refer to illustration 29.16

14   Disable the airbag system, see the **Warning** above.

15   Remove the glovebox (see Chapter 11).

16   Remove the screws and gently pry the airbag unit from the back of the dashboard **(see illustration)**. **Caution:** *The airbag assembly is heavier than it looks; use both hands when removing it from the dash.*

17   Installation is the reverse of the removal procedure.

30 Wiring diagrams - general information

Since it isn't possible to include all wiring diagrams for every year covered by this manual, the following diagrams are those that are typical and most commonly needed.

Prior to troubleshooting any circuits, check the fuse and circuit breakers (if equipped) to make sure they are in good condition. Make sure the battery is properly charged and has clean, tight cable connections (see Chapter 1).

When checking the wiring system, make sure that all electrical connectors are clean, with no broken or loose pins. When unplugging an electrical connector, do not pull on the wires, only on the connector housings themselves.

12

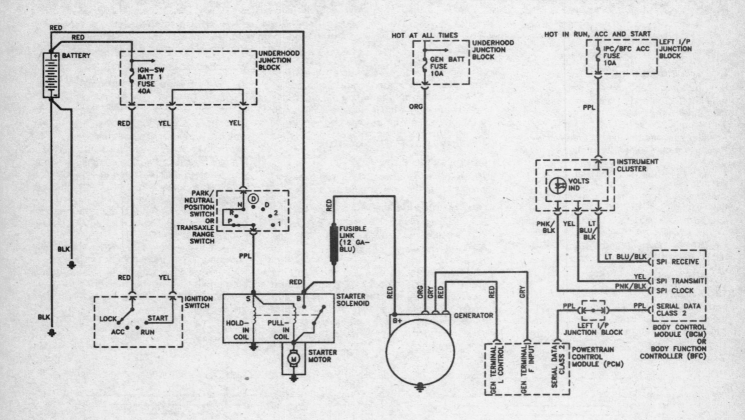

Starting and charging system

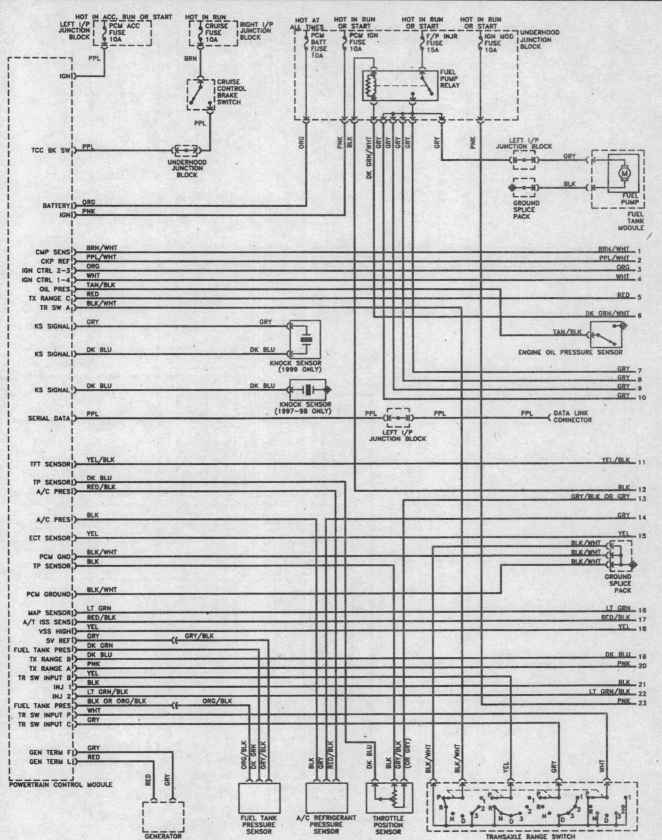

Engine control system, four-cylinder engine (part 1 of 3)

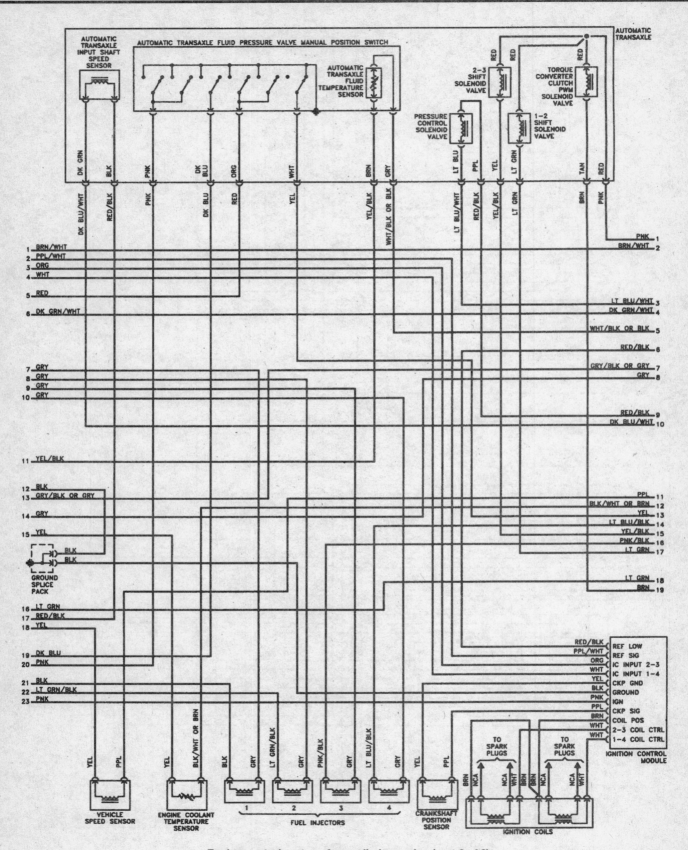

Engine control system, four-cylinder engine (part 2 of 3)

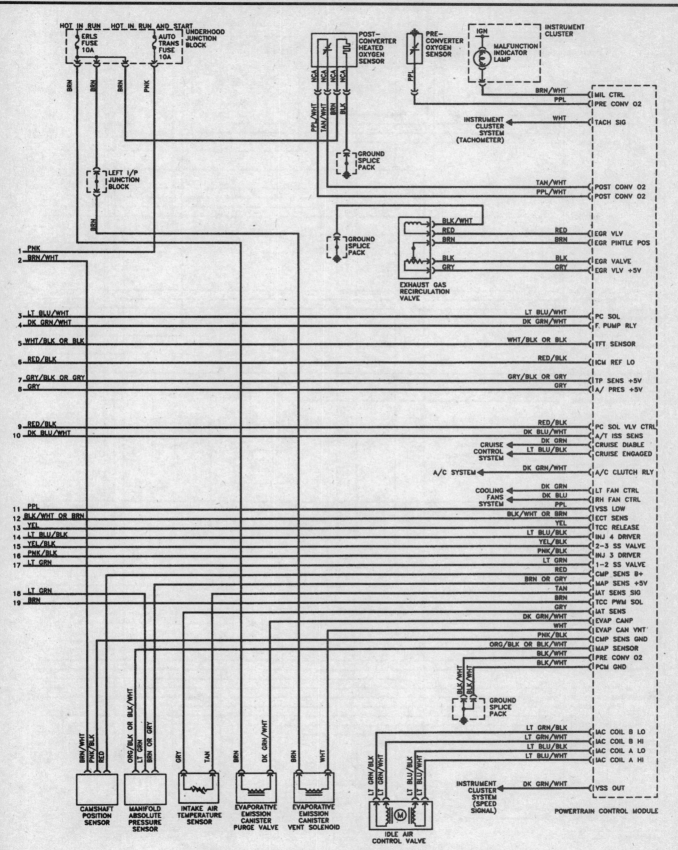

Engine control system, four-cylinder engine (part 3 of 3)

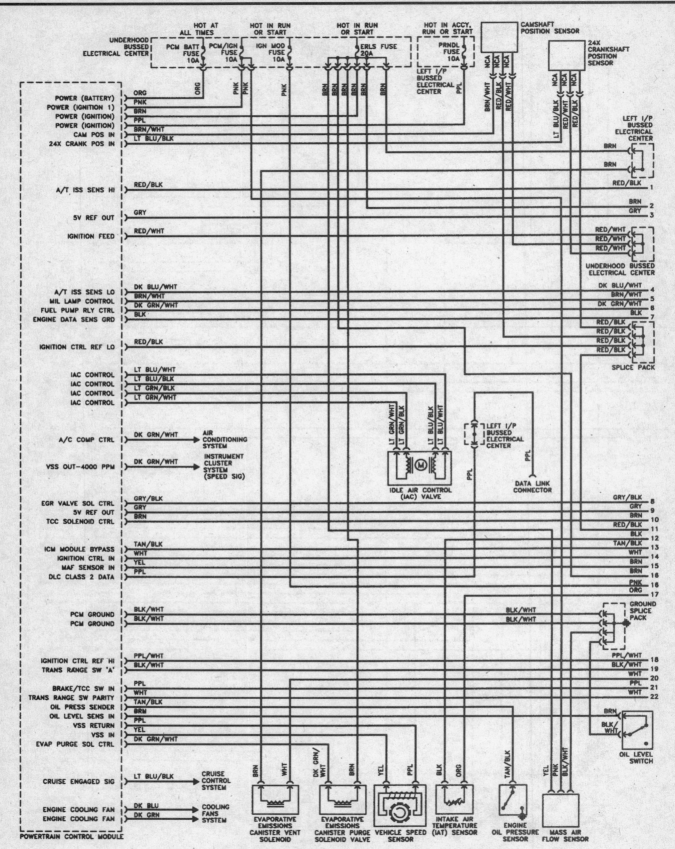

Engine control system, 1997 3.1L V6 engine (part 1 of 3)

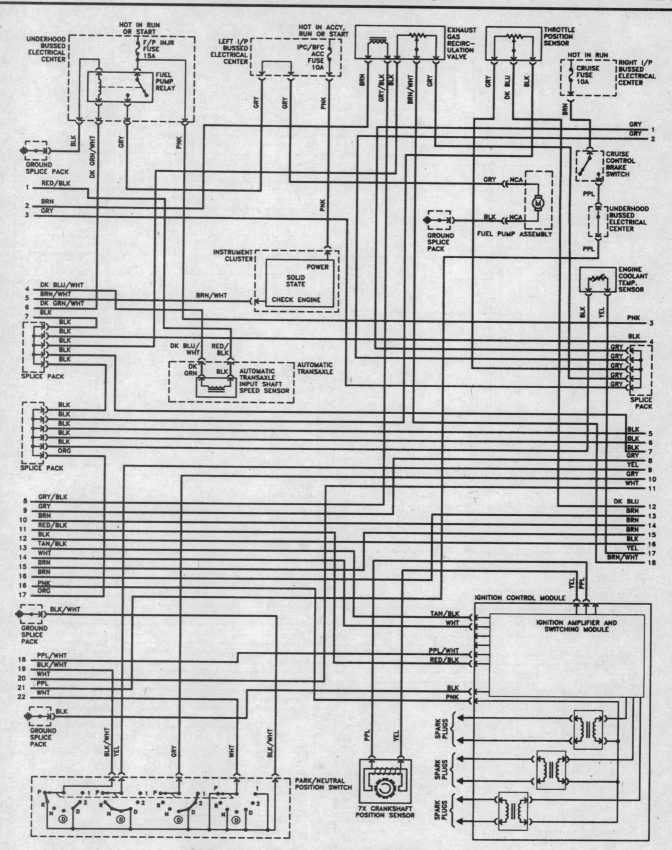

Engine control system, 1997 3.1L V6 engine (part 2 of 3)

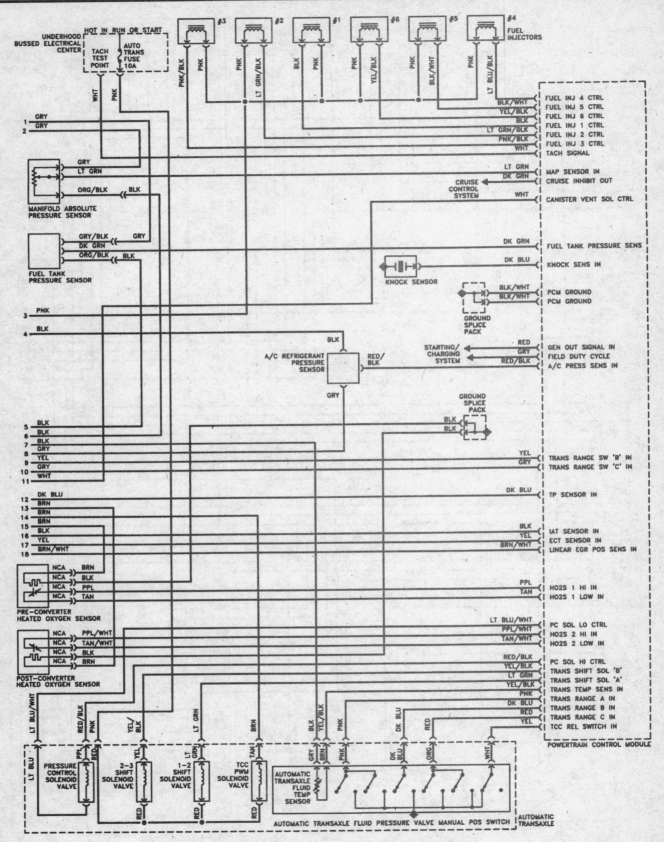

Engine control system, 1997 3.1L V6 engine (part 3 of 3)

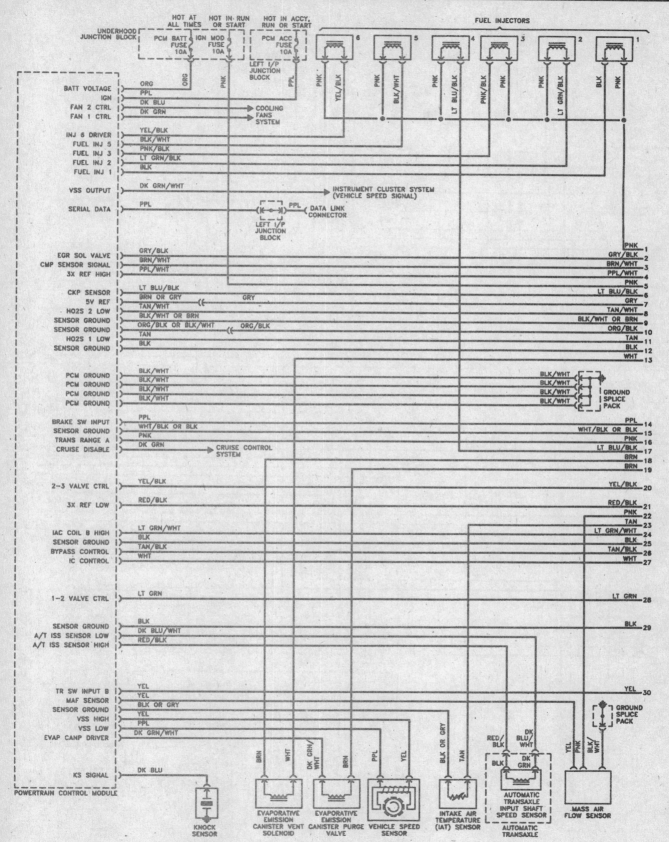

Engine control system, 1998 and later 3.1L V6 engine (part 1 of 3)

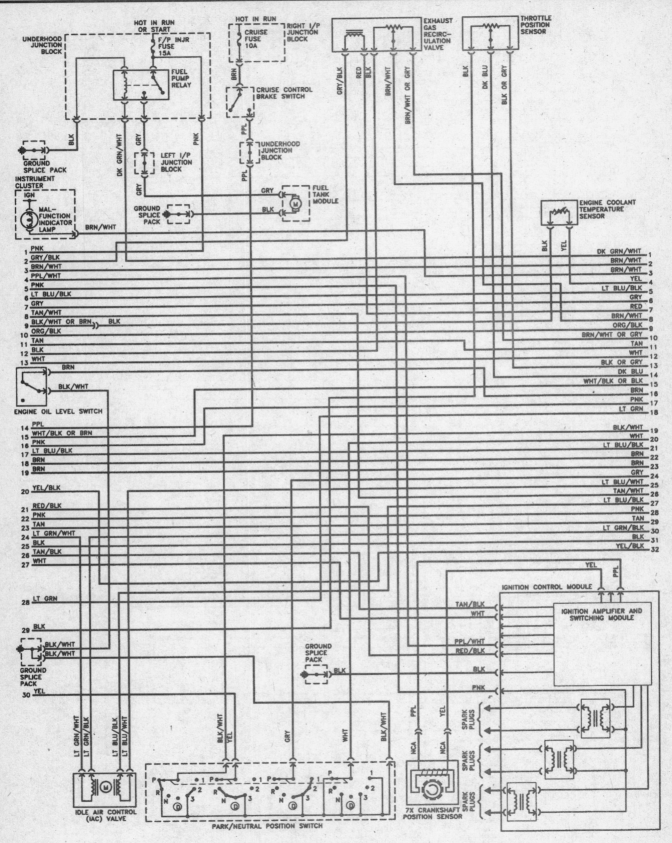

Engine control system, 1998 and later 3.1L V6 engine (part 2 of 3)

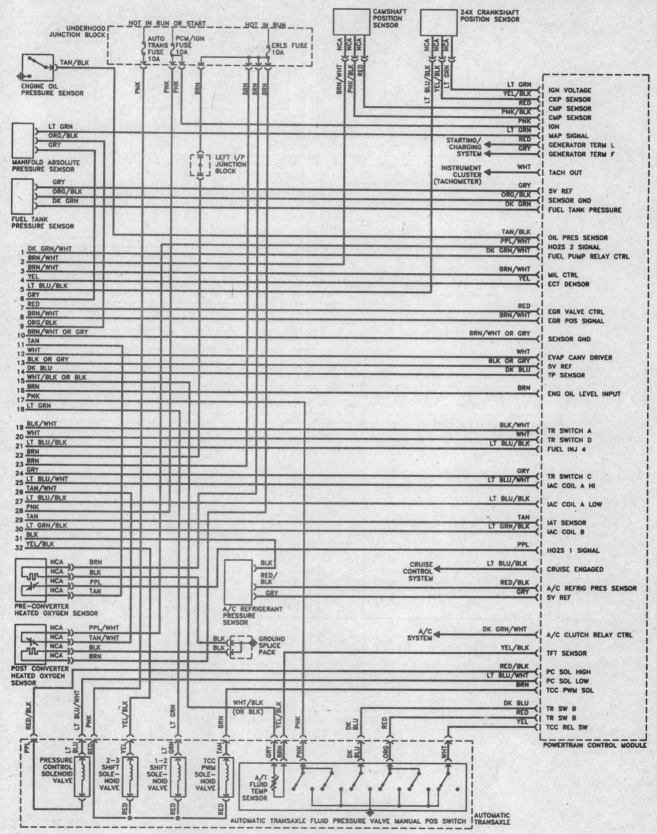

Engine control system, 1998 and later 3.1L V6 engine (part 3 of 3)

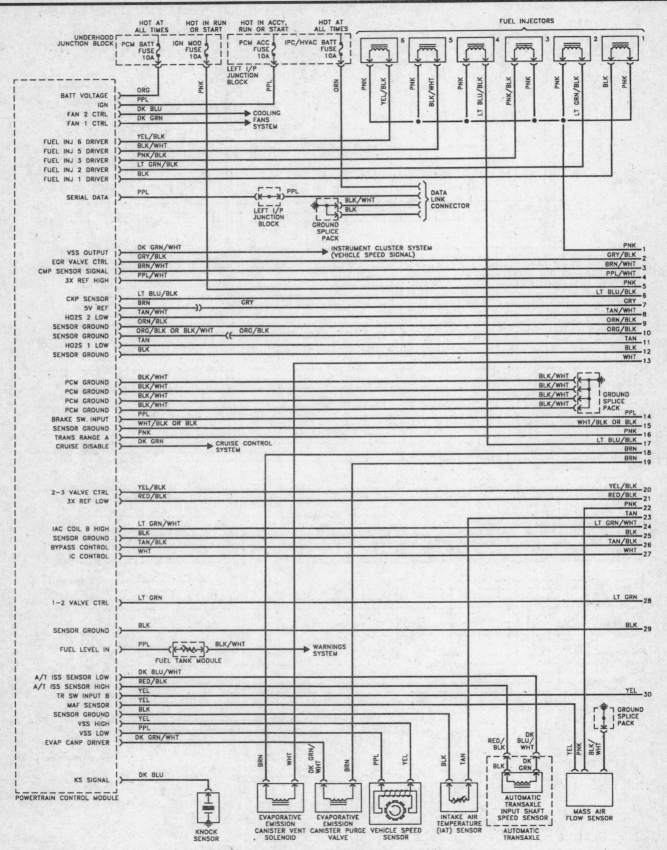

Engine control system, 3.4L V6 engine (part 1 of 3)

Chapter 12 Chassis electrical system

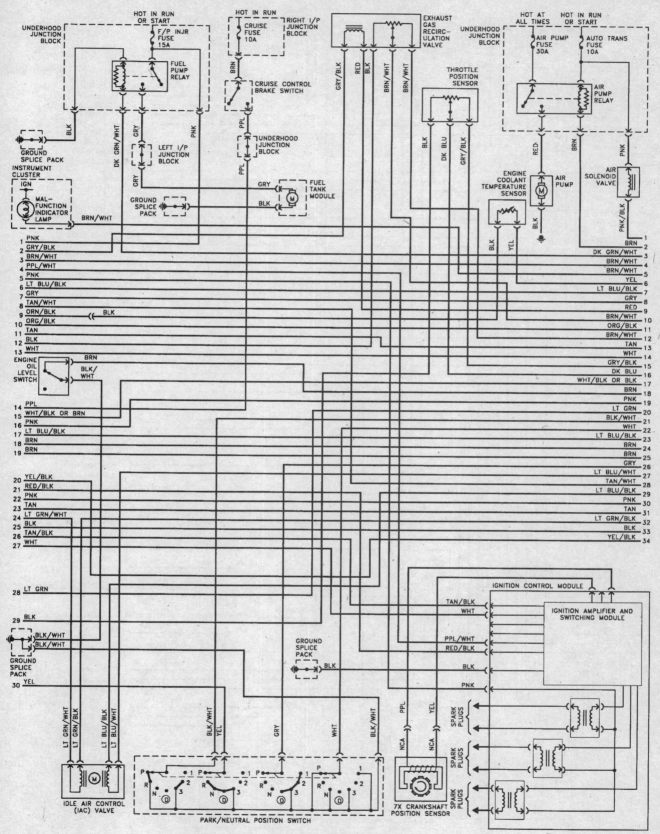

Engine control system, 3.4L V6 engine (part 2 of 3)

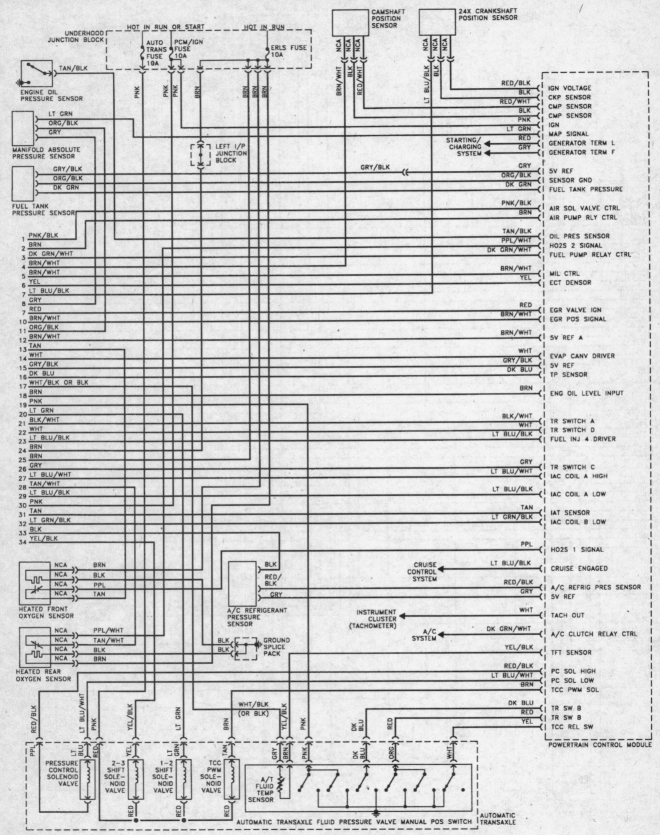

Engine control system, 3.4L V6 engine (part 3 of 3)

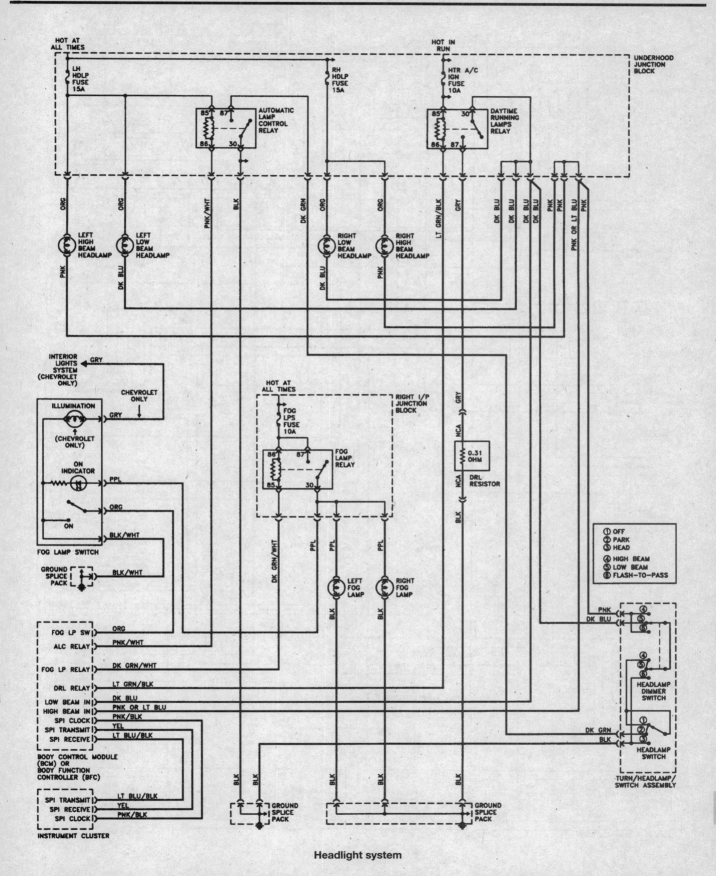

Headlight system

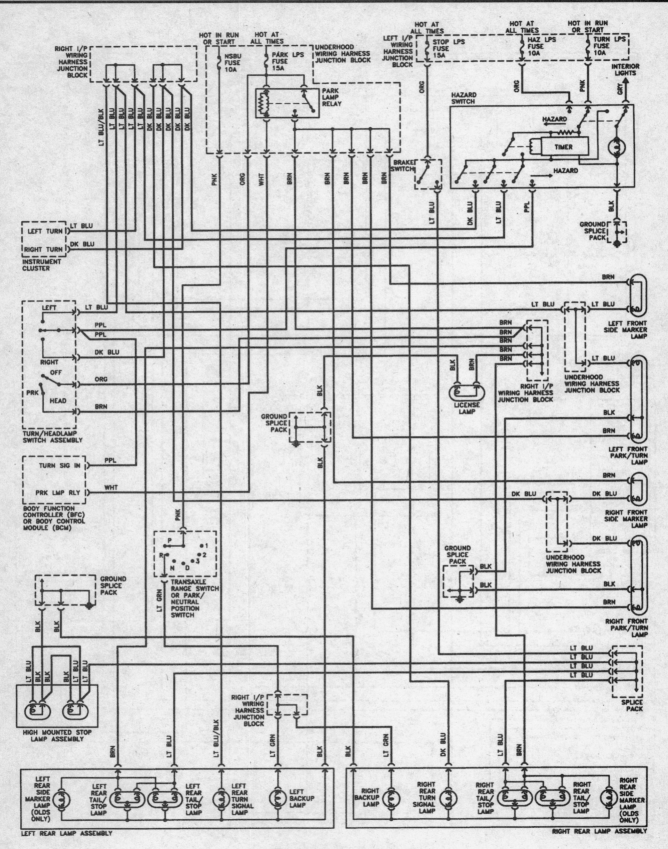

Exterior lighting system

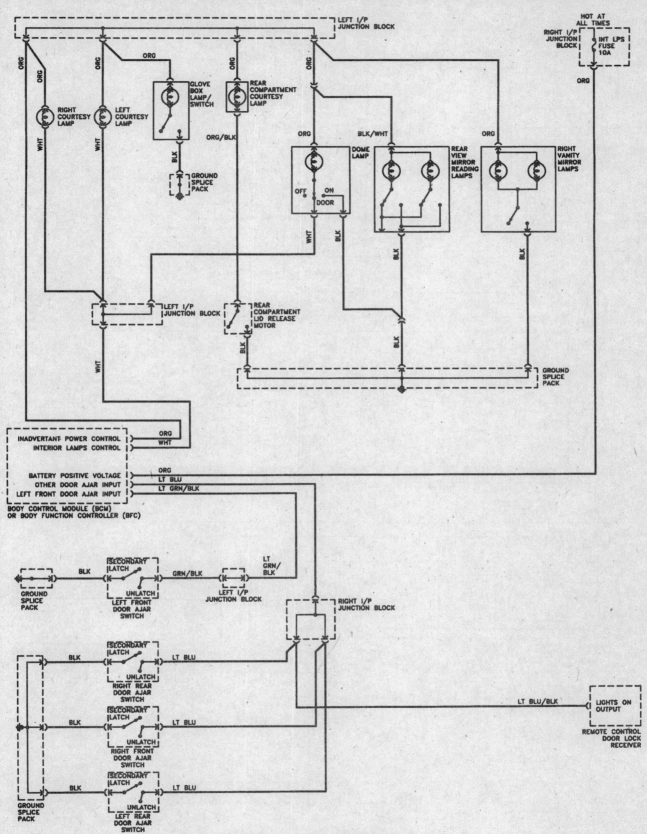

Interior lighting system - courtesy lights

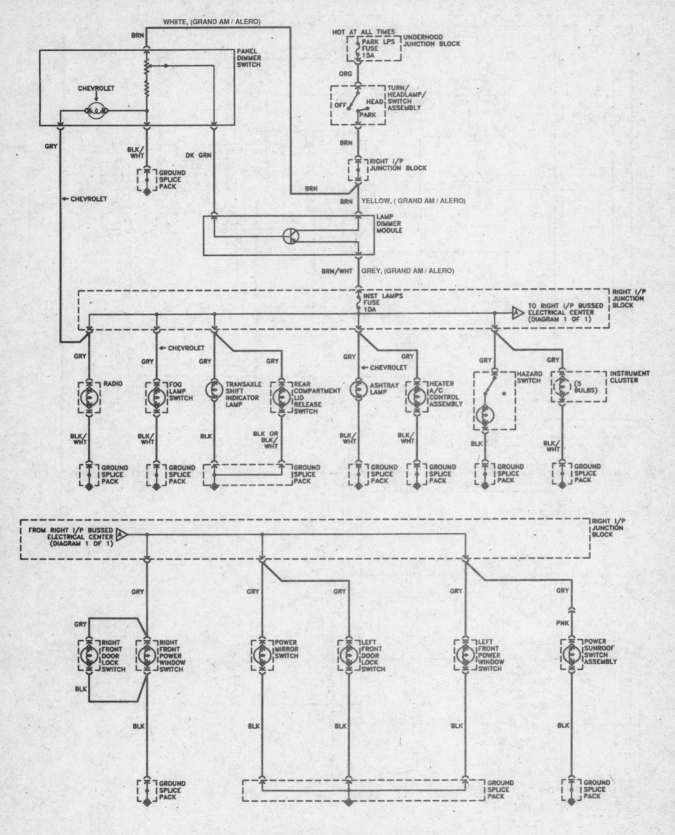

Interior lighting system - instrument panel illumination

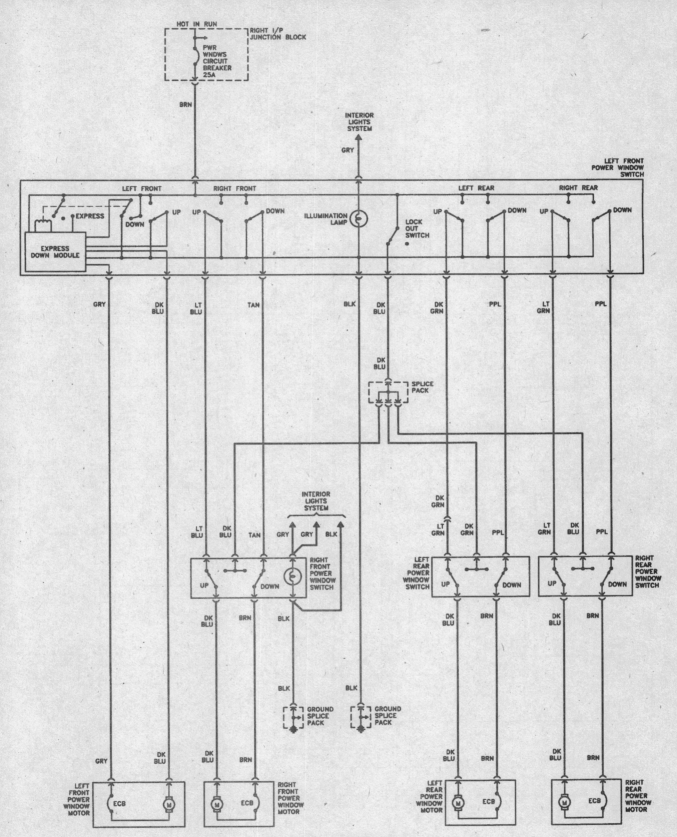

Power window system (sedans)

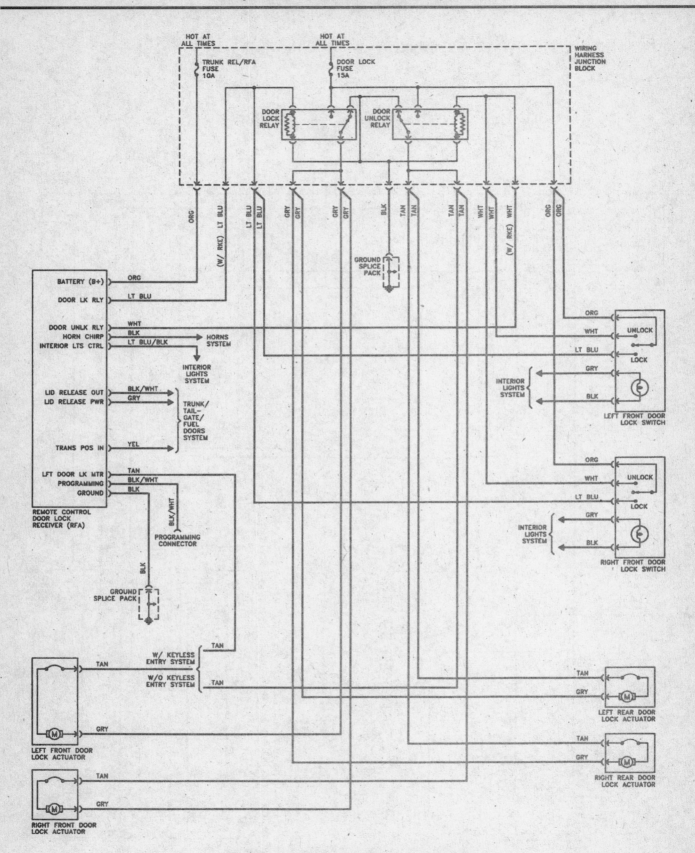

Power door lock system - Malibu/Cutlass models

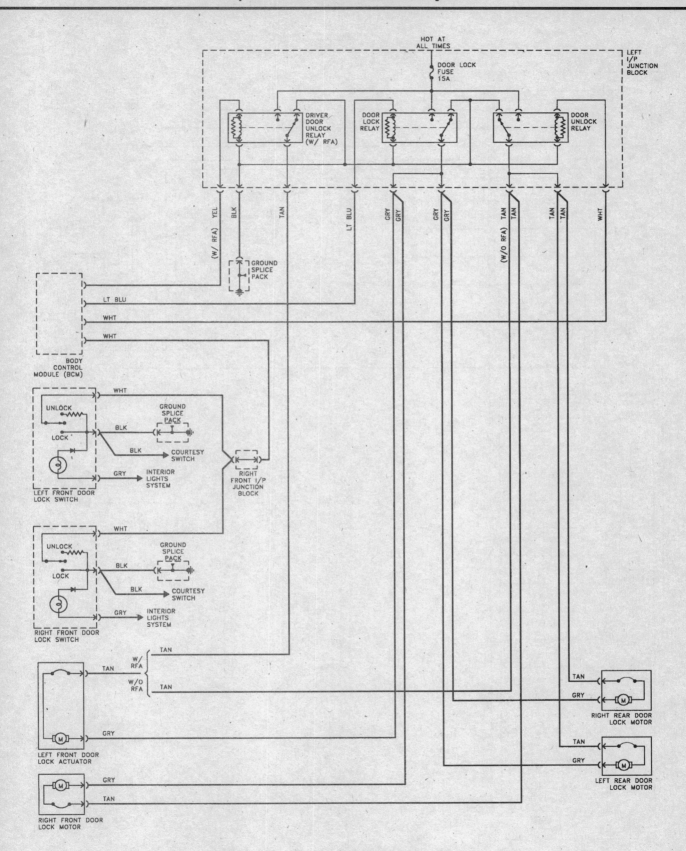

Power door lock system - Grand Am/Alero models

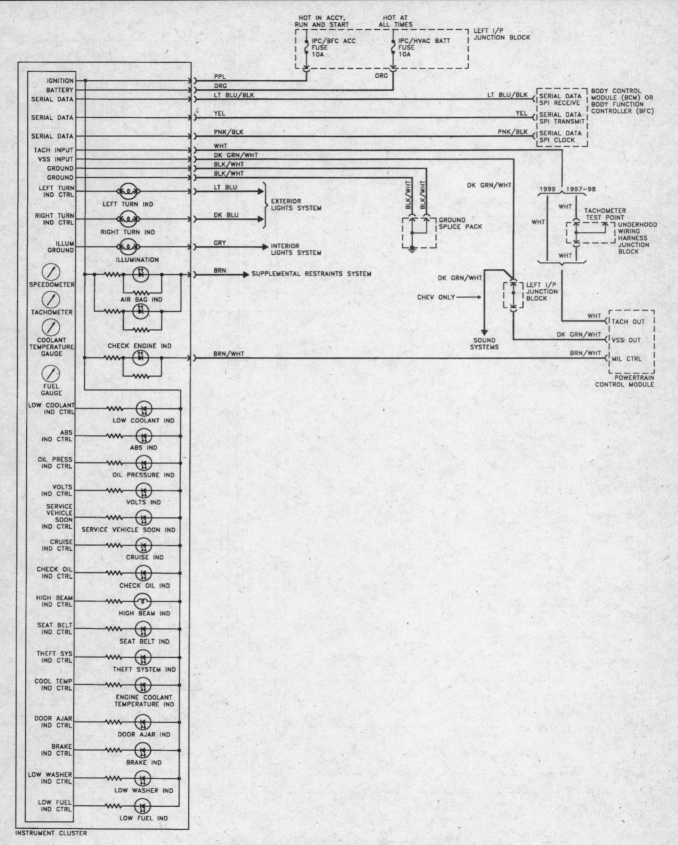

Instrument panel warning light system - 1997 through 1999 models

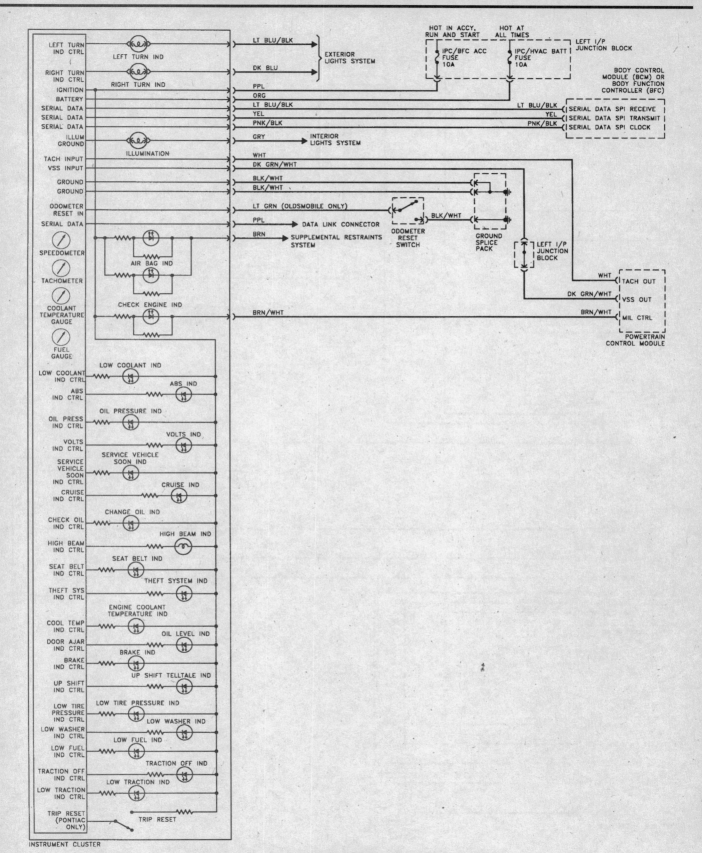

Instrument panel warning light system, 2000 models

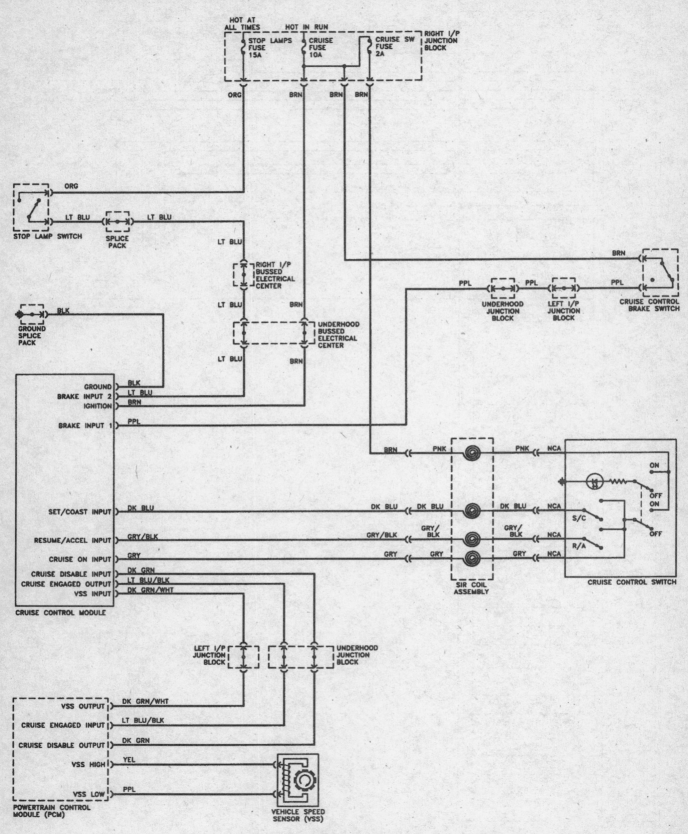

Cruise control system - typical

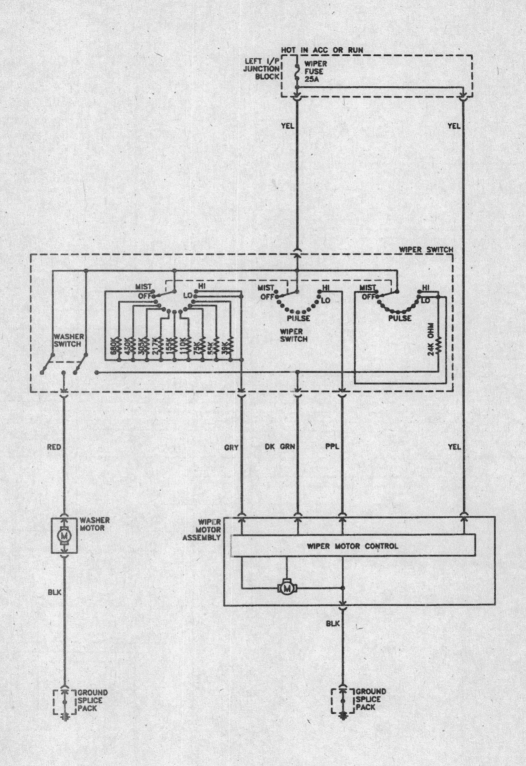

Windshield wiper/washer system

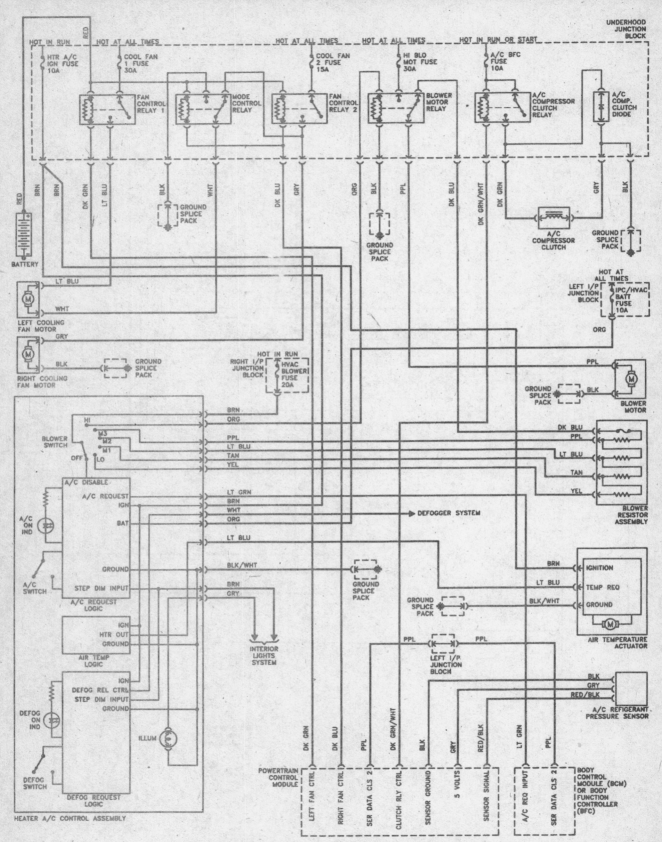

Heating/air-conditioning system (including engine cooling fans)

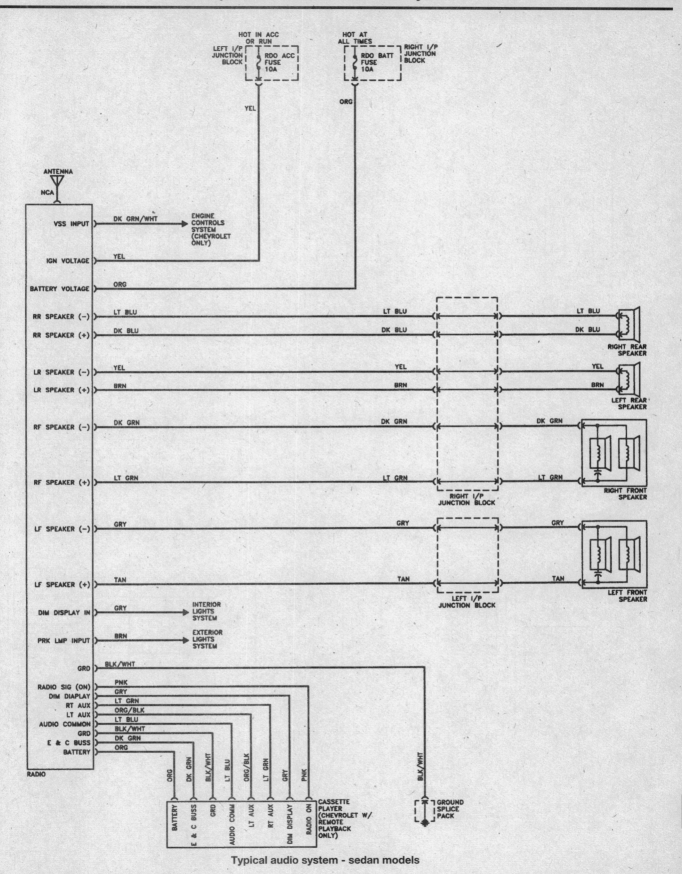

Typical audio system - sedan models

Notes

Index

A

B

Haynes Automotive Manuals

NOTE: New manuals are added to this list on a periodic basis. If you do not see a listing for your vehicle, consult your local Haynes dealer for the latest product information.

ACURA
- 12020 Integra '86 thru '89 & Legend '86 thru '90
- 12021 Integra '90 thru '93 & Legend '91 thru '95

AMC
- Jeep CJ - see JEEP (50020)
- 14020 Mid-size models '70 thru '83
- 14025 (Renault) Alliance & Encore '83 thru '87

AUDI
- 15020 4000 all models '80 thru '87
- 15025 5000 all models '77 thru '83
- 15026 5000 all models '84 thru '88

AUSTIN-HEALEY
- Sprite - see MG Midget (66015)

BMW
- *18020 3/5 Series not including diesel or all-wheel drive models '82 thru '92
- *18021 3-Series incl. Z3 models '92 thru '98
- 18025 320i all 4 cyl models '75 thru '83
- 18050 1500 thru 2002 except Turbo '59 thru '77

BUICK
- Century (front-wheel drive) - see GM (38005)
- *1902Q Buick, Oldsmobile & Pontiac Full-size (Front-wheel drive) all models '85 thru '00
  Buick Electra, LeSabre and Park Avenue;
  Oldsmobile Delta 88 Royale, Ninety Eight and Regency; Pontiac Bonneville
- 19025 Buick Oldsmobile & Pontiac Full-size (Rear wheel drive)
  Buick Estate '70 thru '90, Electra '70 thru '84, LeSabre '70 thru '85, Limited '74 thru '79
  Oldsmobile Custom Cruiser '70 thru '90, Delta 88 '70 thru '85,Ninety-eight '70 thru '84
  Pontiac Bonneville '70 thru '81, Catalina '70 thru '81, Grandville '70 thru '75, Parisienne '83 thru '86
- 19030 Mid-size Regal & Century all rear-drive models with V6, V8 and Turbo '74 thru '87
  Regal - see GENERAL MOTORS (38010)
  Riviera - see GENERAL MOTORS (38030)
  Roadmaster - see CHEVROLET (24046)
  Skyhawk - see GENERAL MOTORS (38015)
  Skylark - see GM (38020, 38025)
  Somerset - see GENERAL MOTORS (38025)

CADILLAC
- 21030 Cadillac Rear Wheel Drive all gasoline models '70 thru '93
  Cimarron - see GENERAL MOTORS (38015)
  Deville - see GENERAL MOTORS (38031)
  Eldorado - see GM (38030 & 38031)
  Fleetwood - see GM (38031)
  Seville - see GM (38030 & 38031)

CHEVROLET
- *24010 Astro & GMC Safari Mini-vans '85 thru '98
- 24015 Camaro V8 all models '70 thru '81
- 24016 Camaro all models '82 thru '92
- *24017 Camaro & Firebird '93 thru '00
  Cavalier - see GENERAL MOTORS (38016)
  Celebrity - see GENERAL MOTORS (38005)
- 24020 Chevelle, Malibu & El Camino '69 thru '87
- 24024 Chevette & Pontiac T1000 '76 thru '87
  Citation - see GENERAL MOTORS (38020)
- 24032 Corsica/Beretta all models '87 thru '96
- 24040 Corvette all V8 models '68 thru '82
- *24041 Corvette all models '84 thru '96
- 10305 Chevrolet Engine Overhaul Manual
- 24045 Full-size Sedans Caprice, Impala, Biscayne, Bel Air & Wagons '69 thru '90
- 24046 Impala SS & Caprice and Buick Roadmaster '91 thru '96
  Impala - see LUMINA (24048)
  Lumina '90 thru '94 - see GM (38010)
- *24048 Lumina & Monte Carlo '95 thru '01
  Lumina APV - see GM (38035)
- 24050 Luv Pick-up all 2WD & 4WD '72 thru '82
  Malibu '97 thru '00 - see GM (38026)

- 24055 Monte Carlo all models '70 thru '88
  Monte Carlo '95 thru '01 - see LUMINA (24048)
- 24059 Nova all V8 models '69 thru '79
- 24060 Nova and Geo Prizm '85 thru '92
- 24064 Pick-ups '67 thru '87 - Chevrolet & GMC, all V8 & in-line 6 cyl, 2WD & 4WD '67 thru '87; Suburbans, Blazers & Jimmys '67 thru '91
- 24065 Pick-ups '88 thru '98 - Chevrolet & GMC, full-size pick-ups '88 thru '98, C/K Classic '99 & '00, Blazer & Jimmy '92 thru '94; Suburban '92 thru '99; Tahoe & Yukon '95 thru '99
- *24066 Pick-ups '99 thru '01 - Chevrolet Silverado & GMC Sierra full-size pick-ups '99 thru '01, Suburban/Tahoe/Yukon/Yukon XL '00 thru '01
- 24070 S-10 & S-15 Pick-ups '82 thru '93, Blazer & Jimmy '83 thru '94,
- *24071 S-10 & S-15 Pick-ups '94 thru '01, Blazer & Jimmy '95 thru '01, Hombre '96 thru '01
- *24075 Sprint & Geo Metro '85 thru '94
- *24080 Vans - Chevrolet & GMC '68 thru '96

CHRYSLER
- 25015 Chrysler Cirrus, Dodge Stratus, Plymouth Breeze '95 thru '00
- 10310 Chrysler Engine Overhaul Manual
- 25020 Full-size Front-Wheel Drive '88 thru '93
  K-Cars - see DODGE Aries (30008)
  Laser - see DODGE Daytona (30030)
- 25025 Chrysler LHS, Concorde, New Yorker, Dodge Intrepid, Eagle Vision, '93 thru '97
- *25026 Chrysler LHS, Concorde, 300M, Dodge Intrepid, '98 thru '01
- 25030 Chrysler & Plymouth Mid-size front wheel drive '82 thru '95
  Rear-wheel Drive - see Dodge (30050)
- *25040 Chrysler Sebring, Dodge Avenger '95 thru '02

DATSUN
- 28005 200SX all models '80 thru '83
- 28007 B-210 all models '73 thru '78
- 28009 210 all models '79 thru '82
- 28012 240Z, 260Z & 280Z Coupe '70 thru '78
- 28014 280ZX Coupe & 2+2 '79 thru '83
  300ZX - see NISSAN (72010)
- 28016 310 all models '78 thru '82
- 28018 510 & PL521 Pick-up '68 thru '73
- 28020 510 all models '78 thru '81
- 28022 620 Series Pick-up all models '73 thru '79
  720 Series Pick-up - see NISSAN (72030)
- 28025 810/Maxima all gasoline models, '77 thru '84

DODGE
- 400 & 600 - see CHRYSLER (25030)
- 30008 Aries & Plymouth Reliant '81 thru '89
- 30010 Caravan & Plymouth Voyager '84 thru '95
- *30011 Caravan & Plymouth Voyager '96 thru '99
- 30012 Challenger/Plymouth Saporro '78 thru '83
- 30016 Colt & Plymouth Champ '78 thru '87
- 30020 Dakota Pick-ups all models '87 thru '96
- *30021 Durango '98 & '99, Dakota '97 thru '99
- 30025 Dart, Demon, Plymouth Barracuda, Duster & Valiant 6 cyl models '67 thru '76
- 30030 Daytona & Chrysler Laser '84 thru '89
  Intrepid - see CHRYSLER (25025, 25026)
- *30034 Neon all models '95 thru '99
- 30035 Omni & Plymouth Horizon '78 thru '90
- 30040 Pick-ups all full-size models '74 thru '93
- *30041 Pick-ups all full-size models '94 thru '01
- 30045 Ram 50/D50 Pick-ups & Raider and Plymouth Arrow Pick-ups '79 thru '93
- 30050 Dodge/Plymouth/Chrysler RWD '71 thru '89
- 30055 Shadow & Plymouth Sundance '87 thru '94
- 30060 Spirit & Plymouth Acclaim '89 thru '95
- *30065 Vans - Dodge & Plymouth '71 thru '99

EAGLE
- Talon - see MITSUBISHI (68030, 68031)
- Vision - see CHRYSLER (25025)

FIAT
- 34010 124 Sport Coupe & Spider '68 thru '78
- 34025 X1/9 all models '74 thru '80

FORD
- 10355 Ford Automatic Transmission Overhaul
- 36004 Aerostar Mini-vans all models '86 thru '97
- 36006 Contour & Mercury Mystique '95 thru '00
- 36008 Courier Pick-up all models '72 thru '82
- *36012 Crown Victoria & Mercury Grand Marquis '88 thru '00
- 10320 Ford Engine Overhaul Manual
- 36016 Escort/Mercury Lynx all models '81 thru '90
- 36020 Escort/Mercury Tracer '91 thru '00
- 36024 Explorer & Mazda Navajo '91 thru '01
- 36028 Fairmont & Mercury Zephyr '78 thru '83
- 36030 Festiva & Aspire '88 thru '97
- 36032 Fiesta all models '77 thru '80
- *36034 Focus all models '00 and '01
- 36036 Ford & Mercury Full-size '75 thru '87
- 36040 Granada & Mercury Monarch '75 thru '80
- 36044 Ford & Mercury Mid-size '75 thru '86
- 36048 Mustang V8 all models '64-1/2 thru '73
- 36049 Mustang II 4 cyl, V6 & V8 models '74 thru '78
- 36050 Mustang & Mercury Capri all models Mustang, '79 thru '93; Capri, '79 thru '86
- *36051 Mustang all models '94 thru '00
- 36054 Pick-ups & Bronco '73 thru '79
- 36058 Pick-ups & Bronco '80 thru '96
- *36059 Pick-ups, Expedition & Mercury Navigator '97 thru '99
- *36060 Super Duty Pick-ups, Excursion '97 thru '02
- 36062 Pinto & Mercury Bobcat '75 thru '80
- 36066 Probe all models '89 thru '92
- 36070 Ranger/Bronco II gasoline models '83 thru '92
- *36071 Ranger '93 thru '00 & Mazda Pick-ups '94 thru '00
- 36074 Taurus & Mercury Sable '86 thru '95
- *36075 Taurus & Mercury Sable '96 thru '01
- 36078 Tempo & Mercury Topaz '84 thru '94
- 36082 Thunderbird/Mercury Cougar '83 thru '88
- 36086 Thunderbird/Mercury Cougar '89 and '97
- 36090 Vans all V8 Econoline models '69 thru '91
- *36094 Vans full size '92 thru '01
- *36097 Windstar Mini-van '95 thru '01

GENERAL MOTORS
- 10360 GM Automatic Transmission Overhaul
- 38005 Buick Century, Chevrolet Celebrity, Oldsmobile Cutlass Ciera & Pontiac 6000 all models '82 thru '96
- *38010 Buick Regal, Chevrolet Lumina, Oldsmobile Cutlass Supreme & Pontiac Grand Prix (FWD) '88 thru '99
- 38015 Buick Skyhawk, Cadillac Cimarron, Chevrolet Cavalier, Oldsmobile Firenza & Pontiac J-2000 & Sunbird '82 thru '94
- *38016 Chevrolet Cavalier & Pontiac Sunfire '95 thru '00
- 38020 Buick Skylark, Chevrolet Citation, Olds Omega, Pontiac Phoenix '80 thru '85
- 38025 Buick Skylark & Somerset, Oldsmobile Achieva & Calais and Pontiac Grand Am all models '85 thru '98
- *38026 Chevrolet Malibu, Olds Alero & Cutlass, Pontiac Grand Am '97 thru '00
- 38030 Cadillac Eldorado '71 thru '85, Seville '80 thru '85, Oldsmobile Toronado '71 thru '85, Buick Riviera '79 thru '85
- *38031 Cadillac Eldorado & Seville '86 thru '91, Deville '86 thru '93, Fleetwood & Olds Toronado '86 thru '92, Buick Riviera '86 thru '93
- *38035 Chevrolet Lumina APV, Olds Silhouette & Pontiac Trans Sport all models '90 thru '95
- *38036 Chevrolet Venture, Olds Silhouette, Pontiac Trans Sport & Montana '97 thru '01
- General Motors Full-size Rear-wheel Drive - see BUICK (19025)

GEO
- Metro - see CHEVROLET Sprint (24075)
- Prizm - '85 thru '92 see CHEVY (24060), '93 thru '96 see TOYOTA Corolla (92036)
- 40030 Storm all models '90 thru '93
  Tracker - see SUZUKI Samurai (90010)

(Continued on other side)

Listings shown with an asterisk (*) indicate model coverage as of this printing. These titles will be periodically updated to include later model years - consult your Haynes dealer for more information.

Haynes North America, Inc., 861 Lawrence Drive, Newbury Park, CA 91320-1514 • (805) 498-6703

Haynes Automotive Manuals (continued)

NOTE: New manuals are added to this list on a periodic basis. If you do not see a listing for your vehicle, consult your local Haynes dealer for the latest product information.

GMC

Vans & Pick-ups - see CHEVROLET

HONDA

- 42010 Accord CVCC all models '76 thru '83
- 42011 Accord all models '84 thru '89
- 42012 Accord all models '90 thru '93
- 42013 Accord all models '94 thru '97
- *42014 Accord all models '98 and '99
- 42020 Civic 1200 all models '73 thru '79
- 42021 Civic 1300 & 1500 CVCC '80 thru '83
- 42022 Civic 1500 CVCC '75 thru '79
- 42023 Civic all models '84 thru '91
- 42024 Civic & del Sol '92 thru '95
- *42025 Civic '96 thru '00, CR-V '97 thru '00, Acura Integra '94 thru '00
- 42040 Prelude CVCC all models '79 thru '89

HYUNDAI

- *43010 Elantra all models '96 thru '01
- 43015 Excel & Accent all models '86 thru '98

ISUZU

- Hombre - see CHEVROLET S-10 (24071)
- *47017 Rodeo '91 thru '97; Amigo '89 thru '94; Honda Passport '95 thru '97
- 47020 Trooper & Pick-up '81 thru '93

JAGUAR

- 49010 XJ6 all 6 cyl models '68 thru '86
- 49011 XJ6 all models '88 thru '94
- 49015 XJ12 & XJS all 12 cyl models '72 thru '85

JEEP

- 50010 Cherokee, Comanche & Wagoneer Limited all models '84 thru '00
- 50020 CJ all models '49 thru '86
- *50025 Grand Cherokee all models '93 thru '00
- 50029 Grand Wagoneer & Pick-up '72 thru '91 Grand Wagoneer '84 thru '91, Cherokee & Wagoneer '72 thru '83, Pick-up '72 thru '88
- *50030 Wrangler all models '87 thru '00

LEXUS

ES 300 - see TOYOTA Camry (92007)

LINCOLN

- Navigator - see FORD Pick-up (36059)
- *59010 Rear-Wheel Drive all models '70 thru '01

MAZDA

- 61010 GLC Hatchback (rear-wheel drive) '77 thru '83
- 61011 GLC (front-wheel drive) '81 thru '85
- *61015 323 & Protegé '90 thru '00
- *61016 MX-5 Miata '90 thru '97
- 61020 MPV all models '89 thru '94
- Navajo - see Ford Explorer (36024)
- 61030 Pick-ups '72 thru '93 Pick-up '94 thru '00 - see Ford Ranger (36071)
- 61035 RX-7 all models '79 thru '85
- 61036 RX-7 all models '86 thru '91
- 61040 626 (rear-wheel drive) all models '79 thru '82
- *61041 626/MX-6 (front-wheel drive) '83 thru '91
- *61042 626 '93 thru '01, MX-6/Ford Probe '93 thru '97

MERCEDES-BENZ

- 63012 123 Series Diesel '76 thru '85
- 63015 190 Series four-cyl gas models, '84 thru '88
- 63020 230/250/280 6 cyl sohc models '68 thru '72
- 63025 280 123 Series gasoline models '77 thru '81
- 63030 350 & 450 all models '71 thru '80

MERCURY

- 64200 Villager & Nissan Quest '93 thru '01 All other titles, see FORD Listing.

MG

- 66010 MGB Roadster & GT Coupe '62 thru '80
- 66015 MG Midget, Austin Healey Sprite '58 thru '80

MITSUBISHI

- 68020 Cordia, Tredia, Galant, Precis & Mirage '83 thru '93
- 68030 Eclipse, Eagle Talon & Ply. Laser '90 thru '94
- *68031 Eclipse '95 thru '01, Eagle Talon '95 thru '98
- 68040 Pick-up '83 thru '96 & Montero '83 thru '93

NISSAN

- 72010 300ZX all models including Turbo '84 thru '89
- 72015 Altima all models '93 thru '01
- 72020 Maxima all models '85 thru '92
- *72021 Maxima all models '93 thru '01
- 72030 Pick-ups '80 thru '97 Pathfinder '87 thru '95
- *72031 Frontier Pick-up '98 thru '01, Xterra '00 & '01, Pathfinder '96 thru '01
- 72040 Pulsar all models '83 thru '86
- Quest - see MERCURY Villager (64200)
- 72050 Sentra all models '82 thru '94
- 72051 Sentra & 200SX all models '95 thru '99
- 72060 Stanza all models '82 thru '90

OLDSMOBILE

- 73015 Cutlass V6 & V8 gas models '74 thru '88 For other OLDSMOBILE titles, see BUICK, CHEVROLET or GENERAL MOTORS listing.

PLYMOUTH

For PLYMOUTH titles, see DODGE listing.

PONTIAC

- 79008 Fiero all models '84 thru '88
- 79018 Firebird V8 models except Turbo '70 thru '81
- 79019 Firebird all models '82 thru '92
- 79040 Mid-size Rear-wheel Drive '70 thru '87 For other PONTIAC titles, see BUICK, CHEVROLET or GENERAL MOTORS listing.

PORSCHE

- 80020 911 except Turbo & Carrera 4 '65 thru '89
- 80025 914 all 4 cyl models '69 thru '76
- 80030 924 all models including Turbo '76 thru '82
- 80035 944 all models including Turbo '83 thru '89

RENAULT

Alliance & Encore - see AMC (14020)

SAAB

- *84010 900 all models including Turbo '79 thru '88

SATURN

- *87010 Saturn all models '91 thru '99

SUBARU

- 89002 1100, 1300, 1400 & 1600 '71 thru '79
- 89003 1600 & 1800 2WD & 4WD '80 thru '94

SUZUKI

- *90010 Samurai/Sidekick & Geo Tracker '86 thru '01

TOYOTA

- 92005 Camry all models '83 thru '91
- 92006 Camry all models '92 thru '96
- *92007 Camry, Avalon, Solara, Lexus ES 300 '97 thru '01
- 92015 Celica Rear Wheel Drive '71 thru '85
- 92020 Celica Front Wheel Drive '86 thru '99
- 92025 Celica Supra all models '79 thru '92
- 92030 Corolla all models '75 thru '79
- 92032 Corolla all rear wheel drive models '80 thru '87
- 92035 Corolla all front wheel drive models '84 thru '92
- *92036 Corolla & Geo Prizm '93 thru '01
- 92040 Corolla Tercel all models '80 thru '82
- 92045 Corona all models '74 thru '82
- 92050 Cressida all models '78 thru '82
- 92055 Land Cruiser FJ40, 43, 45, 55 '68 thru '82
- *92056 Land Cruiser FJ60, 62, 80, FZJ80 '80 thru '96
- 92065 MR2 all models '85 thru '87
- 92070 Pick-up all models '69 thru '78
- 92075 Pick-up all models '79 thru '95
- *92076 Tacoma '95 thru '00, 4Runner '96 thru '00, & T100 '93 thru '98
- 92080 Previa all models '91 thru '95

- *92082 RAV4 all models '96 thru '02
- 92085 Tercel all models '87 thru '94

TRIUMPH

- 94007 Spitfire all models '62 thru '81
- 94010 TR7 all models '75 thru '81

VW

- 96008 Beetle & Karmann Ghia '54 thru '79
- *96009 New Beetle '98 thru '00
- 96016 Rabbit, Jetta, Scirocco, & Pick-up gas models '74 thru '91 & Convertible '80 thru '92
- 96017 Golf & Jetta all models '93 thru '97
- *96018 Golf & Jetta all models '98 thru '01
- 96020 Rabbit, Jetta & Pick-up diesel '77 thru '84
- *96023 Passat '98 thru '01, Audi A4 '96 thru '01
- 96030 Transporter 1600 all models '68 thru '79
- 96035 Transporter 1700, 1800 & 2000 '72 thru '79
- 96040 Type 3 1500 & 1600 all models '63 thru '73
- 96045 Vanagon all air-cooled models '80 thru '83

VOLVO

- 97010 120, 130 Series & 1800 Sports '61 thru '73
- 97015 140 Series all models '66 thru '74
- 97020 240 Series all models '76 thru '93
- 97040 740 & 760 Series all models '82 thru '88
- 97050 850 Series all models '93 thru '97

TECHBOOK MANUALS

- 10205 Automotive Computer Codes
- 10210 Automotive Emissions Control Manual
- 10215 Fuel Injection Manual, 1978 thru 1985
- 10220 Fuel Injection Manual, 1986 thru 1999
- 10225 Holley Carburetor Manual
- 10230 Rochester Carburetor Manual
- 10240 Weber/Zenith/Stromberg/SU Carburetors
- 10305 Chevrolet Engine Overhaul Manual
- 10310 Chrysler Engine Overhaul Manual
- 10320 Ford Engine Overhaul Manual
- 10330 GM and Ford Diesel Engine Repair Manual
- 10340 Small Engine Repair Manual, 5 HP & Less
- 10341 Small Engine Repair Manual, 5.5 - 20 HP
- 10345 Suspension, Steering & Driveline Manual
- 10355 Ford Automatic Transmission Overhaul
- 10360 GM Automatic Transmission Overhaul
- 10405 Automotive Body Repair & Painting
- 10410 Automotive Brake Manual
- 10411 Automotive Anti-lock Brake (ABS) Systems
- 10415 Automotive Detailing Manual
- 10420 Automotive Eelectrical Manual
- 10425 Automotive Heating & Air Conditioning
- 10430 Automotive Reference Manual & Dictionary
- 10435 Automotive Tools Manual
- 10440 Used Car Buying Guide
- 10445 Welding Manual
- 10450 ATV Basics

SPANISH MANUALS

- 98903 Reparación de Carrocería & Pintura
- 98905 Códigos Automotrices de la Computadora
- 98910 Frenos Automotriz
- 98915 Inyección de Combustible 1986 al 1999
- 99040 Chevrolet & GMC Camionetas '67 al '87 Incluye Suburban, Blazer & Jimmy '67 al '91
- 99041 Chevrolet & GMC Camionetas '88 al '98 Incluye Suburban '92 al '98, Blazer & Jimmy '92 al '94, Tahoe & Yukon '95 al '98
- 99042 Chevrolet & GMC Camionetas Cerradas '68 al '95
- 99055 Dodge Caravan & Plymouth Voyager '84 al '95
- 99075 Ford Camionetas y Bronco '80 al '94
- 99077 Ford Camionetas Cerradas '69 al '91
- 99083 Ford Modelos de Tamaño Grande '75 al '87
- 99088 Ford Modelos de Tamaño Mediano '75 al '86
- 99091 Ford Taurus & Mercury Sable '86 al '95
- 99095 GM Modelos de Tamaño Grande '70 al '90
- 99100 GM Modelos de Tamaño Mediano '70 al '88
- 99110 Nissan Camioneta '80 al '96, Pathfinder '87 al '95
- 99118 Nissan Sentra '82 al '94
- 99125 Toyota Camionetas y 4Runner '79 al '95

Over 100 Haynes motorcycle manuals also available

7-02

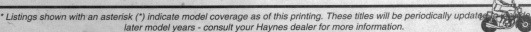

* Listings shown with an asterisk (*) indicate model coverage as of this printing. These titles will be periodically updated to later model years - consult your Haynes dealer for more information.

Haynes North America, Inc., 861 Lawrence Drive, Newbury Park, CA 91320-1514 • (805) 498-6703